T0251301

Petroleum Contaminated Soils

Volume 2

Edward J. Calabrese
Paul T. Kostecki

Technical Editor
Charles E. Bell

CRC Press
Taylor & Francis Group
Boca Raton London New York

CRC Press is an imprint of the
Taylor & Francis Group, an **informa** busines

CRC Press
Taylor & Francis Group
6000 Broken Sound Parkway NW, Suite 300
Boca Raton, FL 33487-2742

© 1989 by Taylor & Francis Group, LLC
CRC Press is an imprint of Taylor & Francis Group, an Informa business

Visit the Taylor & Francis Web site at
http://www.taylorandfrancis.com

and the CRC Press Web site at
http://www.crcpress.com

Preface

Soil contamination has become recognized as a major concern by regulatory agencies during the decade of the 1980s, yet approaches for its assessment with respect to evaluation, fate modeling, risk assessment, and remediation have presented unusually difficult technical, scientific, and regulatory challenges.

The issue of soil contamination, like those of air and water pollution, is often driven by public health concerns. However, there is a major difference that separates soil issues from both the air and water media despite their interrelatedness. This involves the history of solving problems of contamination. For example, the U.S. Public Health Service first established a drinking water standard based on public health concerns in 1914. The Clean Air Act was first passed by Congress in 1963, more than 25 years ago. The span of collective decades of regulatory activity, along with an expanded data base, has helped to mold sophisticated and mature approaches for dealing with both air and water pollution concerns.

In contrast to the experience with air and water contamination, regulatory/public health initiatives with respect to soil contamination are relatively new on the scene. This is reflected in the nature of federal legislation designed to address soil-related problems as well as in seemingly frenetic activity, especially on the state level, to address critical problems associated with the various dimensions of site assessment and remediation.

While there are many types of soil contamination concerns—Superfund sites, agricultural and industrial practices, and household-related lead in soil—the area involving by far the largest number of active sites is that of petroleum contaminated soil, principally as a result of leaking underground storage tanks. Given the mobility in the environment of products such as gasoline, such leaks pose serious threats, especially to nearby groundwater sources.

This book, which represents the proceedings of the Third National Conference on Petroleum Contaminated Soils held at the University of Massachusetts, Amherst, September 19–21, 1988, provides a comprehensive assessment of where the field is with respect to the problem (Chapters 1–4) and research (5–9), as well as specific technical areas such as analysis (Chapters 10–13), environmental fate (Chapters 14–16), remedial techniques (Chapters 17–25 and 27), public health assessment (Chapters 26, 28–34) and regulatory approaches (Chapters 35–37). The expertise from federal and state governmental agencies; affected industries, most notably petroleum and electric utilities; private consultants; and academia has been woven together to provide the reader not only with an historical sense of the problems, but with practical approaches for their solutions within the context of a complex web of regulatory approaches. This book will therefore be of considerable value to those concerned with the challenge of effectively dealing with the problem of petroleum contaminated soils, including the federal and state regulatory community, public health officials, and affected industries, as well as the environmental consultant community, which is deeply involved with developing creative, cost-effective assessments and solutions that can withstand the demands of regulatory requirements. In addition, this book provides students with a broad perspective on the problem by integrating the various dimensions into an intelligible whole so that a proper, balanced appreciation of the problem will be discernible.

Acknowledgments

We wish to thank all the sponsors for their generous support and assistance in presenting the conference. They include:

Agency for Toxic Substances and Disease Registry (ATSDR)/
Office of Health Assessment (OHA)
American Petroleum Institute (API)
Association of American Railroads (AAR)
Department of Energy (DOE)
EA Engineering, Science & Technology, Inc.
Edison Electric Institute (EEI)/
Utility Solid Waste Activities Group (USWAG)
Electric Power Research Institute (EPRI)
EPA/Office of Underground Storage Tanks (OUST)
International Society of Regulatory Toxicology & Pharmacology (ISRTP)
Massachusetts Department of Environmental Quality Engineering (DEQE)
New Jersey Department of Environmental Protection/
Office of Science & Research (NJDEP/OSR)
Roy F. Weston, Inc.
Southern California Edison
The Environmental Institute–University of Massachusetts
University of Massachusetts–Division of Public Health

In addition, we also express our deepest appreciation to the members of the Scientific Advisory Committee who gave of their valuable time to provide guidance and encouragement. They include:

Bruce Bauman . . . . American Petroleum Institute
Ronald Brand . . . . EPA/Office of Underground Storage
Tanks
David Chen . . . . American Petroleum Institute
Peter Conlon . . . . Association of American Railroads
James Dragun . . . . The Dragun Corporation
Rick Helms . . . . Edison Electric Institute/Utility Solid
Waste Activities Group
Steven Jinks . . . . EA Engineering, Science & Technology,
Inc.
Barry Johnson . . . . Agency for Toxic Substances and Disease
Registry

Kathleen Jones.........Roy F. Weston, Inc.
Dorothy Keech.........Chevron
Mary McLearn ....Electric Power Research Institute
Gordon Newell ....Electric Power Research Institute
Robert Tucker ....New Jersey Department of Environmental
                       Protection/Office of Science & Research
Ed Williams...........Department of Energy

Edward J. Calabrese is a board certified toxicologist who is professor of toxicology at the University of Massachusetts School of Public Health, Amherst. Dr. Calabrese has researched extensively in the area of host factors affecting susceptibility to pollutants, and is the author of more than 240 papers in scholarly journals, as well as 10 books, including *Principles of Animal Extrapolation, Nutrition and Environmental Health,* Vol. I and II, and *Ecogenetics.* He has been a member of the U.S. National Academy of Sciences and NATO Countries Safe Drinking Water committees, and most recently has been appointed to the Board of Scientific Counselors for the Agency for Toxic Substances and Disease Registry (ATSDR). Dr. Calabrese also serves as Chairman of the International Society of Regulatory Toxicology and Pharmacology's Council for Health and Environmental Safety of Soils (CHESS).

Paul T. Kostecki, a Senior Research Associate/Adjunct Faculty in the Environmental Health and Science Program at the University of Massachusetts, Amherst, received his PhD from the School of Natural Resources at the University of Michigan in 1980. He has been involved with risk assessment and risk management research for contaminated soils for the last five years, and is coauthor of *Remedial Technologies for Leaking Underground Storage Tanks* and coeditor of *Soils Contaminated by Petroleum Products* and *Petroleum Contaminated Soils,* Vol. I. Dr. Kostecki's yearly conference on petroleum contaminated soils draws hundreds of researchers and regulatory scientists to present and discuss state-of-the-art solutions to the multidisciplinary problems surrounding this issue. Dr. Kostecki also serves as Managing Director for the International Society of Regulatory Toxicology and Pharmacology's Council for Health and Environmental Safety of Soils (CHESS).

Contents

PART I
STATEMENT OF THE PROBLEM

PART II
STATE OF RESEARCH

PART III
ANALYTICAL AND ENVIRONMENTAL FATE

PART IV
REMEDIAL OPTIONS

PART V
CASE HISTORIES

PART VI
RISK ASSESSMENT AND RISK MANAGEMENT

PART VII
REGULATORY CONSIDERATIONS

PART I

Statement of the Problem

CHAPTER 1

Environmental Regulations Place
New Responsibilities on Property Owners

Byron D. Taylor

Since the discovery of the Love Canal's toxic waste problem in the mid-1970s, public and legislative awareness of the hazardous waste issue has grown dramatically. In response to the overall issue, Congress adopted the Resource Conservation and Recovery Act (RCRA) in 1976 to regulate the generation, transport, and disposal of hazardous/toxic wastes. Within four years, however, it became apparent that neither RCRA nor other existing environmental laws adequately addressed the problem of active or inactive hazardous/toxic waste. Therefore, Congress adopted the Comprehensive Environmental Response, Compensation and Liability Act of 1980 (CERCLA), also referred to as "Superfund."

Congress imposed strict liability on a broadly defined group of landowners, transporters, and generators of hazardous waste for the cost of cleaning up hazardous waste sites. The liability includes responsibility for technical studies, actual cleanup costs, and administrative and enforcement costs, as well as damages to certain natural resources. Liability may be imposed on responsible parties either after the government itself has cleaned up the site or by a court order requiring responsible parties to clean the area. Liability is broadly defined to include all persons and is joint and severable, meaning that damages can be recovered from any one or more of the responsible parties.

As originally enacted, the Superfund law raised many questions, but the courts were quick to deal with the law's ambiguities. In short, the courts found that the liability under the act was strict (current owners and operations may be held accountable for the cost of cleanup—actual responsibility for the problem does not have to be shown). Also, the courts stated that the necessary proof of causation linking the defendant to a particular waste was relatively loose.

The Cost of Cleaning Up Your (or Someone Else's) Act

Because the Superfund law made an owner of such a facility liable for cleanup costs for past activity on the site, it raised serious questions for lending institutions and commercial developers. For example, in May of 1986 the federal court in Baltimore held Maryland Bank & Trust Company potentially liable for cleanup costs of more than $550,000 on property for which it was the mortgagee in the amount of $335,000 and which it had purchased following foreclosure for $380,000. Although the act contains an exception for lending institutions whose ownership is related solely to the protection of its security interests, the court concluded that the bank's foreclosure on the loan and ultimate acquisition of the property at foreclosure made it potentially liable as an owner of the property.

A few months after this decision, Congress amended the law by adopting the Superfund Amendments and Reauthorization Act of 1986 (SARA), which became effective on October 17, 1986. The rather complex amendments sought to clarify the obligations of commercial developers and lending institutions under the federal hazardous waste laws. The act makes it clear that real property owners may be held liable for all costs of cleaning up onsite hazardous substances unless they can prove they can satisfy the standards for the "innocent landowner" immunity.

To qualify for the immunity, any party to a land purchase agreement or financing agreement that provides for title transfer in the event of default must be able to show all of the following:

1. The release of hazardous substance was caused by an act of God or a third party.
2. The property was acquired after the hazardous waste had been disposed of.
3. There was no knowledge and no reason to know that any hazardous substance had been disposed of onsite.
4. "Due care with respect to the hazardous substance concerned" was exercised.
5. "Precautions against foreseeable acts or omissions of any third party" were taken.

The legislation links cleanup liability to a broadly defined category of parties. They are:

1. The current owner or operator of the site where the hazardous substances were deposited;
2. The owner or operator of the site at the time the hazardous substances were deposited;

3. Generators of hazardous substances, who arranged for disposal;
4. Transporters who accepted hazardous waste for transport and selected the disposal site.

The definitions combined with the joint and several aspect of the law infers that landlords (owners) and tenants can be equally liable for cleanup costs. Therefore, it stands to reason that a tenant which disposed of a hazardous material, intentionally or not, would create a cleanup liability for itself and the landlord.

The major reason for this increased attention is the national concern over the contamination of private and municipal wells and springs. The EPA estimates that 50 percent of the U.S. population depends on potable groundwater. In rural areas, the estimate is closer to 95 percent. Waste disposal sites and leaking underground storage tanks are the primary sources of such contamination.

This legislation, in combination with new federal regulations governing underground storage tanks containing petroleum and hazardous substances which were issued September 23, 1988, create additional concerns for landowners. The EPA published Underground Storage Tank (UST) regulations in the Federal Register (40 CFR Part 280 and 281). Section 9003 of RCRA requires EPA to establish requirements for leak detection, leak prevention, financial responsibility and corrective actions for all USTs containing regulated substances. The effective date of the regulations was December 22, 1988. These regulations in combination with the required state implementation regulations place ongoing requirements on virtually every owner of a property which contains a UST.

Who Should Be Wary?

Businesses likely to feel the greatest impact of these regulations include industrial operations, industrial and agricultural terminal operations, service stations, and anyone buying or selling property with underground storage tanks.

The EPA estimated there are currently 2.5 to 3 million underground storage tanks that are used to store petroleum and hazardous substances in the United States. Surveys indicate that up to 30% of the existing tanks leak, with the likelihood of leakage increasing as tanks age. About 90% of leaks are caused by corrosion, and more than 50% of those originate in piping associated with the tanks. Corrosion occurs because most of the tanks and piping are fabricated of bare steel, which is prone to corrosion in the subsurface environment.

Based on these statistics, proposed regulations mandate steel tanks either be fitted with cathodic protection or be replaced with fiberglass tanks. The regulations will also govern the removal and closure of existing tanks and the installation and operation of new underground tanks. In addition, they will require substantial upgrades for existing tanks, including leak detection systems within three to five years, and corrosion protection, spill prevention, and overflow protection within ten years.

To real estate owners, developers, lenders, and practitioners, this entire body of regulations and associated state statutes have created and will continue to create obstacles in all forms of transactions. Not only are the primary responsible parties affected but innocent, adjacent landowners are affected as well. As stated earlier, this innocent, adjacent landowner may be excluded from liability; however, the fact remains that his property must be cleaned up prior to finalizing any transaction. In actual practice this innocent party will in all probability undertake the cleanup and initiate action against the source of the contamination.

The litany of examples of the implementation of these regulations is creating serious reservations on the part of developers and major corporations before entering into a real estate contract. Instances where leaking tanks were discovered after a property had been purchased and development commenced, have led to serious financial concerns and bankruptcy in some cases. Additionally, the tentacles from one of these problems are so lengthy and entangled that a solution is usually the resultant of extended litigation and extensive costs.

In many cases, even though the wastes have been removed and the property now shows a clean bill of health, their effects linger. The taint of hazardous waste will be part of the appraisal of such a parcel of property for recorded history.

As the marketability of such contaminated properties is impaired, logically lenders are also understandably aware of contaminated properties. If a bond insuring cleanup costs can be secured, as is the case for many mildly contaminated properties, a major obstacle has been overcome. However, some lenders refuse to advance funds for such a property, bond or not, strictly as a matter of policy. Usually a limited number of lenders can be found for mildly contaminated sites.

There is virtually no chance of obtaining mortgage financing for a seriously contaminated property. Many lenders are still formulating their policies on dealing with contaminated properties.

With the possibility of lenders being held directly liable for cleanup costs, via foreclosure, as well as the subordination of their lien to state and federal "super-liens," mortgagors can expect increased attention, on the mortgagee's part, to potential environmental risks. Environmental risk assessments, or audits, will become as commonplace as title insurance when applying for a mortgage. Loan documents will also reflect this concern to make the borrowing principal fully liable for all costs in connection with environmental hazards. Afterwards, lenders may become involved with the compliance with local environmental laws and regulations as well as whatever financial control is necessary to preserve the value of its collateral.

Protect Yourself!

Buyers and sellers can protect themselves by a relatively few techniques. First, they should state in each and every contract who will pay for and be responsible for an environmental problem. This contractual affixing of responsibility is only as certain as the financial ability of the parties may reflect.

Buyers should also obtain warranties or representations from sellers that establish there is no known contamination, that the seller and its agents have done nothing to contaminate, or if that is not the case, they should disclose the actual situation. Sellers retaining an interest in the property should also ensure that nothing will be done to the property that will make them liable for cleanup in the future. Warranties, covenants, and indemnifications are a way to provide some protection in the contract itself.

The only method of protection which provides a level of "due diligence" that is generally unchallengeable involves the undertaking of environmental audits and risk assessments. These examinations can cost from a few thousand dollars up to $100,000. Environmental checks for contamination should include a review of applicable federal and state laws as well as county and municipal environmental ordinances. In many areas local officials maintain lists of properties that may have contamination problems. A search of those records may warn a buyer if the property is contaminated or located near a contaminated property.

For companies maintaining ownership of sites where there is the potential for inadvertent release of hazardous or toxic materials, continuous monitoring and testing is the only insurance against a significant reduction in land values. Once a property has been determined to have been contaminated, its worth is permanently minimized. Therefore, continuous monitoring and testing and requisite cleanup is the prudent approach for the maintenance of land value and future marketability.

CHAPTER 2

Providing Environmental Impairment Liability Insurance Coverage

John J. Metelski and William P. Gulledge

STATEMENT OF THE PROBLEM

The insurance industry has for many years protected society from the risks associated with both everyday activities and extraordinary events. In recent years, the accountability of individuals and corporations for the consequences of their actions has increased, challenging the insurance industry to create greater capacity while simultaneously reducing the risks of catastrophic loss. The threat to the environment has become one of society's greatest challenges, and once again legislators as seen in this year's political process, have turned to the insurance industry for assistance.

As with other risks in the past, environmental regulators asked the insurance industry to provide financial protection for damages caused by pollution incidents. Unlike other new liabilities, the insurance industry faces two major road blocks to the development of pollution liability insurance coverage:

1. the potential liability for gradual pollution coverage under existing general liability coverages using "occurrence policy forms" intended to cover only sudden or accidental pollution events; and
2. the technical and legal inability of insurers to demonstrate to the courts

when a pollution event started and when it should trigger coverage under an insurance policy.

First, insurance companies issue one of two types of third party damage insurance contracts: an "occurrence" policy or a "claims made" policy. The difference between the two is that a claim can be presented against the occurrence policy at any time as long as the occurrence that caused the damage occurred during the policy period. The advantage to the insured is that it makes no difference when the discovery of damage is identified by the damaged party; there is insurance affordable under the policy that was in force at the time the occurrence took place. In most types of liability situations, the period between the occurrence and discovery of damages is short or at least tolerable from the underwriter's perspective. An example of a "short tail" discovery period is a car accident (the "occurrence") causing bodily injury or property damage (the "damage to a third party"). The discovery period is nonexistent because the two events can be visually associated with one another. An example of an occurrence not clearly associated with third party damages "long-tail," is a gradual leak of an underground storage tank and the effects of the contaminants on property and health of exposed individuals.

A "claims made" policy requires a claim to be reported during the policy period. Some "claims made" policies provide retroactive coverage for events that may give risk to damages but are not known to the insured at the inception of the "claims made" policy. A standard practice among "claims made" policy insurers is to provide coverage for events occurring after the inception date of the first policy.

Insurers of general liability coverage, that up to 1974 provided coverage for pollution occurrences, used the occurrence form exclusively. After 1974, general liability insurers attempted to exclude gradual pollution occurrence and provide coverage for only sudden and accidental pollution occurrences. This limitation of coverage has been interpreted by courts to varying degrees in accomplishing its intent.

The combination of the use of the "occurrence" policy forms that provided large limits of liability (sometimes hundreds of millions of dollars) for unlimited time periods, and the concern over the failure of the pollution exclusion created a reluctance by major insurance companies to provide a pollution liability insurance product.

Secondly, pollution liabilities are ongoing problems that require a high level of technical skill to determine the condition of a property before it is insured. The possibility of having more than one policy limit of liability available on one pollution occurrence is made a reality through the "triple trigger" doctrine adopted by courts. This trigger includes: separate occurrences provided by the start of the contamination, the manifestation of effects within a third party and the discovery of damages by the third party.

Under an occurrence form, an insurance company may pay three limits of liability on one pollution incident for a coverage they never intended to provide.

Compounding the problem, many insurance carriers were and still are, technically incompetent to assist insureds with the most basic pollution incident.

In the late 1970s and early 1980s many insurance companies responded to the increased demand for Environmental Impairment Liability (EIL) insurance. Their efforts to enter this technically challenging insurance field were less than successful. Failure of most EIL programs can be contributed to the following:

- providing blanket coverage for all owned sites and all nonowned disposal sites
- providing unlimited retroactive coverage
- providing unlimited defense cost in addition to limits of liability
- not administrating claim deductibles
- not requiring applications on each insured site
- not requiring the signature of a competent authority verifying technical information disclosed in applications
- not attaching application forms to the policy contract to form a part of the policy
- using inadequate application forms and underwriting criteria providing limited information
- not using environmental risk assessment firms or personnel capable of adequately assessing environmental risk
- not requiring environmental monitoring data
- using property or casualty in-house staffs for environmental risk assessments, or using technical or personnel unfamiliar with insurance practices
- not providing ongoing loss control during the policy period
- using legal services inexperienced in environmental law
- not designing claim defense strategies or developing claims administration capabilities
- insuring extremely high risk industries
- broadly accepting accounts by industry groups rather than on a facility-by-facility basis.

Hindsight is always 20/20 and it is easy today to look back at what appears to be obvious mistakes in accepting risks that caused large losses and the severe contraction of the EIL insurance. But rather than criticize the efforts of others, these efforts can provide great insight into the structuring of future EIL programs that can help them survive the first few critical years of underwriting that are critical to any new insurance program.

A basic requirement for the survival of any new insurance program is profitability. The providers of EIL insurance should not have to make apologies for achieving underwriting profits. The rewards for bearing risks on behalf of others are very unpredictable. Hopefully the company's input to the insured on the following will have helped to prevent future environmental damages:

- establishment of emergency spill procedures
- training of personnel
- limiting quantities of hazardous material stored and property management of these materials
- direct disposal of wastes to proper disposal sites
- modification, if feasible, of manufacturing procedures to limit generation of hazardous wastes or hazardous substances
- recycling and treatment of hazardous wastes
- maintaining proper permits
- identifying waste streams
- establishing inventory controls on hazardous materials
- identification of all above- and underground tanks onsite and their condition
- establishment of clear lines of authority and responsibility for hazardous wastes disposal and hazardous materials handling

A responsible EIL insurer should be able to demonstrate to the insured certain capabilities such as:

- ability to pay claims
- willingness to pay claims
- national cleanup contractor network to respond to pollution incidents
- national environmental law network
- national environmental consultant network for underwriting and claims assistance
- competent in-house technical and claims administration staff
- legislation and regulations tracking
- quality reinsurance support

More and more EIL insurers are recognizing the keys to providing quality EIL insurance services. The Environmental Protection Agency's financial responsibility requirement for owners/operators of underground storage tanks will increase the available EIL options. The problems created by past EIL coverages will most likely continue for large insurance carriers leaving EIL development to specialty insurance entities.

CHAPTER 3

Issues Affecting Contaminated Soils Management: The Railroad Industry Perspective

Peter C. L. Conlon

The railroad industry has a keen interest in research aimed at understanding the behavior of diesel fuel spilled into the soil. We are looking for cost-effective ways to eliminate any health and environmental hazards associated with these fuel residues. About 35 years ago, the railroad industry adopted diesel-powered locomotives as their primary means of motive power. Since then, the industry has used some 85 billion gallons of diesel fuel. Some of that fuel has leaked from storage, distribution, and dispensing systems, and some has spilled in derailments. The fuel crises of the 1970s put a stop to most of the leaks, but derailments continue to occur. As a result, the soil at many fueling facilities and accident sites has been contaminated.

As the railroad industry's trade association, the Association of American Railroads (AAR) conducts research for its member companies. Their interest in finding cost-effective solutions to soil contamination problems led to the establishment of a fuel spill research program at AAR. We recognized that despite the petroleum exemption in the Comprehensive Environmental Response, Compensation, and Liability Act (CERCLA), fuel spills in soil would eventually become regulated.

The issues we were concerned about were (1) what happens to fuel spilled onto the ground, (2) what environmental and health risks does remaining fuel pose after free liquid has been removed, and (3) how can these risks be eliminated with the minimum cost and disruption to the environment.

Early in our research, we found that information about the chemical composition of diesel fuel was scarce. Also, very little was known about what happened to fuel once it was spilled onto the ground. This led us to direct our efforts toward these topics. We asked Dr. Donald Mackay of the University of Toronto to investigate the chemistry of fresh fuel and fuel samples recovered from spill sites. His task was to describe the fuel's chemistry before and after exposure to the natural environment and to deduce what environmental factors affected the recovered fuel.

Studies performed by Mackay indicated that spilled fuel undergoes significant changes in the environment. We have termed this "weathering." Weathering is a complex process involving three major processes: evaporation, dissolution, and biodegradation. Mackay estimated the effects of each of these natural processes acting on fuels by comparing the fresh fuels with the recovered samples. The comparison indicated that these samples showed an average loss of 75% of their straight chain alkane components. This suggests that considerable biodegradation took place because normal alkanes are easily degraded by microorganisms. This was measured by comparing the ratios of pristane and phytane to C_{17} and C_{18}, respectively. It was difficult to distinguish between the effects of evaporation and dissolution, but evaporation seemed to be the more significant process. The amount of fuel that dissolved was very small, probably less than 1%, and was undetectable as a composition change. Fresh oil solubility in water averaged 3 mg/L. Weathered fuels averaged 1 mg/L, largely due to evaporation and dissolution of volatile water-soluble compounds.

A problem with this study was that the recovered samples were of undetermined age and origin, so we could not be sure just how quickly the environmental processes acted on the fuel. This is an unavoidable flaw in conducting research with samples from old spill sites. We tried to correct this with subsequent research on fresh fuels subjected to controlled weathering, but the results of this research were inconclusive due to the design of the experiment. Nonetheless, the results of Mackay's research suggested that diesel fuel in soil could pose a threat to groundwater as long as the water-soluble component remained. We believed the main objective of further research was to examine ways to remove the water-soluble component from the fuel in the soil with a minimum of excavation and offsite soil disposal.

We began a field study in 1986 to try some low-technology biodegradation methods including in situ land farming and subsurface treatment involving the recycling of water with common materials (fertilizer and hydrogen peroxide) through the soil matrix, somewhat like a trickling filter. As of the end of 1988, we have reached some conclusions.

Biodegradation of diesel fuel in soil works. It occurred more rapidly in the plot where we circulated water with fertilizer and hydrogen peroxide through the soil. The simple land farming plots exhibited a loss of fuel as well, but the chromatograms indicate that this occurred more slowly. Our results are inconclusive, however, because we have had some major problems with laboratory analysis. Commercial labs widely differ in their capability to produce consistent, reliable results. Our primary problem was difficulty in getting reproducible results with the C_{17}/pristane and C_{18}/phytane ratios. Though it can clearly be seen in the gas chromatograms that degradation is occurring, as the peaks get smaller, errors in determining the ratios become larger.

There is also apparently very considerable variation in degradation within the $10' \times 10'$ plots. This suggests that the use of these ratios to quantify degradation may not be workable. We are now examining other analytical approaches: total fuel loading in the soil and water solubility. The former is a straightforward analysis but tells us nothing about the effects on groundwater. The latter is valuable because it can help quantify the risk of water contamination from fuel-contaminated soil. Thus, we believe that it is very important to develop a simple, reliable analytical method to determine the water solubility of fuel in a soil matrix.

In summary, we see several issues that are important to our ability to economically manage the problem of fuel-contaminated soils. We need to be able to quantify the rate of evaporation, dissolution, and biodegradation on spilled fuel in soil. We would like to see scientific and regulatory agreement on the risks posed by fuel-contaminated soil. Finally, we would like to encourage further research on simple, low-cost risk-reduction procedures that can be applied at operating facilities such as railroad yards.

CHAPTER 4

Implications of Dealing with Real Estate-Based Cleanup Statutes

John L. Greenthal and Mark P. Millspaugh

STATE HAZARDOUS WASTE STATUTES WHICH IMPACT ON REAL ESTATE TRANSACTIONS

There are several federal statutes which impose requirements for the reporting of spills and releases of hazardous substances and petroleum. These statutes, which include the Resource Conservation and Recovery Act, the Comprehensive Environmental Response, Compensation, and Liability Act ("CERCLA"), Title III of the Superfund Amendments and Reauthorization Act, the Toxic Substances Control Act, and the Clean Water Act, have different scopes as to the materials covered and the extent to which spills or releases of such materials must be reported. As a result, situations can arise where no reporting is required under any federal statute or regulation, and these circumstances can create problems for subsequent owners of the contaminated property inasmuch as they may be unaware of the true environmental conditions at the property.

Partially in response to such gaps in the federal statutes, some states have adopted controls which go beyond those found in the aforementioned federal statutes and their state analogs. There are, basically, three varieties of such state legislation.

The first of these imposes cleanup responsibilities at the time of transfer of industrial property or closure of an industrial facility. Such legislation has been enacted in New Jersey and Connecticut, among others, and has been considered in a number of other states, including New York.[1] Although different in some respects, all such statutes are modeled after New Jersey's Environmental Cleanup Responsibility Act ("ECRA").[2] The purpose of these statutes is to ensure that industrial property is not sold or transferred, or operations terminated thereon, until it is free from serious contamination by hazardous materials and/or petroleum.

In order to attain this goal, the current owner must, before transfer of the property or cessation of operations, submit either a negative declaration (i.e., a formal certification that no unauthorized releases have occurred at the site or that discharges which did occur have been adequately cleaned up) or a compliance plan outlining the cleanup procedure and ensuring its completion. These requirements usually apply not only to the sale of real property, but also to the termination of a lease, transfer of a controlling interest in the corporation owning the property, transfer of title through foreclosure, and the initiation of bankruptcy proceedings.

Penalties for failure to comply with ECRA's provisions may be very stiff. The sale or transfer may be declared void, the purchaser may recover damages, and the seller may be liable for civil penalties and cleanup costs.[3]

A second type of state legislation imposes notice requirements which go beyond the reporting requirements contained in federal legislation. These statutes are designed to afford a means of protecting a purchaser's interest in avoiding the costs of cleaning up latent contamination, but they represent a less drastic approach than that taken by the ECRA-type statutes described above.[4] Rather than imposing cleanup responsibilities, statutes of this kind require that sellers notify purchasers that hazardous materials and/or petroleum are present or have been disposed of at the property, thereby providing the purchaser with the opportunity, before the closing, to negotiate with the seller over responsibility for cleanup obligations.

The notice usually must be in the deed or otherwise recorded in the land records so that all subsequent purchasers are notified of the contamination. Penalties for violations of the notice provisions range from civil and criminal penalties to liability by the violating seller for the costs of cleanup. States which have this type of notice statute include Massachusetts,[5] Minnesota,[6] Pennsylvania,[7] and West Virginia.[8]

The third type of legislation in this area involves "superliens." Six states have superlien statutes: Arkansas,[9] Connecticut,[10] Massachusetts,[11] New Hampshire,[12] New Jersey,[13] and Tennessee.[14] These statutes provide that the state has a first-priority lien on contaminated property which is subject to a cleanup requirement. This "superlien" often has the effect of wiping out previously filed liens, including mortgages, since the property value is usually less than the cleanup costs. The federal lien imposed by the 1986 CERCLA amendments is not as powerful as state superliens because the federal lien does not take priority over previous purchasers and creditors.[15]

MARKET-DRIVEN FORCES

Motivated largely by the anticipated impact of this legislation, as well as by CERCLA and analogous state statutes which impose broad cleanup liability on parties only remotely connected to property, various forces in the economy are beginning to scrutinize any potential relationship to contaminated property.

For one thing, the demands made by prospective purchasers in order to guarantee that they are not buying contaminated land will only intensify in the future. Similarly, sellers may become more demanding so as to ensure that they will not be liable for subsequent acts, or failures to act, by their purchasers. For example, both the buyer and seller may know that the property is contaminated, and because of the timing involved in the transaction, it may not be possible for the seller to undertake remediation before the sale. The seller may, however, want to avoid the potential for ever-increasing remedial costs as the result of the purchaser's failure to undertake timely and effective cleanup measures, a failure which exacerbates the environmental problem. As a result, the seller may insist on staying involved after the sale to supervise remediation by the purchaser. The seller can then ensure that cleanup is done completely and correctly and that he will not, thereafter, be liable.

In addition, under CERCLA, anyone making a secured loan to a past or present owner/operator of a contaminated piece of property may itself be liable for cleanup costs if it exercises enough control to, in effect, be an owner/operator itself.[16] As a result, lending institutions are becoming increasingly cautious about making commercial loans, and are carefully scrutinizing property before giving a mortgage. This is done to avoid being in a position where a choice must be made between writing off a bad loan due to contamination or foreclosing on the property and becoming liable as an owner/operator.[17] Additionally, environmental liability may severely impair the value of the property as security for the loan since cleanup costs, liability for damage to natural resources, and liability to private citizens affected by the contamination may easily exceed the value of the property.

The results of these concerns by banks are a reduction in the number of commercial loans that are given (as banks become less willing to take risks where the potential liability is high), an increase in the cost of obtaining a loan (environmental audits of property to check for contamination can be very expensive), and an increase in the amount of the loan itself if the bank chooses to pass the cost of the bank's investigative and remedial expenses to the mortgagor. Banks may also seek to obtain collateral other than real estate and to impose provisions in mortgages that enable the loan to be called if a risk arises. These forces represent a very potent control on industrial expansion and development as lenders attempt to balance their desire to preserve the collateral with their equally strong desire to avoid becoming a CERCLA owner or operator.

The insurance industry represents another powerful market force influencing industrial actions. Often, parties paying cleanup costs seek to be indemnified by their insurance carriers. However, in light of the vast potential liabilities connected to the cleanup of hazardous material and petroleum contamination, insurers are becoming increasingly hesitant to provide environmental impairment coverage.[18]

Another influential market force arises from the structure of corporations themselves. Although a parent corporation which owned or operated a subsidiary when the subsidiary caused contamination would probably be liable for cleanup costs, a parent corporation of a successor subsidiary which inherits a preexisting problem, but which did not cause it, may not be liable.[19] As a result, corporations may be reorganized and subsidiaries created where property is known to be contaminated and a parent corporation is attempting to shield itself from liability. Such an action, along with the increasing trend toward declarations of bankruptcy in order to avoid large-scale personal injury liability (so-called "toxic torts"), may have a profound impact upon the business dealings of an industry.

PREVENTIVE MEASURES AND RECOMMENDATIONS

In order to avoid the draconian effects of this emerging legislation and the economic forces driven by concern over the high costs of environmental liability, the uppermost management of industrial/manufacturing facilities must set an overall policy requiring proper maintenance of facilities and equipment as well as an overall good "housekeeping" policy, together with proper and thorough documentation of these activities. By doing so, those industries will be in a position to assure themselves an edge over their competition when the time comes to sell or transfer their property or business.

Historically, the organization of industrial and manufacturing facilities has worked against the development and implementation of such corporate programs. Environmental managers, charged with operation of treatment facilities, often have little or no say regarding manufacturing processes and production. Environmental managers must also compete with production departments for available funding.

In most industrial/manufacturing organizations, environmental managers are limited to dealing with matters at the "end of the pipe." However, in today's marketplace, environmental managers have a need-to-know regarding the in-plant uses and handling of petroleum products and chemical raw materials. Environmental managers must also be versed in production processes and by-products so that potential problem areas are identified and addressed before becoming serious liabilities. On the other hand, production managers, who are responsible for optimization of production processes and equipment, are pressured to minimize the cost of production in order to stay competitive in the marketplace. As such, there is a disincentive for production managers to budget funds for maintenance and housekeeping.

The pressures and policies requiring proper facility maintenance and house-keeping must come from uppermost management.

Corporate management must assume a leadership role by pulling together environmental control and production teams under an overall, coordinated policy designed to reduce or eliminate the vast long-term liability associated with contaminated sites.

The following section offers observations on problem areas affecting industries as the result of their routine activities, many of which are not subject to governmental requirements, and provides recommendations on how to alleviate them. At the same time, however, it is essential to bear in mind that adherence to permitting and environmental regulations is equally, if not more, important.

PROBLEM AREAS

Review of site assessment and investigative reports addressing a representative cross section of industrial plants, manufacturing facilities, and other commercial establishments which routinely use petroleum products or hazardous chemicals, indicates several common problem areas.[20] The following are frequently encountered sources of groundwater, surface water, soil, and building contamination.

Sewer Systems

Sewer systems are often overlooked, but have proved time and again to be a very serious source of subsurface contamination. Usually sewer systems are tributary to regulated treatment facilities and permitted point source discharges. However, the proper operation and maintenance of such sewers are rarely required by the treatment/discharge permit. Included in the category of "sewer systems" are the following facilities:

- process sewers which carry away wastewater and chemical process by-products from reactor vessels, air pollution control devices, and/or other sources of chemically contaminated wastewater. The concentration of chemicals found in such sewers is usually very high; percent level concentrations frequently being encountered.
- floor drains, building sumps, foundation drains, etc. Such facilities, especially those encountered in older industrial plants, may have ceased operation before the state and federal regulations governing treatment requirements and standards were put into effect.
- storm sewers, catch basins, storm inlets, etc., particularly those providing storm drainage in the vicinity of chemical and petroleum storage and loading facilities.

A review of several industrial facilities utilizing a wide variety of chemical raw materials, manufacturing processes, and process water and wastewater disposal methods indicates a pattern of environmental problems due to operation of sewer systems.[21] More often than not, underground piping leaks through numerous pipe defects and joints as the result of years of neglect, poor maintenance and repair, substandard construction, and chemical corrosion of the pipes.

It should also be noted that, in most states, sewer construction within a privately owned industrial facility is not subject to the review and approval of state or local government agencies as would be the case with public sewer facilities. Consequently, the original construction may be substandard, particularly for older sewers. Construction plans and as-built drawings are nonexistent for many facilities, and it is not unusual for some sewer construction and repairs to have been performed by plant maintenance personnel with little or no training.

Further, older sewers may not be tributary to currently used treatment facilities. Frequently, older floor drains and combined industrial/sanitary sewers are discharging to onsite, subsurface leachfields or septic systems long thought to be abandoned or unknown to current property owners.

Sewers and storm drainage facilities are often found to be migration pathways for contamination at a facility. Storm drainage facilities frequently receive runoff from parking areas and raw material loading and unloading areas. Dissolved chemicals and suspended materials are carried along with storm runoff. Additionally, it is not uncommon for groundwater infiltration into the storm drains and bedding materials to carry contamination into the pipes or along the sewer trench.

At several facilities contamination was so severe that separate phase product and dense nonaqueous phase liquids (NAPLs) were encountered. The assumed source of such extensive contamination was past onsite dumping and/or unreported bulk spills. After expending many thousands of dollars, the industries learned that their environmental problems were in fact due not to these assumed causes but rather to years of relatively minor releases through process piping, sewers, etc. Had these industrial facilities developed appropriate maintenance programs, such environmental problems and their associated liabilities may never have occurred.

Chemical Storage and Handling Facilities

Raw material loading/unloading and handling areas are often sources of contamination. Frequent, albeit small, losses or drippage can result over time in seriously high levels of soil and groundwater contamination if precautions are not taken to routinely collect such materials.

Chemical permeation of building floors, walls, and even ceilings is frequently encountered due to poor management practices. It is not uncommon to find the surfaces of concrete floors and walls so seriously contaminated that extensive decontamination or removal of walls, floor slabs, and other contaminated structures will be necessary prior to resale or transfer of the property.

Maintenance Facilities

Maintenance facilities such as garages and repair shops are well-documented problem areas, particularly motor vehicle repair centers where oil, engine coolant, and gasoline are routinely drained from the equipment. Indiscriminate use and spillage of cleaning solvents are also frequently encountered.

Many maintenance facilities are equipped with floor drains or sumps. Years of washing down the floors results in mobilization of contaminants to these drains and sumps. In addition, the seepage of oil, solvents, etc., through joints and cracks in floor slabs also results in extensive soil and groundwater contamination.

Parking Areas

Parking areas, particularly those in use over an extended period of time, are frequently found to be contaminated by gasoline, motor oil, and engine coolant leaking from automobiles. While the quantity of material lost per vehicle is fairly small, it is important to consider the number of vehicles and period of time a parking area is in use. Unpaved parking areas can be the source of continuing releases of oil and other chemicals to surficial soil and groundwater, as such materials are mobilized with each rainfall. Storm runoff from paved parking areas often transports contaminated sediments and dissolved chemicals through storm drainage facilities and sewers. Once in the sewer system, such contamination can present the same environmental problems previously mentioned.

RECOMMENDATIONS

Table 1 summarizes those problem areas frequently encountered in an industrial setting which are not traditionally subject to governmental regulations or permits. Table 1 also presents specific recommendations for identifying and minimizing operational releases.

Further, it is imperative to fully evaluate and assess a piece of property before any real estate transaction is finalized. Such an environmental audit should include (1) obtaining information on the site's history, use, and the conditions of the plant and property; (2) reviewing regulatory records, including environmental permits and monitoring, sampling and accident reports; (3) performing a detailed site review; (4) ascertaining the maintenance and repair history and getting a full description of leaks, spills, and any cleanup or remediation efforts; (5) obtaining a list of citations and alleged violations of environmental laws and regulations; and (6) procuring complete records of past, current and potential litigation and administrative proceedings relative to releases, leaks, or spills on the property.

In addition, it would be prudent to consider including clauses in the real estate sales contract which (a) require an environmental audit with the right to terminate the contract after an audit and (b) contain limited representations and warranties regarding the parties' knowledge of actions affecting the environmental character of the property.

Provisions concerning indemnification may also be crucial should cleanup be required in the future. The parties may desire that either the buyer or the seller indemnify and hold the other party harmless. If both buyer and seller are aware that a site is contaminated, it is especially important to include a contractual provision delineating who will pay for acceptable remediation of the site. This could be included as part of an indemnification clause.

At the same time, it is essential that an industry maintain accurate and professionally credible records of all of its activities relative to environmental contamination. When spills or releases of hazardous substances or petroleum occur, their nature and extent should be documented thoroughly as should all measures undertaken to clean them up. Especially important is the performance of an environmental audit by both the seller and the purchaser at the time of transfer of property so that a comprehensive and reliable "snapshot" of environmental conditions can be preserved following the transfer. In this way, both the seller and the

Table 1. Frequently Encountered Problem Areas and Recommended Preventive Measures.

Potential Problem Areas	Recommended Preventive Measures
SEWER SYSTEMS AND UNDERGROUND PIPING	
Leaky pipe joints, Pipe defects, Unknown locations, Unknown connections, Unknown pipe destinations	•Map all underground piping. Investigate and verify sewer connections and destinations (smoke test, dye trace, etc.) •Inspect all manholes, valves, etc., for leaks, sediments, and proper operation •Initiate program to properly clean and maintain sewers •Video inspection of pipes noting all defects, leaky joints, poor flow conditions, etc. •Repair and replace defective pipes, pressure grout leaky pipe joints
Chemical migration through pipe bedding material	•Intercept and collect materials migrating through bedding for proper treatment and disposal
FLOOR DRAINS AND FOUNDATION DRAINS	
Receive contaminated floor wash, spills	•Determine/verify that floor drains are connected to sewers and not dry wells •Reduce/eliminate discharges to floor drains •Possibly connect floor drains to temporary holding tank so collected spills may be properly managed

Table 1. (Continued)

Potential Problem Areas	Recommended Preventive Measures
SEPTIC SYSTEMS, LEACHFIELDS, ETC.	
Older systems may have received chemicals; soils and groundwater around leachfield may be saturated with chemicals resulting in a continuing release of chemicals and petroleum products	• Identify and map all septic systems • Test pits and soil borings to determine if leachfield is contaminated • Remove/eliminate soils found to be contaminated • Investigate/verify possible sewer connections
Systems thought abandoned may still be receiving wastewater flow	• Eliminate any identified connections
CHEMICAL STORAGE AND HANDLING FACILITIES	
Spills and sloppy use cause contamination of building floors, walls, soil, and groundwater	• Establish strict policy for chemical storage, handling and use • Prompt cleanup of spillage • Assure all chemical handling areas are paved and curbed to facilitate cleanup and removal of spills • Seal floors and walls in contact with chemical raw materials • Identify previously impacted areas and implement appropriate remedial measures to prevent further migration of previously spilled chemicals
MAINTENANCE FACILITIES	
Oil and solvents used in maintenance as well as oil, gasoline, and coolant drained from motor vehicles are mismanaged, spilled, etc.	• Investigate all current and former maintenance facilities to evaluate handling and disposal practices • Assure that building sumps and floor drains are not sources of continuing releases of chemicals
PARKING AREAS	
Drippage and leaks of gasoline, motor oil, and coolant can enter storm drainage facilities and/or contaminate surficial soils	• Equip storm drainage facility with oversized catch basins and/or grit chambers so that contaminated sediment can be removed • Segregate storm drainage so that any contamination associated with parking facility is not misinterpreted to be associated with industrial/manufacturing facility • Pave where possible to reduce potential for surface soil contamination
Oiling of road/parking area for dust control	• Identify areas where oil was applied, identify source of oil used • Verification sampling to assure oil was not contaminated

purchaser will be able to account for the contamination attributable to his activities, thereby limiting costly and time-consuming disputes in the future over who is responsible for environmental problems at the facility.

The foregoing suggestions not only apply to real estate sales contracts, but may also be useful for adaptation into lending and insurance agreements. The goal in any such transaction, be it a transfer of property, loan, or insurance, is to protect the parties. In order to effect this, it is vital that steps be taken to make all parties aware of the character of the property and to address all contingencies which may occur should environmental contamination be involved.

CONCLUSION

Environmental contamination is a significant problem in today's world. Government is attempting to control and remedy contamination by enacting statutes and regulations which have a profound and wide-reaching effect. In turn, a number of market forces are acting to impose further controls on industries' actions relative to environmental contamination. Although there are pros and cons concerning both statutory and market controls, it is evident that both will continue to work, probably with increasing force, on business and real estate transactions.

Corporate policies and management practices which stress preventive maintenance and proper environmental housekeeping will substantially reduce, and possibly eliminate, economic and operational disruption when the time comes to sell or transfer an industrial facility. It is also crucial that all parties to transactions involving real estate be aware of the condition and extent (or lack) of contamination at a site, an awareness achieved only through the performance of environmental audits and assessments of property before transactions are finalized. Undertaking and completing these types of protective and preventive measures will minimize the disruption caused by environmental contamination.

REFERENCES

1. Dean, J. P., "How State Hazardous Waste Statutes Influence Real Estate Transactions," *Environ. Reporter*, 18:933–935 (1987) (hereafter "Dean").
2. N.J. Stat. Ann. 13:1K-6 *et seq.* (1984).
3. N.J. Stat. Ann. 13(a), (c).
4. See generally Dean.
5. Mass. Gen. Laws Ann. Ch. 21c.
6. Minn. Stat. Ann. § 115B.16.
7. 35 Pa. Cons. Stat. Ann. § 6018.405.
8. W. Va. Code § 20-5E-20.
9. Ark. Stat. Ann. §§ 82-4708, 82-4720.
10. Conn. Gen. Stat. Ann. § 22a-452a.
11. Mass. Gen. Laws Ann. Ch.21E, § 13.

12. N.H. Rev. Stat. Ann. § 147-B; 10-6.
13. N.J. Stat. Ann. 58:10–23.11f(f).
14. Tenn. Code Ann. § 68-46-209.
15. 42 U.S.C. § 9607(f)(1), (3).
16. Murphy, M., "The Impact of 'Superfund' and Other Environmental Statutes on Commercial Lending and Investment Activities," *Bus.Lawyer*, 41:1133–1163 (1986).
17. Idem.
18. Tasher, S. A., J. P. Dean, S. Oster, and B. F. Kaufman, *Environmental Laws and Real Estate Handbook*, Government Institutes, Inc. (1987).
19. Redburg, M. L., "General Environmental Law Considerations Affecting Business Transactions," The Impact of Environmental Regulations on Business Transactions 1987: Real Property Transfers and Mergers and Acquisitions, pp. 99–133 (1987).
20. See, e.g., "New York State Inactive Hazardous Waste Site Remedial Plan Update and Status Report," New York State Department of Environmental Conservation, October 30, 1987, and "Inactive Hazardous Waste Disposal Sites in New York State: An Annual Report, A Joint Report of the New York State Departments of Environmental Conservation and Health, December 1987.
21. Idem.

PART II

State of Research

CHAPTER 5

Current Issues in Management of Motor Fuel Contaminated Soils

Bruce Bauman

INTRODUCTION

There is little doubt that the issue of petroleum hydrocarbon contamination of soil has become a topic of great interest to many during the past several years. While we all recognize that there is a wide variety of types of petroleum hydrocarbon soil contamination, my comments are directed, as were most of the papers presented at the conference, toward soil contamination caused by fuels such as gasoline, diesel, and heating oils.

Last year, when I shared my thoughts at the conference, my comments emphasized the following points:

- that soil is the most complex and variable portion of the biosphere, an elaborate and intricate physical and chemical environmental media,
- that soil is inherently different than the other two major environmental media, the fluids air and water,
- that several soil properties combine to provide a natural capacity to buffer and ameliorate the impacts of chemical contaminants (e.g., petroleum hydrocarbons),
- that petroleum hydrocarbons in soil can be considered to be relatively nonpersistent contaminants, due chiefly to their biodegradability and, for some compounds, their volatility,

- that decisions regarding cleanup requirements at sites with soil contamination should be based on the presence of risk to human health and the environment, and that the characterization of that risk is a function of the toxicity, mobility of the contaminants, and likelihood of human exposure.

Today those same themes will again be expressed, for they are as valid today as they were last year, and they remain key components to the development of a rational policy concerning soil contamination. I would posit that such a policy would be composed of five main issues:

- accurate, consistent, and effective measurement of site contamination
- characterization of the risks posed by the contaminated soil
- assessment of the need for corrective action
- determination of appropriate and realistic cleanup goals
- selection of a cleanup technology or methodology.

Further, any regulatory policy should reflect:

- an understanding of the interaction of soils and petroleum hydrocarbons
- protection of human health and the environment
- cost effectiveness
- timely response by both industry and regulatory agencies.

Soil contamination has quickly become a considerable environmental issue. This swift development, coupled with a generally inadequate scientific understanding of the interaction of fuels and soils, has created a multitude of inconsistent regulatory approaches. However, a significant volume of recent experimental and applied research has been conducted. Further, with the large number of field sites being addressed on an annual basis, there is much new information being developed that continues to improve our understanding of the nature of this type of contamination in such a rapidly evolving technical subject area. It is important to regularly stop and assess current policies and practices in the context of new research and experiences. This will promote the evolution of more uniform regulatory approaches that can be considered scientifically valid. Toward this desirable goal, in the following paragraphs I will consider two topics of particular interest and debate: characterization of the level of contamination (i.e., soil chemical analysis), and the development of cleanup standards for petroleum contaminated soil.

I would also like to introduce another consideration—that of the need to work toward a policy that recognizes that there is a point of diminishing return regarding expenditures that provide increased environmental protection (i.e., for each site, there is some point at which each additional dollar spent for continued removal of contaminants from soil provides only a marginal reduction of the risk presented by those contaminants). In order to allocate limited resources most effectively, and to simultaneously maximize reduction of risk, we must obtain an approximate understanding of where that point is.

COMPARING AIR, WATER, AND SOIL POLLUTION

Before addressing the principal technical components of this chapter, it is beneficial to place soil contamination in its proper perspective relative to other types of environmental contamination. It is important to recognize the inherent and significant differences in the risks posed by differing levels of soil contamination, in comparison to similar contamination of the other major environmental media, air and water. This difference has its roots in the fact that the soil, by its very nature is not a commodity that is normally breathed, ingested, or brought in contact with skin (i.e., it is not a mobile environmental media, at least in comparison to the fluids, air and water). For example, a concentration of 100 mg/Kg of volatile organic compounds (VOCs) in groundwater may be considered a significant risk to human health and the environment, primarily because of its potential to be consumed directly. Conversely, 100 mg/Kg of VOCs in soil will pose a much reduced risk, primarily because the probability of direct exposure (e.g., ingestion) is much lower.

ANALYTICAL ISSUES

A more basic problem, however, is how to accurately measure the amount of contamination in the soil. It has long been recognized that soil sampling and analysis presents a host of difficult technical obstacles that confound attempts to quantify the concentration of various soil contaminants.[1] This becomes more true when attempting to measure progressively lower concentrations (e.g., parts per million of VOCs). Some of the more important factors that can interfere with accurate soil analysis include:

- matrix effects created by the other nontarget soil compounds (including 'native' or natural organic compounds commonly found in soils),
- the volatile nature of some of the contaminants which can escape from the sample at several points during the sampling, handling, and analytical processes,
- the lack of analytical methods specifically developed for the characterization of soil contamination (e.g., many generic solid waste methods which are commonly used may fail to address the unique characteristics of the soil matrix),
- the ambient soil moisture content.

These analytical shortcomings have contributed to a proliferation of field techniques that provide a qualitative, or at best, semiquantitative analysis of contaminant concentrations. A strong impetus for the development of these field methods has been the need to quickly characterize site contamination, especially when corrective action decisions must be made in the field (e.g., during soil excavations). Among these field methods, many commonly employ some form of analysis of the volatile hydrocarbon vapors given off by the liquid fuel adsorbed to soil particles and pores. Although such data is inherently widely variable it can be of utility for certain qualitative judgments. These methods avoid the potential losses of volatile compounds from samples that may occur during transport, storage, and handling. However, one must be aware of the analytical limitations, and exercise caution in the interpretation of results.

Two of the most common vapor methods are:

- the simple procedure of waving an air sampling device over a mound of soil and analyzing the vapor with a photoionizing detector (PID) or flame ionization detector (FID), and
- headspace methods, which involve placing a quantity of soil in a jar, allowing it to equilibrate for a short period of time, then sampling and analyzing the vapor in the headspace above the soil, again, with a PID or FID.

Obviously there is a need for standardizing these procedures as much as possible, although it is doubtful whether the first procedure ("wand-waving"), can ever provide more than a rough qualitative characterization because it is so error-prone. Headspace techniques are relatively convenient methods, but their accuracy and the true meaning of their results are unknown. There are no published studies that demonstrate a quantitative relationship between a measured headspace concentration and the mass of contaminant in the soil (either total petroleum hydrocarbons, or for BTXE). Insufficient rigorous study has been conducted to determine the ultimate utility of these field analytical methods but given their lack of scientific validation, their results must be subject to cautious and knowledgeable interpretation.

In most cases, the greatest concern at contaminated soil sites is soluble transport of contaminants to groundwater. Perhaps the ideal analytical approach would be leaching studies of the soils to determine how much of the BTXE might eventually reach the water table.[2] However, such studies are lengthy, complex, and costly. As an alternative it may be possible to provide estimates of leachable BTXE based on local soil concentrations. Research which would provide meaningful guidance would likely require correlative leaching studies for many different types of soils. The American Petroleum Institute is currently conducting a study which includes leaching of gasoline contaminated soil cores, which should be completed within the next 6–8 months. The results of this study should provide insight as to the potential for soluble transport of residual hydrocarbon to groundwater for some soil types.

An interesting recent Environmental Protection Agency (EPA) study also serves to provide estimates of possible hydrocarbon leaching. The study used the toxic characteristic leaching procedure (TCLP) to evaluate the potential leaching from fuel contaminated soil cores.[3] The TCLP extract (an acetic acid buffer) is not directly representative of the leaching of such soils by water (it is actually designed to simulate leaching conditions in a municipal landfill). The study results do, however, provide a clear demonstration of how soil properties strongly influence the leaching characteristics of motor fuel contaminated soil, and their potential environmental risks. Tables 1–3 contain the results of these experiments.

In this study, the EPA examined two soil types: a sandy soil (the Kaw), and a clayey soil (the Action). Each was spiked with gasoline concentrations of 10, 100, 1000, 5000, and 10,000 ppm; diesel fuel from 100 to 10,000 ppm; and #6 fuel oil—1000 to 10,000 ppm. The TCLP zero headspace extraction of these spiked soils occurred over an 18-hour period. There was a significant difference in the extractable BTXE between the sandy and the clayey soil. For example, the extract of the clayey soil originally spiked with 1000 ppm gasoline contained only trace amounts of benzene, while the sandy soil contained 241 μg/L (ppb) benzene. No benzene was found in the extract of the clayey soil spiked at 100 ppm gasoline, while the sandy soil extract contained 31 μg/L. Again, these are not definitive values for potential water leaching of BTXE from soils, but are illustrative of the considerable differences that will be present among different types

Table 1.   Target Analyte Concentrations in TCLP Leachates (μg/L) from Unleaded Gasoline-Contaminated Soils.

Analyte	Soil	Contaminated Soil Nominal Concentration (ppm)				
		10	100	1,000	5,000	10,000
Benzene	Action	NA	ND	TR	TR	109[a]
	Kaw	TR	31.6	241[a]	1,660[a]	2,380[a]
Toluene	Action	NA	TR	7.62	417	2,620
	Kaw	15.9	209	1,480	8,420	13,100[a]
Ethylbenzene	Action	NA	ND	6.38	402	1,210
	Kaw	4.88	46.9	329	1,800	2,530
m-Xylene	Action	NA	ND	22.2	1,220	3,460
	Kaw	13.5	129	876	4,520	6,560
o- and p-Xylene	Action	NA	TR	28.6	1,340	3,670
	Kaw	12.7	131	899	4,600	6,720

NA = Not analyzed.
ND = Analyte not detected.
TR = Analyte detected at trace levels (below the quantitation limit).
[a]Analyte concentration in the TCLP extract exceeds threshold level for determination of hazardous wastes.

Table 2.   Target Analyte Concentrations in TCLP Leachates (µg/L) from Diesel Fuel-Contaminated Soils.

Analyte	Soil	Contaminated Soil Nominal Concentration (ppm)			
		100	1,000	5,000	10,000
Benzene	Action	NA	ND	TR	4.04
	Kaw	2.04	51.8[a]	7.45	42.9
Toluene	Action	NA	TR	11.7	62.2
	Kaw	2.09	51.6[a]	36.8	6.4
Ethylbenzene	Action	NA	TR	20.8	52.0
	Kaw	TR	15.0	31.3	47.4
m-Xylene	Action	NA	3.57	47.6	106
	Kaw	TR	32.6	64.5	99.1
o- and p-Xylene	Action	NA	4.30	51.0	110
	Kaw	TR	33.8	66.5	104

NA = Not analyzed.
ND = Analyte not detected.
TR = Analyte detected at trace levels (below the quantitation limit).
[a]Potentially inaccurate results.

Table 3.   Target Analyte Concentrations in TCLP Leachates (µg/L) from No. 6 Fuel Oil-Contaminated Soils.

Analyte	Soil	Contaminated Soil Nominal Concentration (ppm)		
		1,000	5,000	10,000
Benzene	Action	NA	115[a]	17.3
	Kaw	8.96	6.26	11.1
Toluene	Action	NA	52.9[a]	47.9
	Kaw	11.7	13.4	38.0
Ethylbenzene	Action	NA	5.44	13.8
	Kaw	2.60	5.57	10.2
m-Xylene	Action	NA	14.4	39.6
	Kaw	6.80	16.7	30.3
o- and p-Xylene	Action	NA	13.4	40.0
	Kaw	6.66	17.0	31.3

NA = Not analyzed.
[a]Potentially inaccurate results.

of soils. Further, they may be considered 'worst case' examples of potential leaching losses, as they were the result of leaching from 'fresh' fuel contamination. Thus, it was more likely that the soils contained a higher proportion of the more soluble fraction of the fuels (i.e., BTXE) than would occur at some real world leak sites where weathering of the fuels would have resulted in volatile losses of the soluble components.

These results clearly demonstrate that, depending on the properties of the soil, there may be large differences in the potential for groundwater impacts, even among soils with similar levels of contamination. Further, they indicate that the establishment of generic cleanup standards or action levels is inappropriate for motor-fuel contaminated soils, and that there should be a mechanism to allow for site-specific determination of applicable standards.

SOIL CLEANUP STANDARDS

As is clear from the above discussion, soil analysis is performed to provide input to the process of evaluating potential health and environmental risks. The most common risk of greatest concern is that of leaching of contaminants from the soil to groundwater, and eventually to drinking water wells. It is this risk which leads to the need for assessing the relative site contamination, and its potential for contamination (i.e., at what levels of contamination does the risk cease to constitute a significant risk to human health and the environment).

This introduces the second principal topic of this paper, that of 'action levels' or cleanup standards for contaminated soil. Many formal and informal standards have been developed in the last several years by various states, and others are in the process of being developed.[4] These standards vary considerably, both in their numeric values, and in the type of contaminant addressed. Some examples of the standards contained in the EPA survey include: 500 ppm total petroleum hydrocarbon (TPH), 50 ppm benzene, 500 ppm cumulative BTX, etc. As noted in my comments at the 1988 conference, it would seem that regulatory policies or guidelines should avoid blind allegiance to generic numerical standards that lack scientific credibility (i.e., that have been selected without an understanding of the relationship between soil properties and contaminant properties).

The EPA's study of TCLP leachate reinforces the validity of that approach. That study demonstrated the considerable differences in mobility (and thus potential exposure) between a sandy soil and a clayey soil with identical levels of contamination. Thus the risks posed by the two different soils would also be quite different. While it is easy to understand the appeal of having a single number for a cleanup standard, it is difficult to justify the adoption of these numbers, without providing sound technical justification. Standards should, at a minimum, recognize the differences in contaminant mobility in different soil types.

The wide range of the state cleanup standards found in the EPA state survey reflects the uncertainty regarding what substances, and at what levels, are considered potentially harmful. In its evaluation of corrective action standards for petroleum contaminated soil, the EPA elected to forego the establishment of a numeric standard.[5] The Agency recognized that the risk present at a given site is not solely a function of the concentration of the fuel in the soil or backfill, but that other factors contribute to the net risk. The EPA decided to require a site-specific analysis of the potential hazards through a risk assessment that addresses various risk factors; for example, the volume of soil contaminated; its position in relation to the water table, the quality of the local groundwater, its current and potential uses, and the proximity of drinking water wells. Such evaluations are becoming more common as both the methods and benefits of risk assessment become more familiar to the regulatory agencies, environmental consultants, and industry. These assessments can be of varying levels of detail, and serve to provide a reasonable prediction of potential health and environmental risk levels associated with different levels of soil contamination present at a given site.

Perhaps the ideal approach to establishing site-specific cleanup standards for soils would involve the field validation of models that could accurately predict the long-term fate of residual hydrocarbon in the soil. A first approximation of such an approach is in use at the state level. The California State Water Resources Control Board has produced the Leaking Underground Fuel Tank (LUFT) Field Manual to provide guidance for investigations and assessments of motor fuel leaks and spills.[6] This innovative manual employs a risk assessment approach for the determination of acceptable levels of soil contamination. Simply described, the results of a modeling process are provided in tabular form, so that knowledge of the contaminant levels at depth, the amount of annual precipitation, and the depth to groundwater for any given site can be used to decide what the acceptable soil standard should be for that site.

The general approach of the manual is to be commended for its recognition that cleanup standards should be site-specific. However, the models used in LUFT have not been field validated, and a number of conservative assumptions are used to generate the tables which contain the recommended cleanup standards. As a result, for many sites the calculated standards tend to be considerably lower than may actually be sufficient for protection of human health and the environment. Further refinement of the models and assumptions used in LUFT should lead to a more accurate assessment of risks presented by motor fuel contaminated soil.

In general, a key component of risk assessments is the consideration of the roles of soil attenuating mechanisms (such as passive biodegradation and volatilization) in reducing contaminant levels in the soil, and limiting the migration of contaminants from the site. In the LUFT manual, these attenuation processes (both in the soil and groundwater) are estimated very conservatively. As already noted in this chapter, contaminant mobility in some soils (e.g., clayey soils or finer-textured soils) is severely limited, due to adsorption and low permeability. Biodegradation of petroleum hydrocarbons occurs naturally, and can be enhanced through careful manipulation of the soil environment.

Most of us have at least a passing familiarity with the principles of biodegradation, and understand that the soil has an impressive capacity to serve as a medium for the microbial destruction of organic wastes. Again, compared to air and water media, soil is a more 'friendly' environment to microbial communities, and can provide plentiful access to nutrients, water, and oxygen. Certainly we recognize that there is a wide range of soil habitats, and that at the extremes (e.g., deserts, permafrost, saline or alkaline/acidic soils), microbial populations will be limited. However, research indicates that the vast majority of soils provide an environment that sustains microbial activity. Populations of these microbes can transform petroleum hydrocarbons into simpler, less innocuous compounds as demonstrated by numerous studies, both in the laboratory and in the field.[7] Because of the important role that biodegradation and soil attenuation play in the ultimate mobility and persistence of petroleum hydrocarbons, these processes must be incorporated into the development of cleanup standards.

Relative Risk and Risk Reduction

Two concepts of considerable importance in the process of determining a reasonable cleanup objective at a soil contamination site are illustrated in Figures 1 and 2. Figure 1 is a theoretical distribution of the relative risks present among a given population of soil contamination sites. If we assume a normal distribution for this population of contaminated sites, the figure shows that a certain proportion of the sites will have very high risks, and others will have very low risks. A further conclusion can be made for at least some of the lower risk sites, i.e., that risks are low enough that there may not be much to be gained by engaging in a corrective action program. At these sites, 'passive remediation' may provide suitable protection of human health and the environment. At sites with higher levels of risk, there would likely be a need to begin a corrective action program to effect a reduction in potential risks. At those sites with highest levels of risk, corrective action programs might be employed most expeditiously to rapidly lower risk levels.

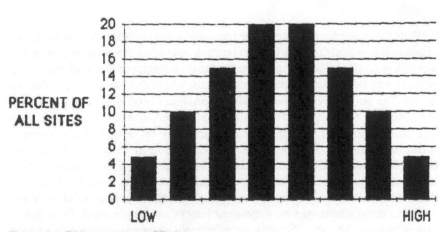

Figure 1. Risk present at UST sites.

Figure 2 provides another theoretical comparison, this one showing the relative risk reduction versus economic costs for three different sites. At Site A, a very high level of risk reduction can be obtained for relatively little cost—the ideal scenario. At such a site, it would be likely that close to 100% risk reduction could be achieved. In contrast, at Site C, there is an essentially linear relationship between risk reduction and costs. This site might require considerable expenditures to achieve an acceptable reduction in risk. Site B presents a particularly interesting and common scenario. At first, a large reduction in risk (perhaps 60–70%) is possible for a relatively small cost, but each further incremental reduction in risk will carry a relatively high price tag (e.g., it may cost as much to obtain an additional 10% reduction in risk as it did to obtain the initial 65% risk reduction).

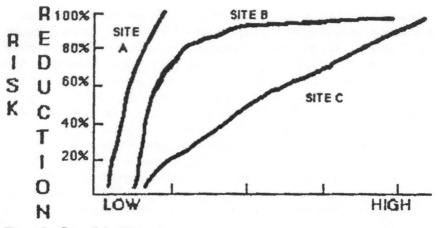

Figure 2. Remedial action costs.

While the examples in Figure 2 are speculative, they serve to demonstrate that, at some sites, a decision may have to be made regarding the relative cost of further risk reduction versus the net benefit of that increment of risk reduction. That is, there may be a point of diminishing returns, where that incremental reduction in risk does not justify the considerable expenditure needed to obtain it. This is not to suggest that this occurs at all sites. However, at those sites where there is a "leveling off" of the risk reduction curve, the net benefit of continuing to spend large amounts of money to provide only minimal reduction in risk must be carefully evaluated. Whether the cleanup funds are coming from industry, from insurors, or from cleanup funds, there are limited resources available to address these incidents of environmental contamination. It is important to identify and prioritize those situations where the most benefit can be obtained. In so doing, expenditures can be most efficiently and effectively allocated to provide the greatest overall reduction in risks.

Given the inherent variability in the types of sites that require remedial action, it is probable that there will be some sites where high costs for relatively minor reductions in risk will continue to be incurred. These decisions will be made case-by-case. However, at many sites it should be possible to determine a more reasonable point of compliance through consideration of the factors discussed above. It cannot be assumed that the most expensive cleanup solution is the best one, or that the level of environmental protection can be inferred by the amount it will cost to conduct the cleanup. Rather, careful considerations of risks must be balanced with reasonable economic expenditures in a policy to maximize the effectiveness of remedial activities.

SUMMARY

The petroleum industry has a strong and proven commitment to cleaning up sites where it has been responsible for the introduction of contaminants into the environment, and it is committed to preventing any future contamination. The active UST replacement programs of the major oil companies, initiated in advance of state and local regulations for USTs, are evidence that the industry recognizes its responsibility to provide the petroleum products our society requires (in excess of 400 million gallons of fuels alone each day), in a manner that minimizes the associated environmental risks. However, given the volume of activity, and the large number of sites where petroleum is stored, it is inevitable that some releases will occur. When they do, it is important to respond in a manner that will provide a high level of protection for both human health and the environment.

In recognizing this responsibility, the industry also recognizes that in responses to releases that occur, the goals of corrective action programs should not be dominated by the need to remove or eliminate every single molecule of contamination as rapidly as can be arranged. Instead, the emphasis should be to manage affected sites in a manner that ensures the protection of water resources and eliminates unnecessary risks and exposures, yet that does not waste limited economic resources.

The above remarks are not intended to imply that the industry is attempting to minimize its responsibilities toward environmental protection, that it feels that contamination of soil and water resources is not a problem, or that there is no need for regulatory involvement. Rather, these comments are intended to transmit the reality that every incident of environmental contamination creates a unique level of risk, and that corrective action requirements should be a function of that risk level. Regulatory policies that recognize this reality, and which utilize it in a responsible manner, will provide the most effective solutions to current problems of soils contaminated by petroleum products.

REFERENCES

1. Sampling and Analytical Methods for Petroleum Hydrocarbons in Groundwater and Soil, American Petroleum Institute, Washington, DC, 1987.

2. Dunlap, L. E., and D. D. Beckman. "Soluble Hydrocarbons Analysis from Kerosene/Diesel Type Hydrocarbons," in *Proceedings of the Conference and Exhibition: Petroleum Hydrocarbons and Organic Chemicals in Groundwater*, National Water Well Association, Dublin, OH, 1988.
3. Romeau, A., C. King, D. Blevins, and N. J. Soulor. "Mobilization of Volatile Toxic Components from Petroleum Product-Contaminated Soils by TCLP," in *Solid Waste Testing and Quality Assurance, Proceedings of the Fourth Annual EPA Symposium*, American Public Works Association, Washington DC, 1988.
4. Survey of State Programs Pertaining to Contaminated Soils, EPA Office of Underground Storage Tanks, Docket #UST2.4.SB.23 Draft report prepared by Midwest Research Institute, 1988.
5. *Federal Register*, September 23, 1988.
6. Leaking Underground Fuel Tank Field Manual, California State Water Resources Control Board, Sacramento CA, 1988.
7. Bossert, I., and H. Bartha. "The Fate of Petroleum in Soil Ecosystems," in *Petroleum Microbiology*, R. Atlas, Ed. (New York: Macmillan Publishing Co., 1984).

CHAPTER 6

EPRI-Sponsored Research on Underground Storage Tanks

Mary E. McLearn

Introduction

Underground storage tanks (UST) are used by the electric utility industry for a number of functions ranging from storage of motor vehicle fuel at service centers to storage of backup diesel fuel at power plants. While no accurate count has been made, it is estimated that the industry operates less than 2% of the UST in the country; this represents, however, more than 15,000 underground tanks. Almost all electric utility companies have some UST, and some have several hundred.

In response to increasing concern about leaking tanks and potential resulting groundwater contamination, the electric utility industry has initiated a wide-ranging research program to address the management of UST. Research projects sponsored by the Electric Power Research Institute (EPRI) and the Utility Solid Waste Activities Group (USWAG) address generic problems associated with UST, while those conducted by individual utility companies address issues specific to their tanks.

This chapter focuses on EPRI research projects, several of which have been conducted jointly with the USWAG Underground Storage Tank Committee.

Both the Coal Combustion Systems Division and the Environment Division at EPRI have active research programs which address UST. Coal Combustion Systems Division has focused on the engineering aspects of UST operations, examining the implementation of options for tank management. Environment Division has studied risk management methods for UST and developed tools to evaluate and facilitate choice among the various available engineering options. In addition, Environment Division has included UST in its integrated project to investigate the environmental behavior of organic substances; this project seeks to develop new information for modeling the release and subsequent dispersion, degradation, and attenuation of organic substances during transport through the environment.

Engineering Options

Research on engineering options focused first on the important issue of remediation of contaminated soil and groundwater from UST leaks. Thirteen techniques, both in situ and non-in situ, were described and evaluated for application at UST sites; these techniques are listed in Table 1. In addition to the engineering aspects (equipment, operation, maintenance), institutional, environmental, and cost considerations in the application of each technique were explored. The results of this project are presented in EPRI technical report CS-5261, "Remedial Technologies for Leaking Underground Storage Tanks." This project was co-sponsored by the USWAG Underground Storage Tank Committee.

Available technologies for leak detection, including tank tightness testing, in-tank monitoring, and external tank monitoring, are assessed in CS-5780, "Monitoring Technologies for Underground Storage Tanks." The capabilities of each technique, factors to be considered in choosing a technology most suitable to site conditions, and practical design and cost considerations are included.

To delineate the trends in state and local regulation of UST which will influence utility decisions on long-range tank management strategies, EPRI explored in depth the regulations imposed in 11 selected state or local jurisdictions; the locations studied are listed in Table 2. Key areas of state or local regulation were monitoring and leak detection, upgrading of existing tanks, construction of new tanks, and tank removal and disposal practices. In addition, a broad summary of state and local regulations, including key state contacts on UST matters, was prepared. The results of this project were published in CS-5520, "Trends in State and Local Underground Storage Tank Regulation."

Table 1. Remedial Technologies for Underground Storage Tanks.

In Situ Technologies	Non-In Situ Technologies
Volatilization	Land treatment
Biodegradation	Thermal treatment
Leaching	Asphalt incorporation
Vitrification	Solidification
Passive remediation	Groundwater extraction
Isolation/containment	Chemical extraction
	Excavation

Table 2. State and Local Jurisdictions Studied.

Kern County, CA	Santa Clara County, CA
Chesprocott, CT	Broward County, FL
Dade County, FL	Cape Cod, MA
Newark, NJ	Nassau County, NY
Suffolk County, NY	Philadelphia, PA
Spokane City and County, WA	

Two additional engineering aspects of UST management are currently being explored: repair and retrofit of UST and closure of UST. Documents presenting the results of these studies should be available in late 1989.

Risk Management

Electric utility companies are strongly motivated by economic considerations to minimize UST leaks, but development of a cost-effective risk management strategy for tank operations is a complex, multifaceted project. To aid utilities in the choice of the optimal options for UST management, EPRI has developed a decision support tool. The model, called TANKS, balances the known costs of leak prevention and mitigation against the uncertain costs of tank leaks, using tank- and site-specific data to address key questions. These questions include how much testing or monitoring to conduct at each site, what actions to take based upon testing or monitoring results, when to replace a tank, and how to design a replacement tank system.

Input data for TANKS include the reliability of the subject tank type, its age, its previous performance history based on testing and monitoring results, hydrogeologic factors which may introduce vulnerability to the tank system, and cost and expected accuracy of the testing and monitoring technologies. For each management option, TANKS assesses outcomes including the expected life cycle cost of the tank system, the expected time until replacement of the system, the likelihood of a leak, and the expected costs of leaks.

TANKS uses sensitivity analysis to identify those variables which critically influence sound tank management, and those variables for which additional data must be collected before sound decisions may be made. The primary output of TANKS is the various components of life-cycle cost. In addition, the model estimates the probability that the tank is leaking and determines whether the tank should be removed. Finally, the model calculates the probability distribution for future tank leaks based upon age of the system, testing and monitoring results, and other tank- and site-specific considerations.

The model was tested by the USWAG Underground Storage Tank Committee, which used it to review alternative tank management policies concerning frequency of testing, replacement of existing tanks, and selection of new tanks. Several thousand tank management scenarios were analyzed, covering alternatives of tank type (bare steel, cathodically-protected steel, fiberglass), tank age, likelihood of leakage, possible site risks, and strategies for testing, monitoring, and replacement. Several important insights resulted from the test. Leak costs, including remediation, are frequently a significant fraction of total life cycle costs. The optimal management strategy is often very sensitive to site characteristics as well as to tank type. Site characteristics determine the optimal testing and monitoring program as well as the value of installing a more expensive and leak-resistant tank. Finally, the model indicates that individual tank management strategies should be developed for each UST site.

Environmental Behavior

The cost and effectiveness of cleanup of UST leaks depend on the extent of migration of the leaked material through the subsurface environment. Design of cost-effective remedial action plans therefore requires a thorough understanding of a tank leak, including the nature, quantity, and characteristics of the leaked material, such as solubility, adsorptivity, attenuation characteristics, and specific gravity. Additional site-specific information is needed concerning the soil characteristics and groundwater proximity. Often, these kinds of information are not available when a site cleanup is initially planned.

One component of EPRI's Environmental Behavior of Organic Substances (EBOS) project is to develop reliable and rapid techniques to sample and analyze contaminated soils to determine the presence or absence of contamination. Effective sampling and analysis provide a reliable map of the contaminant plume, which aids greatly in selection of a cost-effective remedial action plan. The EBOS program has investigated the use of supercritical fluid extraction, which uses high-temperature and high-pressure carbon dioxide to remove semivolatile organic species from soil samples; the extracted fluid is then analyzed by gas chromatography to quantify the amounts of organic compounds present. A portable field unit has been constructed to demonstrate supercritical fluid extraction at sites where real-time analyses are needed. In 1989, the EBOS project will characterize plumes developed at leak sites to develop empirical associations between hydrogeochemical conditions and the extent of migration. The results will be used to develop and validate improved computer-based models of contaminant migration.

Coordination

The three components of the EPRI research program on underground storage tanks are designed to complement one another. The risk management model, TANKS, explores the suitability of various tank management strategies, and the manuals on engineering options describe the implementation of the optimal strategies. The EBOS projects will contribute to an overall understanding of the impacts of petroleum contamination on the environment and to an improved ability to model contaminant migration in addition to providing new techniques for assessment and verification of site cleanup. The USWAG Underground Storage

Tank Committee works closely with the EPRI staff in the development and application of the results of these EPRI research projects.

Acknowledgment

The author wishes to thank Victor Niemeyer and Ishwar Murarka of Environment Division, EPRI, for their guidance and support in the preparation of this chapter.

CHAPTER 7

Health Effects Research Initiatives at the Agency for Toxic Substances and Disease Registry—Applicability to Contaminated Soils

Cynthia M. Harris

INTRODUCTION

The Agency for Toxic Substances and Disease Registry (ATSDR), part of the U.S. Public Health Service, was created to implement health measures mandated under the Comprehensive Environmental Response, Compensation, and Liability Act (CERCLA) of 1980 (Public Law 96-510). ATSDR's mission is to prevent or mitigate adverse human health effects and the diminished quality of life resulting from environmental exposures to hazardous substances. The Superfund Amendment and Reauthorization Act (SARA, 1986) greatly expanded ATSDR's authorities and responsibilities, resulting in the establishment of additional health-related program areas.[1]

The contamination of soils with toxic environmental contaminants is a pervasive problem of potential human health concern to those working and residing near hazardous waste sites. Soil contamination may directly impact human health through the following human exposure pathways: incidental ingestion of surface soil (particularly by children who may gain access to the site or onsite workers exhibiting frequent hand-to-mouth activity [e.g., smokers]), direct dermal contact, and the inhalation of contaminated fugitive dusts by workers and nearby residents during remedial activities or inclement weather conditions. In addition, soil contamination may adversely impact human health by contributing to

49

groundwater and surface water contamination via infiltration and surface runoff, respectively. Chemicals that are commonly found in petroleum products (e.g., polyaromatic hydrocarbons, benzene, lead) are compounds commonly identified as soil contaminants at various hazardous waste sites, and toxicological profiles of several of these substances have been drafted.

CERCLA authorizes ATSDR to support generalized applied health effects research on toxic contaminants found at hazardous waste sites. This health effects research initiative will be applicable to the identification of data needs for all substances on the Hazardous Substances List. Specifically, ATSDR plans to sponsor research to determine acute and chronic toxicity from exposure to hazardous environmental substances and any subsequent short- and long-term health effects, to generate pharmacokinetic data on these substances, and to compile available human exposure and health effects data. In addition, this research will include pilot health studies (e.g., exposure surveys), epidemiological studies, health surveillance, toxicological testing of chemicals, clinical toxicology, human exposure assessments, and studies on occupational safety and worker health. The ATSDR staff will then disseminate the health-related information on specific contaminants and offer health education programs.

This chapter will describe: (a) ATSDR's legislative mandates, particularly those related to health effects research; (b) ATSDR program narratives and proposed general research initiatives, with emphasis on the development of toxicological profiles and the identification of chemical-specific data gaps applicable to constituents in petroleum products; and (c) the sponsorship of a newly formed organization to help develop methodologies for assessing soil risk—the Council for Health and Environmental Safety of Soils (CHESS).

LEGISLATIVE MANDATES RELATED TO HEALTH EFFECTS RESEARCH

CERCLA provided the congressional mandate to clean up abandoned hazardous waste sites and to provide federal aid in toxic emergency situations. ATSDR was created by CERCLA and charged with implementing the health-related sections of that statute. CERCLA identifies a process for ATSDR to follow in coordinating health effects research activities: (1) ATSDR and the Environmental Protection Agency (EPA) should jointly prepare a list of substances most commonly found at National Priorities List (NPL) sites and that pose a significant risk to human health [Section 104 (i)(2)(A)]; (2) Section 104 (i)(3) of CERCLA requires ATSDR to prepare a toxicological profile for each of the substances on the list; (3) Section 104 (i)(5)(A) of CERCLA further requires that for each of the substances on the list, ATSDR, in consultation with EPA and other public health agencies, assess whether adequate information on the health effects of each substance is available; and (4) Section 104 (i)(5)(D) of CERCLA states that it is the sense of Congress that the costs for conducting this research program be borne by private industry, either under the Toxic Substances Control Act (TSCA) (15 U.S.C. 7901 *et seq.*), the Federal Insecticide, Fungicide, and Rodenticide Act (FIFRA) (7 U.S.C. 136 *et seq.*), or cost recovery under CERCLA.

ATSDR has also been charged to conduct Health Assessments at hazardous waste sites and to assist the Environmental Protection Agency in determining which hazardous substances should be regulated and the levels at which they are of concern to human health. A Health Assessment is defined as ". . . the evaluation of data and information on the release of toxic substances into the environment in order to: assess any current or future impact on public health, develop health advisories or other health recommendations, and identify studies or actions needed to evaluate and mitigate or prevent human health effects."

HEALTH EFFECTS RESEARCH INITIATIVES

CERCLA mandates that ATSDR and EPA jointly rank, in order of priority, those substances that are most prevalent at Superfund sites and pose the greatest threat to human health. The primary criteria used in ranking these substances are: (1) overall toxicity of the substance (acute, subacute, and chronic toxicities); (2) frequency of occurrence; and (3) human exposure potential. This effort has resulted in the compilation of lists of the first and second one-hundred most prevalent contaminants at Superfund sites. Twenty-five substances will be added each year to this list and ATSDR will develop a Toxicological Profile for each of these chemicals.

ATSDR has been mandated by CERCLA (as amended by SARA) to develop and maintain up-to-date Toxicological Profiles on chemicals on the NPL. The ATSDR Toxicological Profiles are compilations of the toxicological information available and the health effects resulting from exposure to these substances. The contents of the Toxicological Profiles include a public health statement, health effects summary, data gaps, toxicological review, environmental fate, human exposure information, chemical/physical properties, and regulatory standards and guidelines.

The Toxicological Profiles are used by ATSDR, EPA, and the National Toxicology Program (NTP) to identify significant scientific data gaps in knowledge (e.g., environmental human exposure data, epidemiological studies). ATSDR's responsibilities in the development of the Toxicological Profiles involve adding to the body of knowledge concerning relationships between human exposure to hazardous substances and adverse health effects. ATSDR will initiate a program of sponsored research in an effort to fill any identified data gaps. This research will be coordinated with NTP and other programs of toxicity testing (e.g., private industry, academic institutions) to avoid duplication of research conducted in other programs and under other authorities. The logistics of these procedures are currently being developed by ATSDR, EPA, and NTP. The process of identification of data needs will involve a chemical-specific assessment of the current data, identification of chemical-specific data gaps, and the determination of potential adverse human health effects. These data needs will be prioritized for delineating chemical-specific research needs.

There are several program areas which will sponsor general environmental health effects research, and these include: Health Assessments, Toxicological Profiles, Emergency Response, Exposure and Disease Registries, and Worker Health and Safety. Health effects research, under the various program areas will involve:

1. environmental epidemiology (e.g., health studies on persons living on or near a hazardous waste site);
2. toxicological testing (to address the toxicity of hazardous substances and data gaps on exposure to specific chemicals and environmental mixtures). This research should include studies to determine short-, intermediate-, and long-term health effects; organ-specific, site-specific, and system-specific toxicities; gathering and understanding chemical-specific pharmacokinetic data, and collection of human exposure data;
3. human exposure assessment;
4. implementation of clinical toxicological methodologies (development or improvement of biomedical testing to assess potential adverse human health effects);
5. information dissemination to primary care providers;
6. health education (efficient communication of health risk information to the general public and to health officials and health care professionals);
7. worker safety and health (ATSDR has funded research on personnel protection equipment appropriate for use by remedial workers and emergency responders).[2]

COUNCIL FOR HEALTH AND ENVIRONMENTAL SAFETY OF SOILS (CHESS)

One of the projects ATSDR is helping sponsor (and will provide technical expertise to) is the newly formed Council for Health and Environmental Safety of Soils (CHESS). CHESS was formed to facilitate understanding of current regulatory guidelines on soil contamination and subsequently, to respond to a need for the creation of clear, consistent risk assessment methodologies for use by federal, state, and private agencies. ATSDR has been proactive in the development of CHESS because of the Agency's responsibility to assess health effects from potential exposures to contaminants at Superfund sites. In addition, the goals of CHESS are strongly applicable to the needs of ATSDR in assessing potential health effects from exposure to contaminants in soil.

CONCLUSIONS

In conclusion, the implementation of sponsored health effects research to fill significant data gaps concerning human exposure pathways and the toxicity of substances commonly found at hazardous waste sites is of high priority in accomplishing the global mission of ATSDR—service to the public through the mitigation and prevention of human exposure.

REFERENCES

1. Comprehensive Environmental Response,Compensation, and Liability Act of 1980 as amended by the Superfund Amendments and Reauthorization Act of 1986.
2. Johnson, B. L. "Public Health Effects of Hazardous Waste in the Environment," *Hazardous Waste: Detection, Control, Treatment* R. Abbou, Ed. (Amsterdam: Elsevier Science Publishers B.V., 1988), pp. 1017–1035.

CHAPTER 8

Underground Storage Tank Releases in Arizona: Causes, Extent, and Remediation

Jeanmarie Haney

INTRODUCTION

There are over 17,000 underground storage tank (UST) systems located at over 6000 facilities registered with the state of Arizona. Although UST systems occur throughout the state, the majority occur in the Phoenix and Tucson areas. As of September, 1988, 425 releases from UST systems have been reported to the state.[1] Products released include gasoline, diesel, waste oil, aviation fuels, and kerosene. UST system leaks have occurred due to overfill, product line failure, tank failure, and spillage during tank removal. As expected, the majority of UST system leaks have occurred in the Phoenix and Tucson areas. However, significant numbers of leaks have also occurred in other portions of the state.

Arizona contains three major geologic provinces. These provinces are the Basin and Range province in the southern portion of the state, the Colorado Plateau in the northern portion of the state, and a narrow band, known as the Transition Zone, in the central portion of the state. Phoenix and Tucson are located in the deep alluvial basins typical of the southwest Basin and Range province. Depth to groundwater in these basins typically exceeds 100 feet, and may exceed 300 feet. This provides a relatively thick vadose zone which serves as a "buffer" to groundwater contamination. Although fewer in number, UST system leaks outside of the Phoenix and Tucson area are environmentally significant because groundwater is often at less depth, and groundwater contamination may more readily result.

To assess the environmental impact of UST releases, the Arizona Department of Environmental Quality (ADEQ) requires that the extent and magnitude of soil and groundwater contamination be defined, and the potential environmental impact evaluated. Soil and groundwater cleanup levels have been established for specific petroleum hydrocarbon components. Determining the extent of contamination resulting from a UST release in Arizona usually involves extensive soil sample collection and analysis. Such procedures are problematic, and a conservative approach to soil sample collection and analysis is necessary.

Of the UST releases referred to the ADEQ Groundwater Hydrology Section, 26% have caused documented groundwater impact, 58% have caused significant soil contamination without attendant groundwater contamination, and 16% are currently under investigation to determine if groundwater quality has been affected.[2] To date, UST remedial actions in Arizona have included the installation of soil vapor extraction systems, the excavation of contaminated shallow surface soil, the removal of free product, and the treatment of groundwater to remove dissolved contaminants.

Compilation and analysis of data regarding UST leak investigations in Arizona provides insight into the subsurface movement of petroleum, thereby increasing the efficiency of future UST leak investigations, and facilitating appropriate UST program development.

CAUSES AND EXTENT OF UST LEAKS IN ARIZONA

Owners of UST systems are required to notify ADEQ of the existence of their USTs. The ADEQ UST notification data base contains information regarding 17,726 tanks at 6274 facilities. Figure 1 shows the number of tanks in the major cities of Arizona.

As of mid-September, 1988, 425 UST system leaks have been reported to ADEQ. A dramatic increase in UST leak reports over the past several years has occurred (Figure 2). The number of leaks reported per month is expected to increase even more dramatically since finalization of the federal UST rules. UST leaks have occurred in all portions of Arizona (Figure 3), with the majority of the leaks occurring in Phoenix and Tucson, the major urban areas.

UST leaks have occurred from all portions of the UST system. Approximately 44% of the reported leaks have occurred from the product piping (Table 1). Of these product piping leaks, the majority occurred as the result of faulty installation and/or loose seals at the pump and dispenser connections. The remainder occurred due to corrosion of the piping, especially at the elbows. Approximately 29% of the leaks occurred from the tanks, with the majority of these being due to corrosion, with a fewer number due to faulty installation (Table 1). Approximately 8% of the reported UST leak incidents were due to spills or overflows. The cause of leakage was unknown for the remaining 19% of the reported leaks.

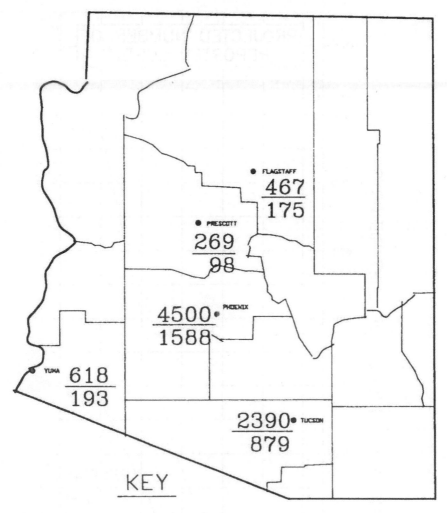

<u>618</u> NUMBER OF REGISTERED TANKS
193 NUMBER OF FACILITIES

Figure 1.   Major cities of Arizona and number of registered USTs and facilities.

In these cases, soil contamination was usually discovered unexpectedly during tank removal, and the source of the leak could not be pinpointed.

In 34% of the reported UST releases, unleaded gasoline was the product lost (Table 2). Fuel oil accounted for 21% of the reported leaks, regular gasoline for 20%, waste oil for 4%, and miscellaneous, including aviation gas and solvents, for 4%. In 17% of the report UST leaks, the type of product lost was unknown.

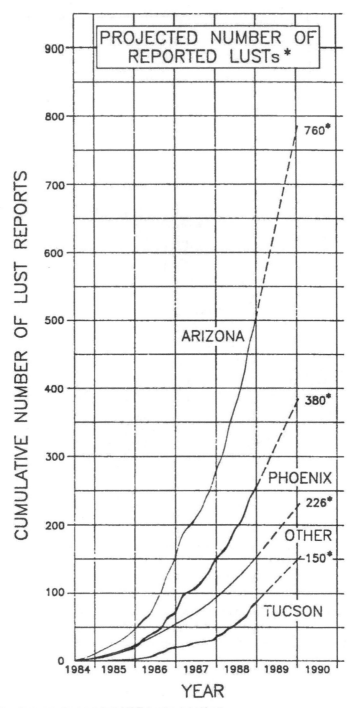

Figure 2. Increase in reported UST leaks over time.

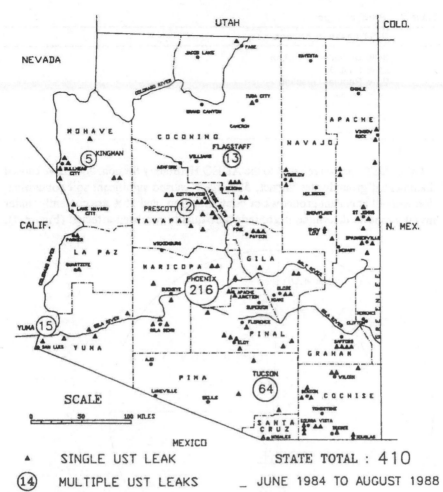

SINGLE UST LEAK   STATE TOTAL : 410

(14) MULTIPLE UST LEAKS   _ JUNE 1984 TO AUGUST 1988

Figure 3. Reported UST leak locations in Arizona.

Table 1.   Causes of UST Leaks.

Hydrology Section UST Data Base

44% Line leaks
    Majority – leaks at loose seals
        faulty installation
    Remainder – corrosion, especially at elbows

29% Tank leaks
    Majority – corrosion
    Remainder – faulty installation

8% Spills or overflows
19% Cause unknown
Total = 215

Table 2. Product Type.

UST Leaks in Arizona
34% Unleaded Gasoline
21% Fuel Oil
20% Regular Gasoline
4% Waste Oil
4% Miscellaneous
17% Unknown
Total = 215

Of the UST releases referred to the ADEQ Hydrology Section, 26% have caused documented groundwater impact, 58% have caused significant soil contamination without attendant groundwater contamination, and 16% are currently under investigation to determine if groundwater quality has been affected (Figure 4).

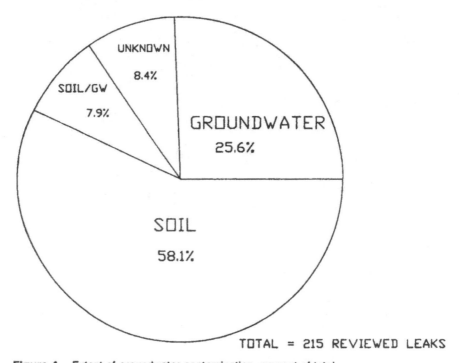

Figure 4. Extent of groundwater contamination, percent of total.

MAJOR GEOLOGIC PROVINCES IN ARIZONA

The three major geologic provinces in Arizona are illustrated in Figure 5. These provinces include the Basin and Range province in the southern portion of the state, the Colorado Plateau in the northern portion of the state, and a narrow band,

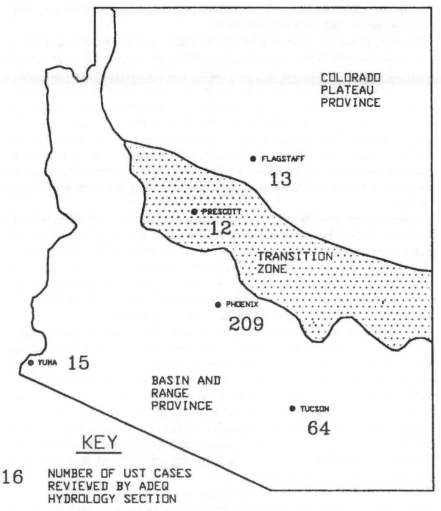

Figure 5. Geologic and physiographic provinces of Arizona showing reported numbers of UST leaks.

known as the Transition Zone, in the central portion of the state. The environmental impact of a UST release, and the method of leak investigation and remediation, varies dependent upon the hydrogeologic setting of the site.

The Basin and Range province includes the southwestern half of the state. The two largest cities in Arizona, Phoenix and Tucson, are located in the Basin and Range province of southern Arizona, and therefore, the largest number of UST systems, and the largest number of leaks, occur in the southern portion of the state. This region consists of elongated northwest-southeast trending mountain ranges separated by broad alluvial basins which are drained by ephemeral rivers.

The alluvial basins are filled with thousands of feet of Cenozoic alluvium, lacustrine sediments, and Cenozoic volcanics.

Phoenix, which lies within the Salt River Valley, utilizes groundwater for approximately one-third of its water supply, with the remainder of the water supply obtained from reservoirs on the Salt, Verde, and Gila rivers. The main source of groundwater in the Salt River Valley is the valley fill deposits that underlie much of the valley. The valley fill deposits are extremely heterogeneous, with discontinuous lenses of silt, sand, clay, gravel, and cobbles. Permeability varies greatly both vertically and horizontally. Depth to water commonly exceeds 100 feet and often exceeds 300 feet.[3] Groundwater is usually unconfined, but semi-confined conditions may exist locally where there is an increase of finer grained materials. Perched conditions also exist, and cascading water in wells occurs locally.

Hydrogeologic conditions in the Tucson basin are similar to those found in the Salt River Valley. However, due to the lack of a surface water supply, groundwater is the only dependable source of water. Tucson is the largest city in the U.S. that is 100% dependent on groundwater. Consequently, the aquifer underlying Tucson has been designated a sole source aquifer. Depth to water commonly exceeds 100 feet, and ranges to a maximum depth of 557 feet below land surface near the eastern foothills of the Rincon mountains.[4]

The Transition Zone in central Arizona is a narrow band characterized by rugged mountains composed predominantly of Precambrian granite and metamorphic rock. Quaternary and Tertiary basalts are common in many areas, and recent alluvial deposits occur along stream valleys. Hydrologic conditions vary widely throughout the area. However, shallow groundwater often occurs in fractures in bedrock. This is true for the town of Prescott, the largest town located in the Transition Zone (Figure 5). Municipal well fields are generally located outside the immediate town area, and normally penetrate the deeper portions of the aquifer. However, shallow private wells in the proximity of UST leak sites within city boundaries is a common occurrence.

The Colorado Plateau in the northern portion of the state is characterized by predominantly horizontal stratified sedimentary rocks that have been eroded into numerous canyons, plateaus, and scarps along which are exposed many colorful rocks ranging in age from Precambrian to Cenozoic. The Plateau province is the least populated portion of the state, and includes vast stretches of sparsely settled land. Cenozoic volcanic activity has created volcanoes and lava flows on top of portions of the Colorado Plateau. Flagstaff, one of the five largest cities in Arizona, sits upon such a lava field (Figure 5). Groundwater occurs in numerous Tertiary sedimentary formations in the Colorado Plateau region, and depth to water varies considerably, but is generally greater than 500 feet, although shallow perched groundwater does occur. As in the Transition Zone, municipal well fields are generally peripheral to the immediate town area, and often are perforated in the deeper portions of the aquifer.

OVERVIEW OF APPLICABLE STATE LAWS AND POLICIES

Arizona regulates UST systems through the state UST statute, effective August, 1986, which requires adoption of a program compatible with the federal UST regulations. State UST regulations have not yet been promulgated. In lieu of regulations, ADEQ has developed general policy and guidelines for UST leak investigation and remediation. These policies and guidelines are consistent with the Arizona Environmental Quality Act (1986). The Environmental Quality Act defines an aquifer as a groundwater bearing zone with a yield of at least 5 gallons per hour, and protects all aquifers in the state for drinking water purposes. Therefore, drinking water standards are aquifer water quality standards.

ADEQ requires that the vertical and horizontal extent and magnitude of contamination be defined in each UST leak incident. Due to the great depth to groundwater which is common in many of the major urban areas in Arizona, defining the extent of contamination almost always involves soil sample collection and analysis. Hollow stem augers are generally utilized for the drilling of soil borings, with soil sample collection performed via drive samplers fitted with split spoon or shelby tubes, with or without liners.

State soil cleanup guidance levels exist for benzene, toluene, ethylbenzene, xylenes (BTEX), and total petroleum hydrocarbons (TPH). Table 3 summarizes these draft guidance levels. The soil cleanup levels are based on the groundwater action levels, assuming a soil attenuation factor of 100 times.

Table 3.   Arizona Soil Cleanup Levels.

Petroleum Hydrocarbon Compounds	
Contaminant	Soil Cleanup Level
TPH	100 mg/kg
Benzene	130 μg/kg
Toluene	200 mg/kg
Ethylbenzene	68 mg/kg
Xylenes	44 mg/kg

SOIL ANALYTICAL METHODS

ADEQ requests that soil samples be analyzed for benzene, toluene, ethylbenzene, and xylenes by EPA method 8020. Method 8015, modified for gasoline, results in quantification of BTEX and TPH. For diesel leaks, the state accepts EPA method 418.1, modified for diesel fuel, or EPA method 8015, modified for diesel. The total petroleum hydrocarbon concentration is usually high in soil impacted by diesel contamination. However, since BTEX occur in much lower concentrations in diesel than in gasoline, BTEX is usually not detected to any significant degree in soils contaminated with diesel fuel. When a diesel leak has migrated to and impacted groundwater, TPH contamination can usually be detected in the groundwater. However, Arizona does not have a groundwater action level for TPH. Groundwater impacted by diesel contamination may have benzene in excess of the benzene action level of 5 μg/L.

Waste oil contamination may affect groundwater quality if groundwater is relatively shallow, and/or the amount of waste oil lost to the environment is large. In the case of waste oil contamination, ADEQ generally requires a solvent scan and metal analysis to be performed on the soil samples, as well as the typical BTEX and TPH analyses.

SOIL ANALYSIS

Defining the extent of contamination using soil sample collection and analysis is problematic. The inhomogeneity of the soil matrix, the complexity of attenuation and partitioning mechanisms, and the chemical complexity of petroleum mixtures, results in extreme variations in concentration values within very small distances. Particle size in typical alluvial basin settings ranges from clay to gravels and cobbles, often in the same sample. Also, where the main compounds of interest are volatile, sample collection procedures may reduce concentrations below that which is actually present. This is especially true when care is not taken to reduce volatile loss during sample collection and transport to the laboratory. Problems also exist with the analytical methodologies utilized for the analysis of soil total petroleum hydrocarbon concentrations. Standard EPA methods are "modified" for quantitative analysis of gasoline, diesel, and waste oil in soil. However, these methods are "modified" in different ways by different analytical laboratories. Therefore, results from different labs, or even results of different sample rounds from the same lab, may not be comparable. Also, extraction methods developed to remove the compounds of interest from the soil are often inefficient, especially if there is a high percentage of organic carbon and/or clay present in the sample. In some methodologies, such as method 418.1 for gasoline, the extraction procedure may result in the loss of up to 90% of the light-end (short-chain) hydrocarbon compounds.

For the reasons cited above, a conservative approach should be utilized when soil sample analysis is used to define the extent of contamination. Major decisions regarding the extent of migration of the leaked product should not be made based on limited data.

SOIL ANALYTICAL RESULTS

Data contained in the UST leak files at ADEQ has been examined in an effort to (1) detect patterns regarding the horizontal versus vertical extent of contamination, (2) determine a correlation between the extent and magnitude of contamination, as defined by soil analysis, and the estimated amount of product lost, and (3) ascertain if a correlation exists between lithology and concentration.

A typical UST leak situation is illustrated diagrammatically in Figure 6. The extent of contamination is outlined by the solid line. The oval (X_{max}) illustrates the maximum vertical extent of contamination if the leak is small and/or groundwater is relatively deep, and the leak is immobilized above the water table. The dashed line represents contamination that has reached the top of the saturated zone. As is illustrated in the diagram, if soil boring B-1 is placed in such a way that X_2 exceeds X_1, the boring will not intercept contamination, and the site will be assumed to be clean. This situation may occur when physical obstacles, such as canopies, prevent drilling at the leak location, or when the leak location is not known. If the leak is large, or groundwater is shallow, product may extend to the top of the saturated zone, and monitor wells will be required to determine if groundwater quality has been affected. If the monitor well is not properly located to intercept groundwater contamination, the groundwater will be judged to be clean.

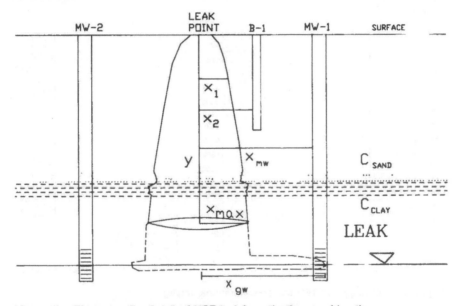

Figure 6. Diagrammatic sketch of UST leak investigation considerations.

Figures 7 through 12 present soil analytical data for a number of leaks that have been investigated in the Phoenix, Tucson, and Yuma areas. The data is presented on cross-sectional diagrams which show lithology, as well as TPH and benzene concentrations in mg/kg. Analysis of such soil analytical data is difficult because, frequently, the leak location is unknown, multiple sources are present, and the amount, and often type, of product lost is not known. Therefore, it is difficult to correlate contaminant migration patterns (e.g., the vertical extent of contamination) to the amount of product lost. This is true even in instances where the subsurface lithology is relatively well known. However, it is possible to comment on the vertical versus the horizontal extent of contamination, and to make some general observations and recommendations regarding investigative techniques.

Figure 7 illustrates a large leak in the Phoenix metropolitan area. Depth to groundwater at this site is 115 ft. Over 400,000 gal of unleaded gasoline were lost. The leak occurred in the vicinity of boring 1. High concentrations of TPH can be observed in soil samples collected from boring 1 down to the water table. The nearest boring, boring 3, is located 43 ft from boring 1. TPH was not detected throughout the unsaturated zone in boring 3. In this case, which represents a major leak, lateral product migration in the vadose zone is less than 43 ft. Also of note in Figure 7 is that 8 ft above the free product layer in borings 2 and 3, no contamination was detected.

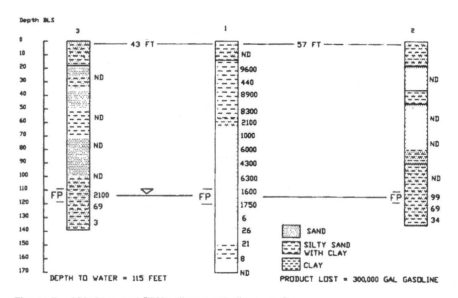

Figure 7. Lithology and TPH soil concentrations, mg/kg.

Figure 8 shows a smaller leak. Approximately 937 gal of regular gasoline were unaccounted for by inventory reconciliation. The exact leak location is unknown. Boring C-3 appears to have intercepted soil contamination, and can be assumed to be located close to the leak point. Boring C-4, located 30 ft from the assumed leak point, and boring C-5, located 20 ft from the assumed leak point, both show nondetect for TPH in all samples collected.

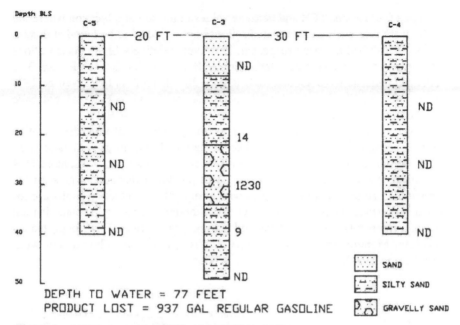

Figure 8. Lithology and soil TPH concentration, mg/kg.

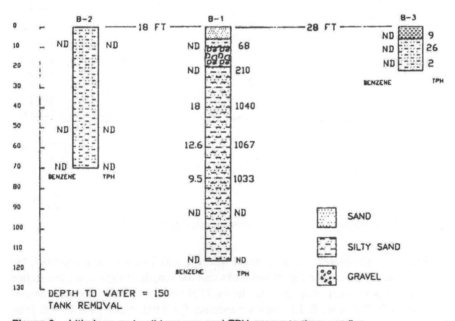

Figure 9. Lithology and soil benzene and TPH concentrations, mg/kg.

Figure 9 illustrates TPH and benzene values in the soil at a location where the USTs were removed. Contaminated soil was not expected to be found at this location; a leak had not been suspected. However, relatively high concentrations of TPH and benzene were detected in boring B-1. In boring B-2, 18 ft from B-1, TPH and benzene were not detected. In B-3, 28 ft from B-1, low levels of TPH were detected. The low level TPH contamination found in B-3 may not have resulted from the same source as did the contamination in B-1.

Figure 10 illustrates a case where an estimated 476 gal of unleaded gasoline were unaccounted for by inventory reconciliation. Depth to groundwater is 31 ft. It was thought that the leak location was near B-1. However, soil sample TPH analytical results are all relatively low, with nondetect throughout most of the unsaturated zone. The only samples with results above detection limits are located at the water table. This indicates that although vadose zone contamination was not intercepted by any of the soil borings, groundwater was impacted by this leak. Monitor well installation was required at this site. The monitor well, installed near B-3, has 4.5 in. of free product.

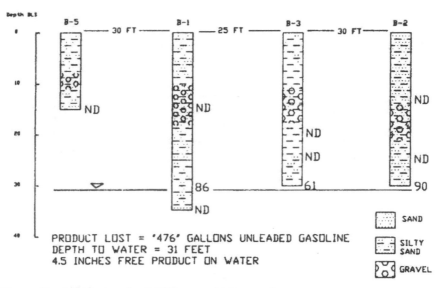

Figure 10. Lithology and soil TPH concentration, mg/kg.

Figure 11 illustrates a tank removal where a leak had not been reported, but was suspected due to the age of the tanks and the length of time since the station had been operational. Boring B-12 shows TPH and benzene soil contamination. Boring B-9, 8 ft from B-12, shows nondetect for TPH. Borings B-7, 4 ft from B-12, and boring B-7, 22 ft from B-12, both show relatively low levels of TPH soil contamination. This contamination may or may not be from the same source that caused the contamination in B-12. Boring B-7 may illustrate that higher concentrations may occur in clay layers, as clay has a higher porosity and a higher retention capacity. Also, the clay layer located between 65 and 75 ft in boring B-12 did not halt the downward migration of product, as a TPH level of 1300 mg/kg occurs in the sand below the clay layer. Soil contamination occurs down to the water table at this site. A number of the monitor wells installed at this location exhibit free product.

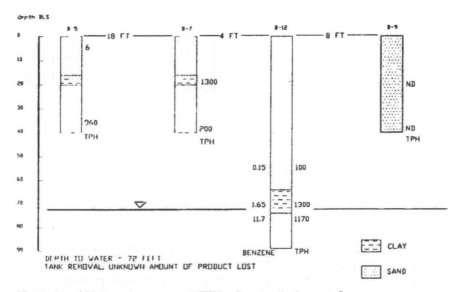

Figure 11. Lithology, benzene and TPH soil concentrations, mg/kg.

Figure 12 illustrates a diesel loss of unknown quantity. The exact leak location is unknown, but is probably near boring B-E. Low levels of TPH contamination were detected in soil samples collected from MW-1, 8 ft from B-E. The caliche layer in MW-1 may be perching a small amount of product, but contamination also apparently occurs below the caliche layer. Contamination appears to occur to within close proximity of the water table at this site. However, thus far, water samples collected from MW-1 have been nondetect for TPH and BTEX. This may be because contamination has not reached the groundwater, or the monitor well may not be properly placed to intercept the groundwater contaminant plume.

Based on the data presented above, the following conclusions may be drawn regarding utilizing soil sample analysis to define the subsurface migration of petroleum products: (1) vertical migration far exceeds horizontal migration; (2) clay and caliche layers do not halt the downward migration of product; (3) contamination is often more extensive than expected, based on the estimated amount of product lost; (4) due to the limitations of the field and laboratory methodology,

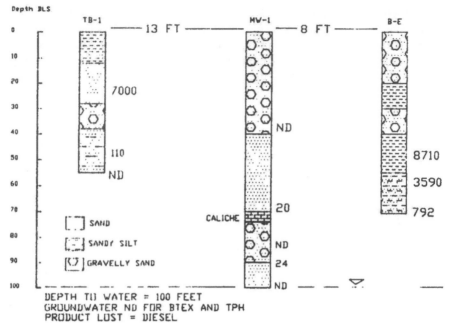

Figure 12. Lithology and soil TPH concentration, mg/kg.

a conservative approach to soil sampling should be utilized. It is always better to take more samples and assure that a problem does not exist than to take fewer samples and not detect what may turn out to be a significant problem.

UST REMEDIAL ACTIONS IN ARIZONA

Remedial actions that have been performed to date in Arizona are summarized in Figure 13. Excavation and treatment of contaminated soil have been performed in approximately 30% of the UST leak cases. Vapor extraction systems have been installed in 15% of the remedial efforts. No remediation has occurred at sites where the leak was small and depth to groundwater was large, and soil sample analyses indicated that only low levels of residual hydrocarbons occurred in the shallow soil. These cases were judged to present no threat to the environment or public health. Free product removal is occurring at 6% of the UST leak sites, and groundwater extraction and treatment is occurring at about the same number of sites.

Disposal of excavated petroleum contaminated soil is a problem at Arizona UST leak sites. Petroleum contaminated soil is not classed as a hazardous waste in Arizona unless testing indicates characteristics of a hazardous waste. On-site aeration of excavated contaminated soil may present a public health hazard in the form of vapors and objectionable odors emanating from the aerating soil.

Precipitation may cause migration of contamination from the excavated soil to other areas above or below the surface. Offsite transport with subsequent treatment involves many county and city ordinances, and may result in increased liability for the responsible party. Excavated contaminated soil may be transported to a hazardous waste landfill, but the cost is often prohibitive. Arizona is developing a petroleum contaminated soil disposal policy to provide guidance to UST owners and operators.

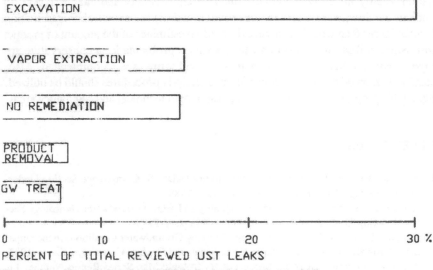

Figure 13.   Remedial actions, UST leaks in Arizona (total = 215).

STATE UST PROGRAM DEVELOPMENT

The state of Arizona will develop UST rules during the upcoming year. In support of these rules, policies and procedures will continue to be developed to help guide the regulated community. Some of the major issues that ADEQ will be addressing include (1) standardization of "modified" EPA tests for petroleum-contaminated soil, (2) in-situ soil remediation for diesel, waste oil, and gasoline contamination, (3) remediation requirements for low level dissolved groundwater contamination, including TPH, BTEX, and other components, (4) petroleum contaminated soil disposal, and (5) tank testing and tank monitoring requirements.

SUMMARY

The UST leak data base was assessed to determine causes, extent, and remediation of UST leaks in Arizona. The majority of UST leaks in Arizona have occurred from the product piping. UST leaks are primarily gasoline and diesel fuel, with lesser numbers of aviation gas, waste oil, and solvent leaks. The environmental impact of UST leaks differs depending on the hydrogeologic setting. Approximately 26% of the UST leaks reviewed by the Hydrology Section have affected groundwater quality, with approximately one-third of these requiring free product removal. The extent of contamination must be defined in all UST leak incidents. Defining the extent of contamination using soil sample collection and analysis is often problematic. An evaluation of the extent of soil contamination at UST leak sites in Arizona demonstrates that (1) the vertical extent of contamination usually far exceeds the horizontal extent, (2) clay and caliche layers do not halt the downward migration of product, and (3) the extent of contamination is often larger than would be predicted based on estimates of the amount of product lost. Since soil sample collection and analysis is problematic, a conservative approach which utilizes a sufficient number of soil samples is recommended. Also, standardization of soil sample collection and analysis procedures should be utilized. State UST program development is ongoing in both technical and regulatory areas.

REFERENCES

1. Arizona Department of Environmental Quality (ADEQ), Compliance Section Underground Storage Tank Leak Report Database, 1988.
2. Arizona Department of Environmental Quality (ADEQ), Groundwater Hydrology Section Leaking Underground Storage Tank Database, 1988.
3. Murphy, B. A., and J. D. Hedley, Maps Showing Groundwater Conditions in the Upper Santa Cruz Basin Area, Pima, Santa Cruz, Pinal, and Cochise Counties, Arizona–1982; Arizona Department of Water Resources Hydrologic Map Series Report Number 11, 1988.
4. Reeter, R. W., and W. H. Remick, Maps Showing Groundwater Conditions in the West Salt River, East Salt River, Lake Pleasant, Carefree and Fountain Hills Sub-Basins of the Phoenix Active Management Area, Maricopa, Pinal, and Yavapai Counties, Arizona–1983; Arizona Department of Water Resources Hydrologic Map Series Report Number 12, 1986.

CHAPTER 9

State of Research and Regulatory Approach of State Agencies for Cleanup of Petroleum Contaminated Soils

Charles E. Bell, Paul T. Kostecki, and Edward J. Calabrese

INTRODUCTION

In light of the recent development of federal underground storage tank regulations and technical requirements, a survey was conducted for the purpose of determining the current state of research and regulatory approach of state agencies with respect to site identification and remediation of petroleum contaminated soils (PCS). The objective of this survey was to provide researchers, state regulators, and other interested parties with a comprehensive summary of approaches and technologies currently in use in dealing with cleanup of PCS. This survey also serves to update and supplement a similar survey conducted by the University of Massachusetts Environmental and Health Science Department in 1984 and 1985.

Although the initial survey provided a great deal of information, results indicated that state regulatory approaches vary widely both among and within states.[1] The inconsistency of cleanup levels between states, and lack of development of adequate standard operating procedures (SOPs) or guidelines for remediation within states, is due in part to the absence of a specific federal regulatory policy for PCS. The inconsistencies and absence of SOPs is also a result of inadequate resources available to an agency to effectively develop and implement regulatory policy.

Although some constituents of petroleum products such as benzene, toluene, and xylene are listed as hazardous waste under the federal Resource Conservation and Recovery Act (RCRA), oil and gasoline are not. From the federal perspective, PCS would not be considered since they are unlikely to exhibit characteristics such as ignitability, corrosivity, and reactivity; or threaten the environment or health as determined by the extraction procedure (EP) toxicity test. State regulators looked to EPA's Underground Storage Tank (UST) program to provide guidance and/or soil standards for cleanup as part of the program's UST Technical Requirements released in September, 1988.[2] However this document falls short of providing specific guidelines and charges state regulatory agencies with developing their own management strategies. The state programs must be as, or more, stringent than current EPA requirements. EPA has suggested several options for the offsite management of PCS: treat PCS as nonhazardous waste that requires disposal into solid waste facilities; as a special waste that requires treatment (such as land farming or incineration) prior to disposal to remove the volatile constituents of the petroleum; or as a hazardous waste that must be disposed of under hazardous waste standards. These and other strategies are already being applied in various states, but questions remain as to their cost-effectiveness and the ability of the responsible party (RP) to comply with regulatory requirements.

METHODOLOGY

Potential respondents from each state were identified from a list of regulatory representatives who participated in the earlier survey. These past participants were contacted by telephone and informed of the nature of the survey and asked whether they could provide the necessary information. In those cases where the past participant was no longer with the agency, could not be contacted, or felt they were no longer the best qualified to participate, a more appropriate individual was suggested by that office and contacted. Frequently, different departments or divisions within a regulatory agency have overlapping jurisdiction with respect to PCS cleanup, and often two or more agency representatives were solicited to participate in the survey.

Once identified, each participant was provided with an overview of the material to be covered in the survey. A phone interview was conducted at a later date, thereby allowing time for participants to collect any material and prepare to respond to the questionnaire. Each participant was read a paragraph at the onset of the phone interview to reiterate background information provided during the initial contact and to further standardize the interview process. With the permission of participants, phone conversations were recorded in order to maximize accurate transcription of answers onto questionnaires. Participants were sent a copy of the questionnaire containing their responses for verification of completeness and accuracy. Respondents were requested to sign and return the questionnaire along with any relevant literature. In those cases in which questionnaires were not returned, survey data were assumed complete.

Participants were asked to respond to a series of questions concerning: the current state of research relating to PCS; the regulatory approach of the agency in terms of a response for cleanup; the types of remedial technologies currently being utilized; any future plans such as pending legislation or policy decisions that could affect selection of particular remedial technologies as preferred cleanup methods; classification or treatment of PCS as a hazardous or nonhazardous waste.

GENERAL FINDINGS

Results are based on 28 states for which survey data were complete as of February 1, 1989. Survey efforts will continue until all 50 states have been contacted, and each state file will then be updated on an annual basis. Representatives from state departments of Environmental Protection, Health Services, Environmental Regulation, Chemical Safety, Spill and Emergency Response, as well as UST and Water Quality Programs, participated in the survey. Phone interviews lasted from 45 to 90 minutes, with as many as six regulatory representatives from different agencies within a state being contacted.

At the time of the survey, 7 of 28 states were conducting or supporting research activities relating to the detection and/or remediation of PCS (Table 1). These investigations primarily focus on analytical methods, public health effects including risk assessment, and application of various remedial technologies. Nearly all of these projects incorporate some air or water quality component as part of the overall scope of work. For example, Florida is currently supporting research investigating microbial anaerobic degradation of gasoline in groundwater and organic contaminant sorption in soil/groundwater systems. The areas of investigation provide an indication as to the concern of state regulators for the continued protection of groundwater resources, and the associated public health risks from exposure to petroleum constituents via air and drinking water sources.

Table 1. State of Research Relating to Petroleum Contaminated Soils.

State	Support Contracted Research	Conduct In-House Research	Nature of Investigations	Incorporate Air/Water Quality Component
AK	No	No		
CA	Yes	Yes	Analytical Public Health Remediation	Air, Water
CT	No	No		
FL	Yes	No	Remediation	Water
HI	No	No		
IL	No	Yes	Analytical Public Health Cleanup Stds	Air, Water

Table 1.   (Continued)

State	Support Contracted Research	Conduct In-House Research	Nature of Investigations	Incorporate Air/Water Quality Component
IN	No	No		
KS	No	Yes	Analytical	Air
KY	No	No		
ME	Yes	No	Remediation	Air
MD	No	No		
MA	No	No		
MI	Yes	No	Remediation	Water
MN	Yes	No	Analytical Remediation	Air, Water
MS	No	No		
NH	Yes	No	Environmental	Water
NJ	Yes	No	Public Health Remediation	Air
NM	No	No		
NY	No	Yes	Remediation	Air
PA	No	No		
RI	No	No		
SC	No	Yes	Remediation	Water
TN	No	No		
TX	No	No		
UT	No	No		
VT	No	Yes	Analytical	Air
WI	No	No		
WY	No	No		

Three states (California, Massachusetts, New Jersey) reported classifying or treating PCS as hazardous waste on a conditional basis. All other states consider PCS nonhazardous unless it fails to meet RCRA criteria for EP toxicity or ignitability. Massachusetts treats soils containing waste oil as hazardous waste, but soils containing virgin product only if it is excavated and exceeds 1800 ppm (volume/volume) headspace analysis or 3000 ppm (mg/kg) total petroleum hydrocarbon (TPH). PCS in New Jersey are considered hazardous waste if they exceed 3% TPH. Other states (Connecticut and Tennessee), although not actually classifying PCS as hazardous waste, may require its disposal as hazardous waste.

Significant variability exists among states regarding the methods used to determine the extent and degree of contamination once a site has been identified (Table 2). Most states, including those that treat PCS as hazardous waste, prefer or require laboratory analysis of soil samples. These analyses may be performed by a contractor or state laboratory, and involve the measurement of many different parameters such as TPH, benzene, toluene, ethylbenzene and xylene (BTEX), lead, semivolatiles, and organo halides. Specific types of analyses may be required or performed by a regulatory agency if the nature of the contaminant is known. Conversely, some states require no laboratory analysis of soil samples, and utilize sensory methods (sight, smell, and touch) as a primary means to identify

Table 2. State Guidelines, Standard Operating Procedures, and Sampling Methodology for Site Investigation.

State	Have Written SOP for Site Investigation	Primary Method Used for Detection	Alternate Methods Used	Provide Specific Guidelines to RPs for Cleanup
AK	No	Lab analysis	Onsite instrumentation sensory	No
CA	Yes	Lab analysis	Onsite instrumentation	Yes
CT	No	Case-by-case	Lab analysis, sensory onsite instrumentation	No
FL	Yes	Onsite instrumentation		Yes
HI	No	Case-by-case	Lab analysis, sensory onsite instrumentation	No
IL	No	Onsite instrumentation	Lab analysis	Yes
IN	Yes	Case-by-case	Lab analysis, sensory on-site instrumentation	Yes, if involving groundwater
KS	No	Sensory for gasoline	Lab analysis, onsite instrumentation	Yes
KY	No	Lab analysis	Sensory, onsite instrumentation	Yes, onsite
ME	No	Sensory		Yes
MD	No	Sensory	Lab analysis, onsite instrumentation	Yes
MA	No	Lab analysis	Sensory, onsite instrumentation	Yes
MI	No	Sensory	Lab analysis	No
MN	Yes	Sensory	Lab analysis, onsite instrumentation	Yes
MS	No	Lab analysis		No
NH	No	Case-by-case	Sensory, lab analysis onsite instrumentation	Yes
NJ	Yes	Lab analysis		Yes
NM	Yes	Lab analysis	Onsite instrumentation for screening	Yes
NY	Yes	Onsite instrumentation	Sensory	Yes
PA	No	Case-by-case	Sensory, lab analysis	No
RI	No	Case-by-case	Sensory, lab analysis onsite instrumentation	Yes
SC	No	Case-by-case	Lab analysis, onsite instrumentation	No
TN	Yes	Lab analysis	Onsite instrumentation	No
TX	Yes	Sensory		No
UT	No	Lab analysis	Sensory, onsite instrumentation	No
VT	Yes	Onsite instrumentation	Lab analysis—water samples only	Yes
WI	Yes	Case-by-case	Sensory, lab analysis onsite instrumentation	No; under development
WY	No	Lab analysis	Sensory, onsite instrumentation	No

and initiate remedial actions. Others prefer the use of onsite instrumentation such as an OVA, H-Nu, or portable gas liquid chromatography (GC). All states will utilize one or more of these techniques to determine the extent of contamination and, more often than not, handle each site on a case-by-case basis.

To better understand the process by which PCS sites are characterized, regulators were asked what chain of events would take place in terms of regulatory response for cleanup, once a site had been identified. Eleven of 28 states contacted have written standard operating procedures or guidelines that are followed during investigation and cleanup of PCS sites. All states with or without formal procedures handle each site on a case-by-case basis and initiate similar steps: outline RP's obligations under state regulations; recover free product/emergency response where applicable; determine extent and degree of contamination, and potential immediate public health or environmental hazard; request voluntary regulatory compliance and submission of corrective action plan from the RP or contractor; issue administrative order as necessary; review and approve corrective action plan; and monitor, verify, and validate remediation. Eight of the 11 states that have SOPs for cleanup also provide specific guidelines and/or recommend procedures to RPs for cleanup. Overall, 16 of 28 states provide some form of guidance document to RPs during the course of cleanup efforts.

Only six of 28 states reported having formal established levels for cleanup of PCS (Table 3). Formal standards were considered those supported or enforceable through state regulations and/or consistently applied from site to site. Fifteen additional states reported they had established levels for cleanup, but that they were informal in nature. Informal standards were considered guidance levels used as a basis for corrective action but result in cleanup goals that may vary from site to site. The majority of informal cleanup standards are based on water quality

Table 3. Existing Established Levels for Cleanup of Petroleum Contaminated Soils.

State	Yes	No	Formal	Informal	Level or Explanation
AK	X			X	cleanup to background or apply assessment matrix in CA LUFT manual to justify alternative level
CA	X		X		leaching potential analysis/LUFT manual: 10–1000ppm TPH—gasoline; 100–10,000ppm diesel; benzene (.3–1ppm), toluene (.3–50ppm), xylene (1–50ppm), ethylbenzene (1–50ppm) (BTXE) for gasoline or diesel
CT	X			X	level necessary to protect groundwater resources
FL	X		X		5ppm TPH, "clean," sum of BTEX <100ppb
HI		X			function of water quality stds—no soils stds
IL	X			X	function of biol/health considerations/ case-by-case basis interim stds—benzene

Table 3.   (Continued)

State	Yes	No	Formal	Informal	Level or Explanation
					5ppb, toluene 2ppm, xylene 440ppb, ethylbenzene 13,600ppb (BTEX)
IN	X			X	100ppm TPH in some cases
KS		X			function of water quality stds—no soils stds
KY	X			X	to background or detection limit
ME	X			X	20–50ppm TPH, depending on water quality factors
MD		X			case-by-case basis
MA	X			X	100ppm TPH or 10ppm total organic volatiles as benzene
MI	X			X	background
MN	X			X	background of <1ppm TPH
MS	X			X	function of water quality stds—no soils stds
NH	X		X		<10ppm TPH and BTEX each <1ppm for gasoline, <100ppm TPH and BTEX each <1ppm for diesel
NJ	X		X		100ppm TPH
NM	X			X	function of water quality considerations
NY		X			under development
PA		X			case-by-case basis
RI	X			X	50–100ppm TPH, depending site-specific factors
SC		X			case-by-case basis, function of water quality—no soil stds
TN	X		X		100ppm TPH for hydrocarbons except gasoline, 10ppm total BTX in soil for gasoline
TX		X			function of water quality stds—no soil stds
UT	X			X	function of water quality considerations, formal soil stds under development
VT	X			X	<10ppm TPH or apply water quality stds
WI	X			X	<10ppm TPH used as guidance level for further corrective action and/or apply groundwater stds for benzene and toluene
WY	X		X		satisfactory cleanup achieved when oil and grease content of groundwater <10mg/L; <10ppm TPH for soils where groundwater <50ft; <100ppm TPH where groundwater >50ft

considerations, and therefore establish levels for cleanup on a case-by-case basis. Very often RPs are expected to clean up all PCS to background levels. One state may apply another state's standards or protocols at a particular site, or use another state's approach as a basis for development of regulatory policy. For example, Alaska has accepted the use of the leaching potential analysis from the California Leaking Underground Fuel Tank Manual[3] to justify an alternative cleanup level other than background for a particular site. At the time of the survey, New York and Utah were in the process of establishing formal cleanup levels specifically for soil. These efforts are indicative of the overall trend toward development of more formalized protocols to address PCS problems.

Participants were asked what types of remedial technologies were most commonly applied in their states (Table 4). Selection or preference of a particular remedial option within a state was observed to be a function of demographic, geographic, and site-specific environmental concerns. For example, soil venting may not be the preferred remedial option at a site adjacent to an elementary school, but may be used at a site located in a rural area with warm and dry climatic conditions. Volatilization or soil venting is the primary in situ remedial option in use, and includes such processes as vacuum extraction, vapor extraction, or air stripping. Nine of 28 states also indicated biodegradation as an option that had

Table 4.   Remedial Options and Technologies Most Frequently Applied Within States.

State	Primary or Preferable In Situ Treatment Option	Alternative In Situ Cleanup Methods Used	Primary Non-In Situ Treatment Option	Alternative Non-In Situ Cleanup Methods Used
AK	volatilization		excavation to landfill	land trt, thermal trt asphalt batching
CA	volatilization	biodegradation	excavation to landfill	thermal trt and asphalt batching as demos
CT	volatilization (L)	biodegradation (L)	excavation to landfill	land treatment
FL	volatilization		excavation to landfill	land treatment, asphalt batching, thermal trt
HI	none		excavation to landfill	none
IL	volatilization	biodegradation	excavation to landfill	thermal treatment
IN	no preference	volatilization biodegradation	excavation to landfill	land treatment
KS	volatilization		land treatment	excavation to landfill
KY	containment		excavation to landfill	
ME	volatilization	passive	land treatment	
MD	volatilization		excavation to landfill	thermal treatment, asphalt batching
MA	volatilization	biodegradation	asphalt batching	excavation to landfill
MI	volatilization		excavation to landfill	
MN	no preference		asphalt batching	land treatment, thermal treatment
MS	volatilization		excavation to landfill	land treatment
NH	volatilization	biodegradation	excavation to landfill	asphalt batching
NJ	volatilization	biodegradation	excavation to landfill	land treatment, thermal treatment

Table 4.   (Continued)

State	Primary or Preferable In Situ Treatment Option	Alternative In Situ Cleanup Methods Used	Primary Non-In Situ Treatment Option	Alternative Non-In Situ Cleanup Methods Used
NM	volatilization	biodegradation containment	land treatment	excavation to landfill
NY	volatilization	biodegradation	excavation to landfill	asphalt batching (L)
PA	no preference	volatilization passive, biodegrad.	excavation to landfill	asphalt batching (L)
RI	volatilization	biodegradation (L) containment	excavation to landfill	land treatment, asphalt batching
SC	volatilization (L)	biodegradation (L)	excavation to landfill	land treatment, asphalt batching (L)
TN	volatilization		land treatment	excavation to landfill, thermal treatment
TX	biodegradation		land treatment	excavation to landfill
UT	no preference		land treatment	excavation to landfill
VT	biodegradation	volatilization	land treatment	excavation to landfill
WI	volatilization	passive containment	excavation to landfill	land treatment, thermal trt (L), asphalt batching (L)
WY	volatilization		land treatment	excavation to landfill

(L) = Limited.

been, or is, currently being utilized. However, it is the non-in situ (either onsite or offsite) remedial options that are most frequently selected. Nineteen of 28 states reported that excavation to a landfill was the primary disposal option applied to date. Seven additional states reported excavating the contaminated material, followed by a land treatment process such as land spreading. Asphalt batching is a third remediation process which is the preferred option in two states. This includes both onsite and offsite processes producing either cold batch or hot mix asphalt. Asphalt batching is a remedial option that has generated considerable interest among regulators looking for alternatives to excavation as the number of potential landfill sites diminishes and tighter controls are placed on those still willing to accept PCS. Seven states reported current regulations in place that prohibit or limit the use of a particular remedial option (excavation to a landfill, chemical extraction), most often due to air or water quality considerations. New York reported pending legislation that may affect selection of a particular PCS site cleanup method, and three states reported pending legislation or policy decisions that could change the current classification of PCS as hazardous or nonhazardous waste.

STATE BY STATE SUMMARY

The ability of each state to effectively develop and implement a regulatory policy is often complicated by finite budget resources, the number of PCS sites reported, and diversity of site-specific circumstances under which corrective actions must be initiated. As a result, significant variation exists with regard to the present status and focus of state regulatory programs. What is commonly recognized is that the approach an agency takes to address this issue must be flexible in order to apply to a wide range of site-specific conditions, yet formalized in order to provide consistency. The following section highlights some interesting and important aspects of individual state regulatory programs and policies.

Alaska

Although not conducting research-related activities at present, the Alaska Department of Environmental Conservation is very active in the development of regulatory policy. Interim standards have been established in an effort to standardize cleanup levels to be applied by districts and regions. Soils are to be cleaned up to background, but alternative cleanup levels are permitted if it can be demonstrated that such levels will not lead to groundwater contamination. Once having outlined the RP's obligation under the Alaska Pollution Statutes, regulators request site-specific information, including the level and extent of contamination, and recommend an environmental consultant be hired to develop a contamination assessment report and remediation plan. These are to be submitted to the department for review, at which time a compliance order by consent may be negotiated, or notice of violation/request for corrective action may be issued. Alaska permits the use of sensory or onsite instrumentation as a screening technique for soil contamination, but usually requires laboratory analysis of soil samples as part of the assessment report. Excavation is the recommended remedial option; however, volatile extraction, landspreading, thermal treatment, and asphalt batching have been applied and considered acceptable methods.

California

Perhaps the most active state in terms of research, California currently administers nearly 1.5 million dollars per year through a grant program to fund demonstration projects and promote new remedial technologies. At the time of the survey 89 demonstration projects involving PCS treatment technologies were being funded by the grant program and/or supported by staff participation. The California Leaking Underground Fuel Tank Manual (LUFT) is widely recognized as the most comprehensive document of its type to date. It was intended to provide guidance to regulatory agencies, including county and municipal offices, to address leaking fuel tank problems from site investigation to remediation. At present it only applies to gasoline and diesel fuel spills, but may be supplemented at a later date to include waste oils, solvents, and other mixtures. California treats PCS as hazardous waste if it exceeds 1000 mg/kg (ppm) TPH concentration. The chain of events that takes place in the event of a spill or leak is based on the selection of a site category as outlined in the field manual. Field hydrocarbon vapor (HV) analyzers are recommended for use only as screening tools. Laboratory analysis of soil samples for BTXE and TPH are required as part of the site assessment protocol. California requires licensing and certification of analytical laboratories and also maintains a permitting system for handlers and disposers of PCS. Nearly all available technologies for cleanup of PCS have been applied either as demonstration projects or for mitigation of contaminated sites. These include vacuum extraction, biodegradation, landfarming, thermal treatment, asphalt batching, and chemical extraction. There are some limitations on the use of particular options such as asphalt batching, due to air pollution control requirements.

Connecticut

The Water Compliance Unit of the Connecticut Department of Environmental Protection takes an indirect approach to regulating the cleanup of PCS. Both cleanup standards and the nature of the regulatory response are a function of groundwater considerations. Cleanup levels are based on groundwater classification or use and Health Department drinking water standards. An agency representative will visit a site to determine the potential or extent of pollution to groundwater. An order for investigation will be issued based on the degree of contamination observed, followed by a recommended timeline for further assessment and remediation. Methods to determine the extent of contamination are handled on a case-by-case basis; however, laboratory analyses of soils samples are required to remove material offsite. The state must approve the remedial action plan for each site, and relies on excavation of material to a landfill as its primary disposal option. However, this material must have a total volatile hydrocarbon level less than, or equal to, 50 ppm to be placed in a landfill; otherwise it must be handled as hazardous material and shipped out of state. In situ technologies such as volatilization and biodegradation have seen limited use, land treatment is not viewed as a viable option, and asphalt batching is considered a promising potential application. In the event an RP fails to respond to a request by the agency for cleanup, an emergency response spill fund is available to remediate a site and subsequently bill the RP.

Florida

The Toxicological Research Program of the Florida Department of Environmental Regulation (DER) supports a number of research projects on the topic of remedial techniques related to cleanup of gasoline and other organic chemicals. The emphasis of these investigations focuses on the groundwater component of the contamination problem, which is understandable, given Florida's extensive dependence upon groundwater for its water supply. The DER Office of Technical Support has released operating procedures and guidelines for assessment and remediation of PCS.[4] The document is part of a comprehensive policy that reflects current regulatory requirements and existing regulations. These guidelines allow for initial remedial actions (IRA) to be taken at sites with free product or soil contamination in excess of 500 ppm for gasoline and 50 ppm for diesel-using flame ionization vapor detector (FID). This approach provides a means to eliminate the source of contamination while a more detailed site assessment and remedial action plan can be developed. RPs complying with DER's IRA notification policy may be reimbursed for cleanup expenses, provided disposal and treatment options have been identified prior to excavation. Excavation to a landfill, incineration, and landfarming are the most commonly used treatment methods, followed by in situ vacuum extraction, while biodegradation has seen limited use. Each treatment option has corresponding maximum allowable concentrations of volatile aromatics and total recoverable hydrocarbons in soil for its disposal.

Hawaii

PCS is not considered hazardous waste in Hawaii, but may be regulated or disposed of as such if removed from a site, or if it poses a threat to public health or the environment. There are presently no written SOPs of guidelines that are followed for site assessment or remediation, and no established levels for cleanup. The Hazardous Evaluation and Emergency Response Program of the Hawaii Department of Public Health currently relies on federal guidelines (RCRA) and state water quality standards to evaluate site-specific environmental and public health risks and determine cleanup requirements. The agency will review plans submitted by RPs for cleanup, and may require field sampling (FID, OVA) or analysis of soil samples from contractors or licensed laboratories. Aeration or volatilization, excavation to a landfill, and the passive (or do-nothing) approach are the three options that have been applied to date. Relatively few spills or leaks have been reported to the agency as yet, perhaps due in part to the fact that most commercial and industrial complexes are located away from inland drinking water sources.

Illinois

The Illinois EPA is not supporting any outside research projects at present but is conducting an in-house demonstration project involving soil gas measurements, and has an ongoing literature search and review program to determine potential health risks from exposure to PCS. It considers PCS as hazardous waste only if it fails RCRA EP toxicity or ignitability tests, and otherwise handles PCS as a nonhazardous, special, or industrial waste. Illinois has released interim cleanup objectives for gasoline, diesel fuel, kerosene, and jet fuel, which are to be used by risk managers as guidance levels when establishing cleanup criteria for a particular site. Excavation is the primary disposal option, but many other treatment options including soil venting, leaching, biodegradation, and incineration have been applied. Land treatment is not a preferred option due to air quality considerations, and in the case where groundwater contamination exists, the pump and treat method is commonly used. Regulators will often require laboratory analysis of soil samples or rely on measurements from onsite instrumentation to determine the extent and degree of contamination. State funding is available for cleanup, and contingency plans submitted by RPs are subject to agency approval.

Indiana

PCS are classified as a special waste in Indiana, for which there are presently 12 landfills in the state licensed to accept this material. The Office of Environmental Response (OER) prioritizes spills based on land use and potential adverse effects on groundwater. Indiana has adopted the cleanup standards of Illinois for disposal of PCS. Excavation to a landfill is used almost exclusively as a disposal option, and state cleanup funds are available for emergencies or when a RP cannot be found.

Kansas

Although not supporting outside research, the Department of Health and Environment reported it is conducting in-house investigations into various methodologies of headspace analysis for field sampling of soil. There are no written SOPs for site assessment and cleanup, other than general guidelines explicitly for service station tank releases. The sensory method (sight, smell) is the primary tool for detecting gasoline spills, while onsite instrumentation and laboratory analysis of soil samples may be required in other cases. Cleanup standards have yet to be established, and site-specific cleanup goals are a function of water quality considerations. Landfilling PCS is the most common disposal option in terms of volume of material; however, land treatment prior to its disposal in a landfill is the preferred method.

Kentucky

The agency does not consider PCS as hazardous waste within the context of RCRA guidelines, and handles this material as a special waste based on the constituents present and disposal option selected. No research activities are presently supported, but in-house SOPs and guidelines are under development for cleanup. The Department of Environmental Protection performs site inspections utilizing sensory and onsite instrumentation methods (H-Nu), or conducts laboratory analysis of soil samples, particularly if the contaminant is unknown, in which case a full screening of volatile organic carbons (VOCs) is performed. The agency requires a site to be returned to background or to detection limit and conducts an onsite verification. Excavation is the primary remedial option, but new landfill regulations on groundwater may result in rejection of this material and promote the application of alternative technologies. Kentucky allows for open burning of free product in some cases.

Maine

At the time of the survey, the Maine Department of Environmental Protection was soliciting a Request for Proposal (RFP) for contracted work to be awarded for research on testing of onsite aeration techniques for remediation of PCS sites. Regulatory response is handled on a site-by-site basis as a function of potential threat to groundwater and public health concerns. Sight, smell, and touch are the primary methods for site investigation. Onsite instrumentation (H-Nu) and laboratory analysis of soil samples are less frequently used, usually in cases where groundwater contamination is present. Cleanup orders are issued to RPs, which may include general guidelines for remediation. The state maintains a fund used in emergency response cases, for which the RP is responsible for reimbursement. Informal soil cleanup standards of 20 to 50 ppm TPH as measured by H-Nu meter are being applied. Site-specific cleanup goals are based on water quality considerations. Excavation followed by onsite land treatment is the most common remedial action taken. Air stripping, biodegradation, and occasionally the passive (do-nothing) approach are other options selected.

Maryland

No outside research activities are currently supported by the Department of the Environment, and PCS are considered hazardous waste only if they exhibit RCRA characteristics of EP toxicity or ignitability. There are no established cleanup levels for soils, as each site is handled on a case-by-case basis. The agency requires 48-hour notice prior to UST removal, followed by an inspection of the backfill area for contamination. Sensory methods are often applied using a "stain rule" as a basis for the presence or absence of PCS. An H-Nu meter may be utilized as a secondary procedure, and laboratory analyses of soil samples are normally performed if a site is located within a well-head protection area. Samples are analyzed for BTEX and TPH, as well as MTBE. As with most states, excavation is the primary non-in situ remedial option applied. Air quality considerations have limited the use of asphalt batching as a disposal option.

Massachusetts

The Massachusetts Department of Environmental Quality Engineering (DEQE) has drafted an interim policy for the management of PCS containing virgin product. PCS containing used waste oil or sites contaminated with gasoline having a volume/volume concentration in excess of 1800 ppm volatiles, or oil residuals with weight/weight concentration greater than 300 ppm TPH, are regulated as hazardous waste. The recently released Massachusetts Contingency Plan consists of a multi-phase process for the site investigation, transport, treatment, and disposal of PCS.[5] Cleanup goals are 100 ppm TPH and 10 ppm for volatiles as benzene; however, site-specific alternative levels may be set at the discretion of agency personnel. Laboratory analysis of soil samples are normally required for soil contaminated with oil residuals, and volatiles are analyzed using headspace analysis methods measured by Photo Ionization Detector (PID). Excavation to a state-approved landfill is the most common, yet least-preferred remedial option used at sites where less than 50 cubic yards of material are involved. The amount of PCS a landfill is permitted to accept is determined by its contaminant concentration, whether it is used as daily cover material, and whether the landfill is lined. Usually gasoline contaminated soil will be permitted in an unlined landfill only if no other disposal options are available. Asphalt batching is the preferred non-in situ remedial option for disposal of PCS. Asphalt batch plants are issued a regulatory permit which incorporates an air quality component as part of the approval process. Volatilization, biodegradation, and isolation/containment methods have seen moderate use. Treatment methods involving onsite disposal of PCS are permitted, based on land use and potential environmental impact.

Michigan

The Department of Natural Resources is interested in gathering information on alternative remedial technologies and is currently supporting research on biodegradation of PCS. Responsible parties are required to clean up to background. A combination of sensory, field sampling (H-Nu, OVA), and laboratory analysis of soil samples is utilized, both for detection and verification of cleanup. RPs must demonstrate contamination removal through a site visit by agency representatives. The agency recommends excavation as the preferred remedial option and reported limited use of other currently available treatment technologies.

Minnesota

The Minnesota Pollution Control Agency currently supports both in-house and contracted research on the topics of analytical and remedial techniques for cleanup of PCS. These investigations take into account both air and water quality considerations as part of the overall scope of work. There are written guidelines and procedures available for cleanup of PCS. Laboratory analysis and field vapor sampling of soils may be required subsequent to initial detection by sensory methods. PCS are to be cleaned to background, or less than 1 ppm TPH. Although not making specific recommendations, the agency encourages the use of asphalt batching as the preferred disposal option for PCS. Incineration and land treatment are favorable alternatives, whereas soil venting and excavation to a landfill are no longer popular. Current regulations in fact serve to limit the use of land-filling as an option for disposal. A state trust fund is available to provide up to 75% reimbursement of corrective costs from $10,000 to $100,000 for onsite cleanup.

Mississippi

The Department of Natural Resources recently released interim cleanup standards for PCS associated with UST removal and replacement. Also available are general guidelines and requirements by the agency for an Environmental Assessment Plan to be submitted within 60 days from site identification. RPs must sample for BTEX and perform EP toxicity testing for lead. BTEX levels greater than 100 mg/L require aeration. If soil concentrations cannot be aerated below 100 mg/L, the Bureau of Pollution Control may require an alternative treatment method or recommend a disposal option. PCS with BTEX levels less than 100 mg/L may be excavated and used for cover material at a landfill. Soil with EP toxicity levels for lead greater than 5.0 mg/L are classified as hazardous waste. Cleanup standards for sites with groundwater contamination are handled on a case-by-case basis. Mississippi has also enacted recent legislation to establish an UST trust fund for costs associated with site assessment and cleanup.

New Hampshire

The Department of Environmental Services is not conducting PCS-related research, but is supporting work involving in situ biodegradation of gasoline in groundwater. The agency does not consider PCS as hazardous waste, and has no licensing or permitting system for handlers, laboratories, or disposers of PCS. No written SOPs or guidelines are currently available for site assessment or cleanup; however, remediation goals have recently been established for gasoline and diesel fuels. Development of these levels were based in part on standards and approaches of other state agencies, such as the leaching potential analysis of the California LUFT Manual. A state trust fund was authorized in May 1988 for costs related to UST cleanup, and provides for up to 1 million dollars per site, with the RP paying for the first $5,000. Any or all methods for detection and verification of cleanup of PCS are acceptable in New Hampshire and handled on a site-by-site basis. Sites are prioritized, based on the potential threat to groundwater resources and public health. Excavation to a landfill is the most frequently used disposal option, followed by volatilization, and the pump and treat method for sites with groundwater contamination.

New Jersey

The Department of Environmental Protection is very active in the areas of research and regulation of PCS. A pending RFP seeks to investigate various remedial options as part of recycling efforts in the state, while other investigations focus on public health effects, including dermal absorption of compounds from contaminated soils. New Jersey treats PCS as hazardous waste if levels exceed 3% TPH; otherwise, this material is handled as an industrial waste. A further action limit of 100 ppm must be observed if the material is to be removed from a site. PCS may be left in place if below this level, depending on land use. Excavation is the primary disposal option; however, the agency prefers to recycle as much of this material as possible. Consequently, air stripping, biodegradation, thermal, and land treatment have also been applied. There are written SOPs for cleanup in the form of a flow chart. The regulatory response that takes place for a given site is a function of who is performing the work, the remedial option selected, and the proposed use of the treated material. Further legislation calls for voluntary cleanup by the RP before ownership of the property can be transferred. The Environmental Conservation and Recovery Act (ECRA) requires certification by an industrial site evaluation group that the site is deemed clean or cleaned up prior to any real estate transaction. Analysis of soil samples are required for detection and verification of removal by state certified laboratories. The agency is also in the process of developing standardized analytical protocols for laboratory analysis of soils.

New Mexico

The Health and Environment Department provides general procedures for disposal of PCS. Hydrocarbon saturated soils are defined as "materials having observable free hydrocarbon product" or "fuel that will drain freely when the soil is suspended on filter paper." Saturated soils must be removed and disposed of in a landfill approved by the Ground Water Bureau. Nonsaturated soils may be left onsite or in place, depending on the potential threat to public health based on: fire hazards; short- and long-term exposure to carcinogenic vapors; and consideration of imminent damage to public utilities and drinking supplies. Use of a PID as a field screening device, followed by laboratory analysis of soil samples from the most and least contaminated locations, is the standard procedure for site assessments. The pump and treat method is applied at those sites with groundwater contamination. Given the arid climate, excavation followed by land treatment is the most commonly used non-in situ treatment option. Material must be spread to a thickness no greater than six inches, and regularly tilled for a minimum of three months at a location where depth to groundwater exceeds 100 feet. Other technologies applied include biodegradation and carbon absorption in conjunction with air strippers. Cleanup levels are informal and a function of water quality concerns.

New York

The Department of Environmental Conservation is not supporting outside research, but does utilize information from field evaluations of new technologies taken from existing sites. At the time of the survey, guidance documents on site assessment and cleanup procedures were undergoing major revision and consolidation. Formal established levels for cleanup are also under development. The agency currently relies on sensory methods and onsite instrumentation (H-Nu), as well as the "jar" or "sheen test" for detection of PCS. PCS are routinely handled as an industrial waste in New York. Current regulations encourage excavation of PCS to a landfill, since once removed from a site PCS falls under industrial waste regulations. The pump and treat method or "muck and truck," as it's sometimes referred to, is most often applied in soil and groundwater contamination situations. Isolation or containment is considered an unacceptable alternative and volatilization is most often used in the case of gasoline contamination. Further legislation is pending in the form of an amendment to solid waste regulations that would waive current handling restriction of PCS when treatment and cleanup is performed onsite.

Pennsylvania

At present there is no state-wide coordinated regulatory effort in regard to cleanup at LUST sites. As a result, cleanup requirements are handled on a site-by-site basis. PCS is handled as hazardous waste only if it exhibits RCRA-defined characteristics of ignitability, corrosivity, reactivity, and EP toxicity. No written guidelines or procedures are currently available from the Department of Environmental Resources for investigation and cleanup of PCS. Sensory and analytical methods have been used to detect and evaluate the extent of contamination at PCS sites. There are no established levels for cleanup; however, the agency does make recommendations as to the type of corrective action and treatment option selected. Various technologies, including asphalt batching, soil venting, and biodegradation, as well as landfilling, have been applied, with no preference for any one approach.

Rhode Island

PCS is not classified as hazardous waste, but rather as a special waste, under the state's solid waste regulations. Each release is handled on a case-by-case basis. The regulatory response is dictated by site location, land usage, and the steps necessary to contain the release to prevent further contamination. The Department of Environmental Management has established informal cleanup standards of 50 ppm to 100 ppm TPH, based on site-specific conditions, and does provide some guidance for cleanup in the form of performance standards. These performance standards take into account both the way in which mitigation proceeds to operate, and the timetable for completion. A variety of methods for detection of PCS include sensory, onsite instrumentation (H-Nu, gas chromatography), and laboratory analysis of soil samples. Soil samples may be analyzed for TPH, polynuclear aromatics (PNAs), and VOCs. PNA concentrations in soil may be cleaned down to 1 ppm and VOC concentrations to 250 ppb, also depending on site conditions. Landfilling is the primary disposal option, but many other treatment methods, including volatilization, biodegradation, asphalt batching, and land treatment, have been applied.

South Carolina

No outside research is currently supported by the Department of Health and Environmental Control; however, the agency is conducting a literature search and developing guidelines for bioremediation as part of a cooperative agreement with the U.S. Geological Survey. The standard regulatory response requires the RP to perform a site assessment through field vapor sampling (H-Nu, PID) or laboratory analysis of soil samples. Samples are analyzed for BTEX and TPH, depending on the nature of the contaminant. Analytical laboratories require state certification and licensed hazardous waste haulers are normally used to transport PCS offsite. The agency reviews the remedial action plan prepared by the RP, and monitors the progress of cleanup activities. The state maintains a tank trust fund for cleanup of USTs, and also administers federal LUST trust funds through a cooperative agreement with EPA. Landfilling is the primary disposal option, followed by land and thermal treatment. Volatilization technologies have seen little use, and asphalt batching has yet to be applied, but would be allowed. Current underground injection control regulations may affect selection of the pump and treat method as a method of choice.

Tennessee

The Department of Health and Environment supports no outside research, but has drafted extensive documentation, guidelines, and policies for the cleanup of PCS. Written procedures are made available to RPs for the development of environmental assessment and remedial action plans. Separate formal cleanup standards have been established for soil contaminated with gasoline and other petroleum hydrocarbons. Soils contaminated with gasoline at a concentration greater than 10 ppm must be removed or treated in place until BTX concentrations are less than 10 ppm. Soils removed from a site with BTX values between 10 ppm and 100 ppm may be taken to an approved solid waste landfill, asphalt batched, or land farmed. Soils exceeding 100 ppm BTX may be handled as hazardous waste and disposed of as such or asphalt batched, provided air quality emission limits are not exceeded. Other petroleum hydrocarbons such as diesel fuel, kerosene, and jet fuel are regulated at the 100 ppm TPH action level. Soils between 100 and 500 ppm TPH may be handled in a similar manner, as soils contaminated with 10 ppm to 100 ppm BTX for gasoline. Laboratory analysis of soil samples are required for a site assessment with onsite instrumentation (PID) used as a field screening method. Excavation followed by land treatment is the most common remedial option. Soil venting as well as asphalt batching and biodegradation have also been applied.

Texas

The Railroad Commission of Texas has no established cleanup levels or standards for PCS. Current guidelines available for oil spill cleanup focus on the removal of free-standing product. This applies to oil on the ground surface and PCS saturated to the extent that runoff or heating from the sun could result in oil being released. The agency has no preference for a particular remedial option as long as site cleanup goals (no free product) are met. Sensory methods are the principal means of detection of PCS.

Utah

The Department of Health applies drinking water standards as a basis for cleanup of PCS. Actual soil cleanup levels are under development. The agency does not handle or treat PCS as hazardous waste unless it applies to RCRA guidelines and maintains an "approval only" role in the development of site-specific cleanup procedures and options. Volatile hydrocarbon screening of soil samples by state health laboratories is the detection method of choice. State investigations into a PCS site normally seek to establish a regulatory handle through application of RCRA guidelines or the EPA Clean Water Act. Each site is handled on a case-by-case basis. Land treatment is the most frequently used remedial option, followed by landfilling, and the pump and treat method for soil and groundwater contaminated sites. Application of any in situ remedial technologies in Utah has been minimal or nonexistent.

Vermont

By definition under state law, PCS is a hazardous waste in Vermont. However, each year the legislature renews a special exemption to handle PCS as nonhazardous waste, and future legislation aims to permanently classify PCS as nonhazardous waste. Current research by the Agency of Natural Resources centers on the investigation of various remedial techniques and their potential application pending reclassification of PCS by the legislature. Vermont uses onsite instrumentation (PID) almost exclusively as a means of detection of PCS, but may also require laboratory analysis of soil samples. The agency can clean up sites and bill the RP if the RP fails to comply with state requirements. A state trust fund for cleanup has been established, with the RP responsible for the first $10,000 and the state pays the balance up to $990,000 per site. Regulators may make recommendations as to appropriate procedures and disposal options for a particular site. There are no regulations in place that limit the use of a particular option or treatment as long as air and water quality regulations are met. PCS may be excavated, treated, and placed in a landfill if concentrations are less than 100 ppm TPH (PID). Soil venting and biodegradation are commonly applied and asphalt batching is currently being permitted for use.

Wisconsin

The Wisconsin Department of Natural Resources reported no outside research activities, but is investigating currently available remedial technologies such as thermal treatment and asphalt batching to offset the ever-decreasing number of landfills willing to accept this material. Recent legislation in the form of new air toxic discharge rules may affect the availability, cost, and ultimate selection of these technologies. A cleanup guidance level of 10 ppm TPH has been used to address the question as to whether further investigation or correction action is needed, what remedial option is most appropriate, and when treated material can be placed "back in the hole." There are written SOPs and guidelines that are followed by the agency during site investigations, and recommended procedures to RPs for cleanup are under development. Although onsite instrumentation (H-Nu, OVA) and sensory methods are used, laboratory analysis of soil samples will likely be required as the Department of Industry, Labor and Human Relations has site assessment criteria requiring contaminant characterization or TPH analysis of soils at site closure. Wisconsin has a certification program in place for analytical laboratories. Solid and hazardous waste regulations cover the transportation of PCS. Most of the commonly used remedial technologies have also been applied in Wisconsin; however, there are current regulations placing restrictions on underground injection which may limit the use of bioremediation.

Wyoming

The Department of Environmental Quality has established formal cleanup action levels for PCS which are consistent with current state water quality regulations. Soil contamination levels 100 ppm TPH or greater in an area where depth to groundwater is greater than 50 feet must be mitigated. Sites where depth to groundwater is less than 50 feet with soils containing 10 ppm TPH or greater must also be remediated. Unlike UST cleanup, there are no written standard operating procedures to be followed during investigation of emergency response cases. Specific guidelines or recommended procedures for cleanup of PCS are made by the agency on a case-by-case basis. Wyoming does make site-specific recommendations as to the type of remedial technology to be applied, both for spill and UST sites. Soil vapor extraction is commonly applied at gasoline contaminated sites, and land treatment most often used at sites with large volumes of material. However, excavation to a landfill is the most common disposal option.

SUMMARY

The results of the survey suggest state regulators are concerned about the problem of cleanup of PCS and its potential effect on public health and the environment. Most states have recently developed, or are in the process of developing, SOPs, guidelines, and cleanup standards to address the PCS problem. States vary in the type of analytical methodologies applied to determine the extent of contamination, and often handle sites on a case-by-case basis. Few states have developed formal standards for cleanup; however, many states have indicated a need for development of formalized standards, as evidenced by the number of states applying informal cleanup levels to address the problem. Many states could benefit significantly from information collected by other states' research activities and experience, through implementation of alternative management strategies. Conventional treatment options such as excavating to landfills may be replaced in the future by new technologies specifically developed to deal with the PCS problem.

REFERENCES

1. Kostecki, P. T., E. J. Calabrese, and E. Garnick, *Soils Contaminated by Petroleum: Environmental & Public Health Effects* (New York: John Wiley & Sons, 1988), p. 415–433.
2. "Underground Storage Tank Technical Requirements and State Program Approval; Final Rules," 40 CFR Parts 280 and 281, Environmental Protection Agency, September, 1988. 40 CFR Parts 280 and 281, *Federal Register*, Vol. 53. No. 185, 37082–37247.
3. "Leaking Underground Fuel Tank (LUFT) Field Manual," State Water Resources Control Board, State of California, Sacramento, CA, May, 1988.
4. "Guidelines for Assessment and Remediation of Petroleum Contaminated Soil," Florida Department of Environmental Regulation, Office of Technical Support, January 1989.
5. Massachusetts Department of Environmental Quality Engineering. 310 CMR 40.00 Massachusetts Contingency Plan, October 1988.

PART III

Analytical and Environmental Fate

CHAPTER 10

Analysis of Petroleum Contaminated Soil and Water: An Overview

Thomas L. Potter

INTRODUCTION

Spills, leaks, and other releases of gasolines, diesel fuels, heating oils, and other petroleum products often result in the contamination of soil and water. Analyses required to evaluate the extent of such releases and the threat they present to public health and the environment take a variety of forms. Analytical objectives are also diverse and often poorly specified. They range from a simple assessment of "presence or absence" to determination of the concentration of certain toxic substances these products contain.

The least specific and most general analytical approach to the problem usually involves some form of "total petroleum hydrocarbon" measurement. In contrast are analyses which are focused on selected target compounds. In either case analytical methods which are commonly used were developed by the U.S. Environmental Protection Agency (EPA). These methods have their origin in techniques specified by the agency for compliance monitoring in certain regulatory programs. It also should be noted that the methods were not specifically developed for the analysis of petroleum contaminated soil and water nor have they been systematically evaluated for this purpose.

In this review, the "state of the art" of petroleum contaminated soil and water analysis is discussed with emphasis on the U.S. EPA Methods.[1,2,3,4,5] Various examples demonstrate that the methods can be used effectively for this purpose but that problems do arise. The need for specific method development in this area is emphasized.

THE PROBLEM DEFINED: PETROLEUM PRODUCT CHEMISTRY

The source material for nearly all petroleum products is crude oil. Initial processing involves distillation into a series of fractions characterized by distillation temperature ranges and pressures. In general, the lighter fractions (lower boiling) represent gasoline-range material. The intermediate or middle distillate fractions represent feedstock for diesel, jet fuels, and "light" heating oils, and the residuum in this process serves as residual fuel oils. The trend from gasolines to the residual fuels is from the highly volatile to the nonvolatile, recognizing that in this case volatility is functionally defined.

Beyond distillation, numerous refinery processes are utilized to optimize yield of certain products and to achieve desired product characteristics. The result is that some products may have little resemblance to the distillate fractions obtained in the initial crude oil processing. Gasolines are probably the best example. Typically these products are blended and various additives are used to meet engine performance criteria.

Regardless of production modes or producers, most products are exceptionally complex materials with a wide range of physical properties. Gasolines, diesel fuels, and related products may contain hundreds or even thousands of individual constituents with boiling point distributions on the order of hundreds of degrees Celsius. Further, several chemical classes are usually represented, including paraffins, olefins, aromatics, heteroaromatics, and various polar hydrocarbons containing oxygen, nitrogen, and sulfur. In turn, each class of compounds is characterized by various homologous series within which structural, enantiomeric, and other types of isomerism are exhibited. The higher alkyl substituted homologs also predominate.[6]

Another significant characteristic of the products is that their composition is variable. This is primarily in terms of the relative amounts of the various hydrocarbons the products contain. Relative product composition may also change dramatically after release into the soil environment. Processes responsible include volatilization, dissolution, and biotic and abiotic degradation. Each process influences to greater or lesser degree certain compounds or groups of compounds and the rates of change are a function of environmental conditions.

These factors and others make petroleum product residue analysis in soil and water a formidable analytical challenge. They require that analytical methods be broad in scope. Where target compound analyses are involved there is also need for very high degrees of analytical selectivity and specificity. Compounds must be able to be detected in the presence of numerous potential interferences and considering the toxicity of many petroleum constituents, high sensitivity is needed. Detection limits in the 1 to 10 μg/L per component must be routinely achieved.

TOTAL PETROLEUM HYDROCARBONS

In light of the physical and chemical complexity of petroleum products, and the associated analytical difficulties, the analytical process is often reduced to the measurement of indicator parameters. Measurements of this type focus on determination of "total petroleum hydrocarbon content." EPA methods include Method 413.1: "Oil and Grease" and Method 418.1: "Total Recoverable Petroleum Hydrocarbons."[1]

Methods Description

These methods, which are similar to other well-known procedures, involve the extraction of hydrocarbon residues from soil and water using an organic solvent or solvent mixture. In the EPA methods, use of trichlorotrifluoroethane (FREON 113) is specified. After extraction the sample is discarded and the solvent concentrated using rotary thin film evaporation and other techniques.

Measurement of the "total hydrocarbon" content is performed using an infrared spectrophotometer. Total hydrocarbon concentration is expressed relative to the detector response to a standard mixture containing aromatic and aliphatic hydrocarbons or a petroleum product reference sample. Selected wavelengths in the 3200 to 2700 wave numbers range are monitored. They reflect absorbtion of vibrational energy by carbon to hydrogen bonds of the two classes of compounds.

In the "Oil and Grease" method, determination of the total hydrocarbons in the solvent extracts may be also be performed gravimetrically. In this case the solvent is completely evaporated and the residue weighed.

Applications

A distinct advantage of this approach is that instrumentation costs are modest and extensive technical training of analysts is not required. This translates to low cost. Unfortunately, there are some important limitations. Precision and accuracy vary widely depending on the products involved and the extent of "weathering." Even more significant is the uncertainty that the use of data obtained from analyses of this type introduces into risk assessment and management schemes.

Specific problems relate to the fact that a significant portion of the more volatile compounds in gasolines and light fuel oils may be lost in the solvent concentration step. This is especially so with gravimetric techniques. With residual fuels and other "heavy distillates," low recoveries often result for another reason. This is because many of their constituents are poorly soluble in Freon and are not effectively extracted.

Another problem, at least with the infrared procedures, is in the selection of standards. The relative response of the IR spectrophotometer to a hydrocarbon mixture is a function of the relative amounts of aromatic and aliphatic hydrocarbons it contains. Hydrocarbon mixtures that are routinely used for instrument calibration with this method have constant composition, whereas the relative composition of petroleum products and their residues are highly variable. This may introduce substantial uncertainty in the measurement.

Attempts have been made to compensate for this standardization problem by using samples of petroleum products as standards and in some cases by artificially weathering them.[7] However, there has been no systematic evaluation of the relative effect of either approach on method precision and accuracy. Some improvement is expected, but the choice of product "standards" and the extent to which laboratory "weathering" should be carried out are complex variables. Various data show that the aromatics content of products may vary by at least a factor of two and after release the relative composition of residues is highly variable, depending on numerous environmental factors.[8,9]

TARGET COMPOUND METHODS: THE 500, 600, 8000 SERIES METHODS

At the opposite extreme of indicator parameter monitoring is the direct measurement of specific constituents in petroleum contaminated soil and water. The U.S. EPA 500, 600, and 8000 series methods are probably the most widely used methods for this purpose.[2-5] Each of these methods has an associated list of target compounds for which it was specifically developed and evaluated.

Methods Description

With the passage of amendments to the Clean Water Act in 1972 and the subsequent consent decree settlement in 1976 the EPA developed a list of compounds termed the "Priority Pollutants." The agency also responded to the need for analytical procedures which could be used to detect these compounds as residues in water and wastewater by development of the 600 series methods.[10] Use of these methods is now nearly universal in public and private sector laboratories.

Related methods developed by the agency for drinking water and solid waste analysis include the EPA 500 and 8000 series methods, respectively.[3-5] In most cases, there are few if any conceptual or procedural differences between corresponding methods in these and the 600 series methods.

A common feature of the methods is that nearly all utilize gas liquid chromatography (GC) and for the most part "packed column" technology is specified. The methods also depend on highly selective detectors. GC/MS techniques in which gas chromatography columns are interfaced to mass spectrometers are emphasized.[10] These instruments are well known for their excellent sensitivity and ability to specifically detect organic compounds. An alternate detector used with methods involving monoaromatic hydrocarbons is the photoionization detector (PID). It has a relatively high selectivity for aromatics over aliphatic hydrocarbons.[11]

Only one series of methods (610, 8100), uses the flame ionization detector. This detector gives nearly universal response to hydrocarbons and offers no selectivity.[12] Identifications are based on chromatographic separations alone.

Another key aspect of the methods is they may be categorized as either a "volatiles" or "semivolatiles" method depending on the relative volatility of their target compounds. The "volatiles/semivolatiles" approach was taken in the development of the methods so that chromatographic separation, sample preconcentration, and injection techniques could be optimized.

A functional definition of the "volatiles" is those compounds which can be effectively recovered from soil or water using "purge and trap" techniques. This involves purging of the sample with an inert gas at room temperature and trapping volatile compounds stripped from the sample with a porous polymer adsorbent. The trapped compounds are desorbed directly into the inlet of a gas chromatograph by rapidly heating the trap after the column carrier gas has been diverted to flow through it.

Alternate chromatographic and sample preconcentration techniques were developed for the higher boiling "semivolatile" compounds. These methods involve liquid/liquid and liquid/solid solvent extraction with accompanying pH adjustment for recovery of "acidic" and "base/neutral" compounds and as such are often termed the "extractables." Solvent extracts are concentrated and aliquots injected directly into gas chromatographs.

Applications

The "volatiles" methods (602, 503.1, 8020, 524.1, 624, 8240) are routinely used to investigate gasoline releases. With other products such as diesel fuels, kerosene, and #2 fuel oil, the "semivolatiles" methods (610, 625, 8250, 8270) are relied upon. This reflects the relative volatility of the various products.

An example of the use of the "volatiles" approach to gasoline contaminated water is shown in Table 1. In this experiment, water was equilibrated with an unleaded gasoline. The water was then analyzed using conditions equivalent to EPA Method 8240. Key compounds detected included benzene, toluene, ethylbenzene and xylenes (BTEX). These compounds are target analytes in this and related methods and are found at relatively high concentration in most gasolines. BTEX also have a relatively high aqueous solubility and at least in the case of benzene, are considered very toxic. The maximum contaminant level (MCL) for this compound in drinking water is only 5 μg/L.[13]

Based on these results, it is clear the use of "volatiles" methods in investigation of gasoline contamination is a reasonable approach. Application of various "semivolatiles" methods where diesel fuels and related products are involved is similarly effective in that key compounds like napthalene and phenanthrene are targeted. Nevertheless, both types of applications encounter numerous problems. Troublesome issues include detection of "nontarget" compounds, the limited range of the methods relative to the composition of the various products, and poor chromatographic resolution with the packed and capillary columns specified.

Returning to Table 1, note that among the compounds detected, the one present at highest concentration was a nontarget compound. Indeed, the concentration of methyl-tertiary-butyl ether (MTBE) exceeded the sum of the concentrations of all other compounds combined. MTBE does not appear on any of the EPA target compound lists and unless identification and quantitation of nontargets are specifically requested, the presence of this compound in contaminated samples may be overlooked.

In fact, "false negative" results for MTBE may be quite common. This is alarming, considering that it is now the octane booster of choice in unleaded gasolines and is used at relatively high concentration.[14] It also is apparently transported in ground water at much faster rates than other gasoline hydrocarbons.[15] In addition, the compound is perceived to be relatively toxic. Interim drinking water standards set in various states are in the 5 to 100 μg/L range.[16]

Table 1. "Volatiles" Analysis of Water Equilibrated with an Unleaded Regular Gasoline.

Compound	Concentration (milligrams per liter)
"Target Compounds	
Benzene	29.5
Toluene	42.6
Ethyl-benzene	2.4
Xylene isomers[b]	14.7
"Non-target Compounds"	
Methyl-tert-butyl ether	116.0
1,3 pentadiene[a]	.1
2-methyl-2-butene[a]	.1
2-methyl-1-butene[a]	.1
methyl pentadiene isomer[a]	.1
2-methyl-thiophene[a]	.1

[a]Indicates tentative identification based on tabulated spectra in Reference 17. The reported concentration is an approximate result.
[b]Analysis using analytical conditions equivalent to U.S. EPA Method 8240, Reference 3.

One of the reasons why MTBE may not be detected is that the potential for this compound to occur as groundwater contaminant is apparently not widely appreciated. A closely related factor is economics. In turn, this is often the reason why the GC/PID volatiles methods (502,602,8020) are used instead of the more complex GC/MS methods (524, 624, 8240). In most commercial laboratories per unit charges for GC/PID analyses are typically less than half that of corresponding GC/MS methods. Both types of methods have BTEX in common as target compounds and can be effectively applied to petroleum contaminated soil and water. The problem is that the GC/PID methods have limited capability to specifically detect MTBE and other nontarget compounds.

MTBE is perhaps the best example of a significant nontarget compound. There are probably many others. Failure to consider this can apparently lead to management decisions which may not comprehensively address contamination problems.

An interesting case history in this regard involves a site where an underground gasoline tank leak had resulted in soil and groundwater contamination. In response to the leak, the tank was removed, free product recovered, and a "packed tower" air-stripping device installed to treat contaminated groundwater. Over several years of operation the influent of the "packed tower" was routinely monitored for "volatiles" using U.S. EPA methods 602 and 624. When the concentration of benzene, toluene, and related compounds in the groundwater had fallen below detection limits of 2 μg/L it was concluded that the established objectives of the remediation program had been met. The water treatment equipment was dismantled and removed from the site.

Several years later, a sample from a monitoring well on the site confirmed that the concentration of the "volatile" hydrocarbons in the groundwater was close to the detection limit (Figure 1a). However, this was not the case for the "extractable" hydrocarbons (Figure 1b). The total concentration of these compounds was in the parts per million range. The initial response to these data was that a fuel oil leak was responsible. This was a reasonable conclusion based on the "volatiles"/"extractables" analytical scheme. However, it did not take into account the fact that many of the more water-soluble compounds found in middle distillate fuel oils are represented in the "heavier ends" of gasolines. Ultimately, it was determined that there had been no recent releases of fuel oils on the site. The compounds that had been detected in the groundwater were gasoline residues from the original spill.

In this case what occurred is that the more soluble "volatiles" (BTEX) were leached relatively rapidly from contaminated soil at the site. Left behind were a significant portion of the "heavier ends," which apparently continue to be leached into groundwater. Further remediation is under consideration and it is not surprising that there are several legal and financial complications.

This situation developed in part because it was not recognized that while the "volatiles" methods are effective procedures for BTEX monitoring, they are not applicable to the entire range of compounds found in gasolines. The heavier ends of these products are in the "semivolatile" range. Typically, the volatile hydrocarbons in gasoline contaminated water exceed the concentration of the semivolatile compounds by an order of magnitude. But, as the product is weathered, the relative distribution can change dramatically.

The corollary to this is that the "semivolatiles" methods are not applicable to the entire range of compounds which occur in middle distillate products like diesel fuel. In fact the data presented in Table 2 show that significant quantities of "volatiles" (BTEX) can be leached from diesel and even from residual fuels.

The picture that emerges is that most petroleum products are not distinct entities but rather represent a continuum over broad ranges. This is clearly shown

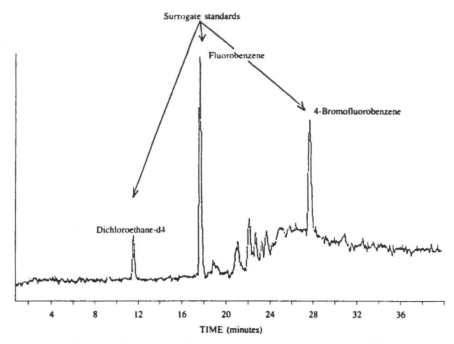

Figure 1a.   Total ion current chromatogram from the "volatiles" analysis, U.S. EPA Method 8240 (Reference 3).

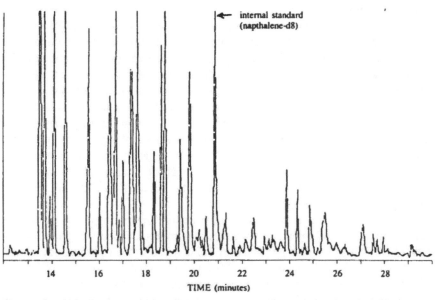

Figure 1b.   Total ion current chromatogram from the "semivolatiles" analysis, U.S. EPA Method 8270 (Reference 3).

by the total ion chromatograms of a gasoline and three middle distillate fuel oils shown in Figure 2. Considering this situation, it is probably appropriate in many circumstances to analyze samples for both "volatiles" and "semivolatiles," regardless of the product type. The solution, however, is not as simple as it may sound. It is limited by chromatography problems, especially with the "volatiles" methods. The "heavier ends" in gasolines and most of the compounds in middle distillate fuels elute very slowly from the GC columns used in the "volatiles" methods. The result is that post-analysis column conditioning is usually required. This can substantially increase analysis time.

With the "semivolatiles" methods, chromatography is also a problem. In this case it is because the very large number of compounds these products contain cannot be resolved on a single chromatography column. This why the "humps" are observed in the total ion current chromatograms of the middle distillate shown in Figure 2. "Humps" is a common characteristic of chromatograms obtained from the gas chromatographic analysis of these products even when high resolution capillary GC conditions are used. The "humps" represent an "envelope" of unresolved compounds which makes identification of target compounds which elute in this region very difficult.

To some degree the "humps" and their accompanying resolution problems are alleviated when analyzing dissolved hydrocarbons in water. This is one of the reasons why acceptable results were obtained in the analysis of the water-soluble fraction of the diesel fuel described in Table 2. In effect what occurs when a product dissolves in water is a group type separation with the aqueous phase strongly favoring the lower molecular weight aromatics. Contaminated soils, however, typically retain the complexity of the products and analytical problems are correspondingly difficult.

One other point regarding the chromatography problems associated with petroleum products can be made by examining Figure 2. Notably, chromatograms of "heavy distillate" products like motor lubricating oils and residual fuels are not presented. Their omission from the figure is not an oversight. It is due to the fact that most of the compounds in these products are not amenable to analysis by gas chromatography.

Table 2.   Benzene, Toluene, Ethyl-Benzene, and Xylenes (BTEX) Concentration in Water Equilibrated with Various Petroleum Products.

Product	Concentration (milligrams per liter)			
	Benzene	Toluene	Ethyl-Benzene	Xylenes
Gasoline	29.5	42.6	2.4	14.7
Diesel fuel	.13	.41	.18	.70
#6 Fuel oil[a]	.01	.03	.007	.05
Drinking water standards[a]	.005 (MCL)	2.0 (MCLG)	.66 (MCLG)	.44 (MCLG)

[a]#6 Fuel oil data and drinking water standards from References 18 and 13, respectively.

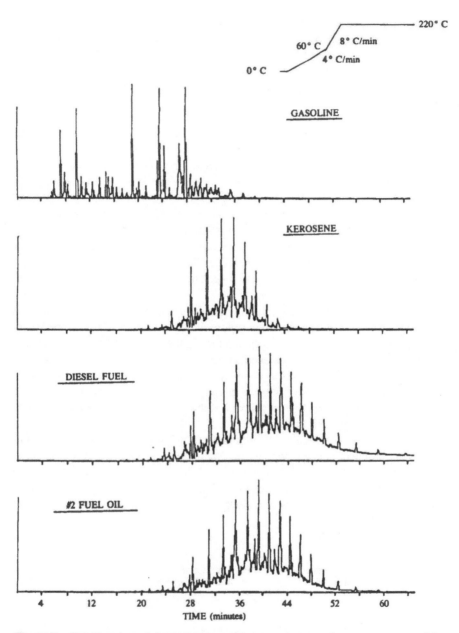

Figure 2. Total ion current chromatograms of four petroleum products on a 60 meter DB-1 (J + W Scientific) fused silica capillary column.

METHOD DEVELOPMENT NEEDS

There are clearly many potential problems in the application of the various U.S. EPA methods to petroleum contaminated soil and water. In part this is because the methods were not specifically developed nor have they been systematically evaluated for this purpose. The modification of existing methods and/or the development of new methods is needed.

With the "total petroleum hydrocarbon methods," identification of better solvents or solvent mixtures is necessary. The objective should be enhanced recovery of the higher molecular weight aromatics. However, problems will remain with low recoveries of the "lighter" hydrocarbons and with standardization. There is apparently no simple way of predicting the relative amount of aromatic or aliphatic hydrocarbons which will occur in a product or its residues recovered from soil and water.

One potentially effective alternative to the solvent extraction methods described is flash thermal desorbtion. This technique has been shown to be applicable to the analysis of a wide range of hydrocarbons in solid samples.[19] When used with a flame ionization detector, it is likely that superior precision and accuracy can be obtained.

For the target compound methods, the recent advances in capillary gas chromatography, information processing, and coupled and multidimensional chromatography techniques are expected to play an important role. The value of large diameter capillary columns with thick films has already been recognized.[20] They can accept high desorbtion flows from "purge and trap" devices and provide enhanced resolution over packed columns. The capillary columns also offer chromatography conditions which are applicable to a much broader range of compounds.

To obtain the full value of these columns in the analysis of petroleum contaminated soil and water, advances in the "purge and trap" methodology are necessary. Studies have shown that desorbtion kinetics and other factors may limit the chromatographic improvement that capillary columns offer.[21] Traps are needed which have high breakthrough volumes for broad ranges of hydrocarbons and which allow quantitative and rapid thermal desorbtion. It is realistic to expect that heated "purge and trap" techniques can be developed which would allow recovery of the higher boiling compounds in gasolines and the various middle distillate products without sacrificing recovery of the more volatile compounds.

Another improvement with the non-GC/MS volatiles methods would be use of coupled detectors. For example, the advantages of operating PID and FID in series have been shown.[11] With this approach the number of compounds detected in a single analysis can be greatly expanded and the detection of nontarget compounds facilitated.

Capillary columns have been in use for a relatively long period of time with the semivolatiles methods. These columns offer many advantages over packed columns when applied to compounds in this category. The capillary columns offer much higher resolution, inertness, and they can be directly coupled to mass spectrometers. But significant limitations remain. This is symbolized by the aforementioned "hump" in the chromatograms of middle distillate products (Figure 2).

Complete separation of all the compounds in highly complex mixtures such as diesel fuel, kerosenes, and related products is probably beyond the limits of a single gas chromatography column.[22] What is needed is group type separations prior to gas chromatography. That is, separation of the compounds into more homogeneous groups by liquid chromatography. Unfortunately, group type separations are time-consuming and labor intensive. Advances in combined microbore liquid chromatography/high resolution gas chromatography offer the potential for performing this "on-line."[23]

One final point that requires emphasis is that target compound lists need to be modified to more comprehensively address the contaminants which may occur in petroleum contaminated soil and water under varying conditions. This must be done with the recognition that numerous factors determine information and ultimately analytical needs. Experiences with MTBE should also tell us that petroleum product chemistry is not constant. As sources of crude oil, other raw materials, and economic factors change, significant changes in product composition are expected. Analytical methods must change accordingly.

REFERENCES

1. "Methods for the Chemical Analysis of Water and Wastes,"EPA-600/4-79-020, U.S. Environmental Protection Agency, 1979.
2. "Guidelines Establishing Test Procedures for the Analysis of Pollutants Under the Clean Water Act; Final Rule and Interim Rule," U.S. Environmental Protection Agency, 1984. 40 CFR Part 136. *Federal Register* 49(209):1–210.
3. "Test Methods for Evaluating Solid Waste," 3rd ed., Doc. No. SW-846, U.S. Environmental Protection Agency, 1986.
4. "Method 524.1 Volatile Organic Compounds in Water by Purge-and-Trap Gas Chromatography/Mass Spectrometry," U.S. Environmental Protection Agency, 1986. Environmental Monitoring and Support Laboratory, Cincinnati, OH.
5. "Method 503.1 Volatile Aromatic and Unsaturated Organic Compounds in Water by Purge-and-Trap Gas Chromatography," U.S. Environmental Protection Agency, 1986. Environmental Monitoring and Support Laboratory, Cincinnati, OH.
6. Speight, J. G., "The Chemistry and Technology of Petroleum," (New York: Marcel Dekker Inc., 1980).
7. DeAngelis, D. "Quantitative Determination of Hydrocarbons in Soil," in M. Kane, Ed. *Manual of Sampling and Analytical Methods for Petroleum Hydrocarbons in Groundwater and Soil*, American Petroleum Institute Publication 4449, Health and Environmental Sciences Department, Washington, DC, 1987.
8. Ury, G. B., "Automated Gas Chromatographic Analysis of Gasolines for Hydrocarbon Types, *Anal. Chem.* 53: 481–485 (1981).
9. Edgerton, S. A., R. W. Coutant, and M. V. Henley, "Hydrocarbon Fuel Dispersion in Water," *Chem.* 16(7):1475–1487 (1987).
10. Telliard, W. A., M. B. Rubin, and D. R. Rushneck, "Control of Pollutants in Wastewater," *J. Chromatog. Sci.* 25:322–327 (1987).
11. Driscoll, J. N., and M. Duffy, "Photoionization Detector: A Versatile Tool for Environmental Analysis," *Chromatography* 2(4):21–27 (1987).
12. Tong, H. Y., and F. W. Karasek, "Flame Ionization Detector Response Factors for

Compound Classes in Quantitative Analysis of Complex Organic Mixtures," *Anal. Chem.* 56:2124–2128 (1984).

13. "National Primary Drinking Water Regulations: Volatile Synthetic Organic Chemicals; Final Rule and Proposed Rule," U.S. Environmental Protection Agency, 1985. *Federal Register* 50(219):46882–46933.

14. Anderson, E., "MTBE Strengthens Hold on Octane Booster Market," *Chem. Eng. News*, October 13, 1986, p.8.

15. Garrett, P., M. Moreau, and J. Lowry, "Methyl Tertiary Butyl Ether as a Ground Water Contaminant," in *Proceedings of Third National Conference on Petroleum Hydrocarbons and Groundwater*, National Water Well Association, Dublin, OH, 1986.

16. Personal communications: Maine Department of Environmental Protection, Connecticut Department of Environmental Protection and Florida Department of Environmental Management.

17. *Eight Peak Index of Mass Spectra*, 3rd ed., Royal Society of Chemistry, The University, Nottingham, UK, 1983.

18. Burchette, G., "Number Six Fuel Oil and Ground Water," *Ground Water Monitoring Review* 6:32 (1986).

19. Crist, W. A., J. Ellis, J. de Leeuw, and P. A. Schenck, "Flash Thermal Desorbtion as an Alternative to Solvent Extraction for the Determination of C8 to C35 Hydrocarbons in Oil Shales," *Anal. Chem.* 58: 258–261 (1985).

20. Reding, R., "Chromatographic Monitoring Methods for Organic Contaminants Under the Safe Drinking Water Act," *J. Chromatog. Sci.* 25: 338–344 (1987).

21. Mosesman, N. H., L. M. Sidsky, and S. D. Corman, "Factors Influencing Capillary Analyses of Volatile Pollutants," *J. Chromatog. Sci.* 25:351–355 (1987).

22. Pitzer, E. W., "Contributions of Stereoisomerism to Peak Shapes of Branched Paraffins in the High-Resolution Gas Chromatographic Analyses of Jet Propulsion Fuels," *J. Chromatog. Sci.* 26:223–227 (1988).

23. Duquet, D., C. Dewaele, and M. Verzele, "Coupling Micro-LC and Capillary-GC as a Powerful Tool for the Analysis of Complex Mixtures," *J. High Res. Chromatog. and Chromatog. Commun.* 11: 252–256 (1988).

CHAPTER 11

Field-Screening Techniques: Quick and Effective Tools for Optimizing Hazardous Waste Site Investigations

Lynne M. Preslo, Walter M. Leis, and Raphe Pavlick

GOALS OF SITE INVESTIGATIONS

The goals for site cleanup at hazardous waste sites can be quite varied. These goals can range from complete cleanup and site restoration of the environment, along with controlling offsite migration of contaminants, all the way to minimal actions focusing on monitoring the dissipation of contaminants. In order to achieve the goals for a particular site cleanup, the engineering team requires data with which to identify remedial alternatives, assess risks, compare these risks and alternatives against ideal situations and to each other, and, not least of all, to evaluate the costs that would be involved with site actions.

The generation of this data base to characterize the site is the function of field investigations. In the past, these investigations often have become a goal in themselves; the goal was to fully characterize a site without any questions or ambiguities remaining. This method often was carried out whether the resultant data were significant to site cleanup or not. Usually such comprehensive investigations generate more data than actually are needed for the design of remedial actions for the site, no matter what goals have been established for cleanup. In addition, significant costs are incurred while conducting these studies, and work schedules are modified to accommodate further fieldwork.

To develop better, more optimal approaches to site characterization and remediation and to control unnecessary expenditures, we must recognize a few basic rules:

1. All sites are different. These differences manifest themselves in variables of investigation and cleanup design. Therefore, a degree of flexibility in approaches must be exercised to avoid formula or "cookbook" approaches that often provide reams of unnecessary, costly data.
2. Engineering feasibility and site investigation phases are joined inexorably, to the extent that the needs for engineering feasibility studies (FS) for remedial actions should guide the remedial investigation (RI) effort.
3. All available tools and methods should be considered and the appropriate ones should be applied at the proper time to provide a flow of necessary data to solve the problem at hand efficiently.

This chapter proposes the application of some available but lesser used field techniques to increase the efficiency of collecting the necessary field data needed for remedial design.

In accordance with the RI/FS guidance documents and the National Contingency Plan (NCP), the RI/FS processes are an integrated work effort. This process produces a fully integrated approach that efficiently combines both the site characterization and engineering design efforts.

The efficiency imposed by such an integrated geological/ engineering approach limits the amount of chance and/or excessive data collection that often is unnecessary, and will focus data collection efforts to answer only those critical questions for remedial design, risk assessment, and ultimately the selection of a final remedial option.

To limit random data collection in an RI/FS program, we advocate the application of a sequence of activities that includes:

1. Conduct a field program driven by the needs of the engineering design program.
2. Exhaust all available site data before collecting new site data.
3. Conduct less intrusive studies which do not disturb the site and possibly cause more risk of exposure before conducting more site destructive evaluations.

As stated in Item 1, this sequence of activities requires a statement of project philosophy which should be followed throughout the program. One form of the application of this philosophy has been discussed above, while the following sections will devote closer attention to Items 2 and 3.

USE OF PREEXISTING SITE DATA

In the preparation of a preliminary definition of a site problem, it is necessary to have adequate information from which to project requirements, if any, of future site investigations. Often, especially at sites in urban and suburban areas, it is possible to glean a high quality data base from past records, and remote sensing data from historical or time sequence aerial photographs.

Such photography usually is available through commercial air photo services, state planning and regulatory agencies, U.S. Department of Agriculture archives, and many times from industries. Photos of historical overflights are compiled and when compared at equivalent scales can provide a wealth of initial data in the following areas:

- *What was the appearance of the site prior to deposition of wastes or before development?* Such information is critical to assess predevelopment drainage, topographic changes, and natural soils and geologic data.
- *What were the modes and times of deposition?* The areal growth patterns of waste areas at a site provide critical information on the deposition sequence and potential migration of contaminants.
- *What are the volumes and surface areas of wastes?* Initial information on the site size and volume is provided so that the scope of site cleanup can be conceptualized even before field work begins.

When evaluated by the entire RI/FS team (scientists and engineers), these photogenerated data can provide a fair degree of quantitation of the problem at hand, and can direct the team to propose initially remedial options to be considered. For example, if these photographs were available to initially review along with local groundwater data, the remedial action team should be able to define, with a high degree of accuracy:

- possible source areas
- potential contaminant migration directions
- candidate remedial actions based on the estimated volumes at the site

All this is possible prior to beginning any field activity at the site.

Additionally, in concert with time sequence photography, a wealth of site data exists from site records, recollections of long-time employees, and informal data provided by workers, officials, residents, and regulatory agencies.

NONDESTRUCTIVE SITE INVESTIGATION TECHNIQUES

Remote-sensing data and searches of site records will allow a fairly accurate assessment of site problems with an initial indication of possible remedial options. The mobilization and field plan preparation phases then would proceed to collect only necessary field data to support engineering options. Such an approach therefore would include sufficient data for characterization of the site, and collection ideally would begin first with minimally intrusive techniques, and then utilize the more intrusive techniques such as monitor well installation, soil borings, or test trenching only after the initial source information is compiled.

The principal categories in the minimally intrusive techniques include shallow geophysics and site-screening data using onsite chemical analyses. Both categories contain a variety of techniques that may be applied to large and small sites in a manner that is both efficient and cost-effective.

SHALLOW GEOPHYSICAL TECHNOLOGIES

Shallow-sounding geophysical techniques include electromagnetic conductivity (EM), shallow and deep resistivity, magnetometric surveys, VLF surveys, IP surveys, ground penetrating radar (GPR) and in certain applications acoustical, gravity, and radiological surveys.

These basic techniques are applied routinely in site investigations and have proved their worth for initial site surveys. The costs associated with running single unit or multiconfiguration geophysical surveys are relatively low, and the system responses for almost every type of buried waste are quite acceptable, provided experienced personnel design and interpret the surveys. Further, signal enhancement and filtering techniques using computer enhancement codes are readily available for microcomputer systems, so that data analyses can proceed even in the field.

A major inhibiting factor to the greater general use of surface geophysical techniques is that in the past, choices of inappropriate techniques often were made due to lack of understanding of the limitations and responses of the particular system. Also, even with appropriate systems and good data reduction, the vertical resolution of geophysical anomalies is fairly shallow and can often be plus or minus 10 feet. However, with proper system selection and operation by trained personnel with a full understanding of anomaly resolution, geophysical techniques are very powerful initial survey tools that can refine the preliminary phase of problem definition (often by orders of magnitude from remote sensing data). By using surface geophysics, the level of understanding of a site and refinement of preliminary remedial options can increase by 10- or 100-fold without incurring massive project costs associated with highly incursive techniques such as monitor well drilling.

ONSITE CHEMICAL FIELD SCREENING

In addition to remote sensing and surface geophysics, onsite chemical analytic techniques are being used increasingly as field screening techniques. Chemical screening in the field consists of a number of techniques which are minimally intrusive, but still require field efforts which cost an order of magnitude or more less than intrusive techniques such as monitoring wells or soil borings. The field screening techniques provide information which streamlines data collection efforts by optimizing:

• the use of intrusive techniques (soil borings and monitoring wells)
• the number of samples sent to the laboratory for confirmatory chemical analysis

These site-screening techniques are showing growing favor with regulatory agencies for assessing source areas or "hot spots" within the soils and the aeral extent of volatile organic chemicals (VOCs) in groundwater through the monitoring of diffused gases in the overlying vadose zone. The analytical approaches for detection of soil gas reported in the literature vary from qualitative relative indicators such as organic vapor analyzers (OVA) to more quantitative soil gas screening applied in field survey conditions[1,2]. At these sites the authors utilized organic screening for groundwater plume tracking, as well as for the identification of waste sources within the vadose zone. Further, applications of field screening of VOCs in soil vapor are applied routinely to check the integrity of underground storage tanks by both mobile and installed in situ monitors. As described in Table 1, the less qualitative meters (OVA, HNu and IR detectors) are subject to interferences, false positives, and false negatives. The more quantitative methods, gas chromatography and gas chromatography/mass spectrometry (GC/MS), are compound-specific and produce much more reliable data.

Field screening has been applied to the detection of less volatile organics on a specific analyte basis (Table 1). For example, uptake of PCBs by glabram (sticky sap) plants has resulted in the identification of a depressed air plume of these chemicals at old landfills in the upper Hudson Valley. Another example is site

Table 1. Field Screening Techniques.

Instrument	Quantification	Limitations
SOIL GAS METHODS (Volatile Compounds)		
OVA or HNu	~1 ppm	-False (+) and (−) -Interferences from other compounds, H_2O vapor -High d.l.
IR Detectors (e.g., Miran)	~1 ppm	-False (+) and (−) -Interferences -High d.l.
Field-mounted GC (with various detectors)	Can be ppt, depending on sample size and detector	-Tentative IDs -Time-consuming -Interferences (other target compounds or extraneous substances)
Field-mounted GC/MS	Can be ppt, depending on sample size and detector	-Operator Training

NON-VOLATILE ORGANIC TECHNIQUES HAVE BEEN DEVELOPED FOR:

-PCBs
-PNAs
-Oil and grease
-Micro-extraction followed by UV, GC/EC, GC/PID, or GC/MS

Table 1. (Continued)

Instrument	Quantification	Limitations
INORGANICS/METALS		
Colorimetric	~1 ppm to %	-Metal group specific -Interferences
Gravimetric	Qualitative	-Interferences
Specific ion probes	~1 ppm to %	-Ion-specific -Interferences
X-ray fluoresence	~1 ppm: maybe lower	-McGovern/Spittler exploring this technique in Region 2

Adapted from: Preslo & Leis, 1985; *AEG Annual Meeting.*[3]

screening of polynuclear aromatic compounds using ultraviolet light detection of PNAs, techniques which have been applied for years by petroleum exploration geologists to detect hydrocarbons in drill cuttings.

Field-screening techniques are not limited to organic chemicals. The technique of trace metal analyses in shallow rock and in plants overlying ore bodies has long been an accepted technique to focus on a commercial metal deposit. In field screening for metals, a classic technique had been to use organic chelating agents to create an organic-metallic complex, the analytical determination of which is accomplished by a color produced by the specific complexing agent and the metallic salt. These techniques have been used in a semiquantitative fashion to estimate the concentrations of metals in a waste source or in soils.

SUMMARY

Even with the limitations listed with each of the methods, these screening techniques represent an efficient investment in time and project funds, since they can in many cases provide the necessary data upon which site cleanup can be based. This is an idea that is in concert with the U.S. Environmental Protection Agency's desire to expedite remedial action at abandoned waste sites.

In summary, the ability to rapidly conduct site characterizations and provide necessary baseline data for remedial action engineering is enhanced by the use of the less intrusive field techniques itemized in this chapter. Further, by conducting field studies in a support role to the remedial engineering, the efficiency of the total project operation is maintained and costs are minimized.

REFERENCES

1. Preslo, L., "Soil-Gas Screening as a Tool to Optimize Site Investigations and Cleanups," *Weston's Way,* Roy F. Weston, Inc., 1987.

2. Lappala, E., and G. Thompson, "Detection of Groundwater Contamination by Shallow Soil Gas Sampling in the Vadose Zone: Theory and Applications," in *National Conference on Management of Uncontrolled Hazardous Waste Sites*, HMCRI, 1985, pp. 20–28.
3. Preslo, L., and W. Leis, "Field Screening Techniques for Hazardous Waste Site Investigations," Association of Engineering Geologists Annual Meeting, December, 1985.

CHAPTER 12

Onsite Analytical Screening of Gasoline Contaminated Media Using a Jar Headspace Procedure

John Fitzgerald

INTRODUCTION

Portable analytical instrumentation is now widely used to investigate and document conditions of environmental contamination by volatile organic compounds, including media contamination by complex mixtures such as gasoline and other light petroleum distillates. In addition to providing substantial cost-savings over traditional laboratory procedures, the utilization of such equipment provides immediate onsite data for use in assessment or remedial response actions.

A majority of portable analytical units currently in use contain or consist of a photoionization detector (PID) or a flame ionization detector (FID). Most units are designed or operated solely as PID or FID response meters, without chromatographic separation, and provide quantitative data on "total organic vapor" (TOV) concentration.

Due to simplicity and expediency of operation, virtually all portable field units are designed to analyze gaseous samples at ambient temperatures. Multimedia analysis of soil and water samples is accomplished by a "headspace" technique, which involves the mass transfer or partitioning of volatile contaminants from an aqueous or bulk soil matrix to an overlying, confined gaseous phase.

Despite widespread usage, very little information has been published on the applications and limitations of such field procedures.

Without qualitative (chromatographic) definition, TOV headspace methodologies are intrinsically limited to the analysis of samples containing a known contaminant or contaminant mixture. Even within this limited universe, additional questions arise on the interpretation, accuracy, and significance of resultant headspace data. Indeed, aside from the broader interpretative issues, the evaluation and even comparison of TOV headspace data has traditionally been hampered by the lack of a standard, universally accepted procedure.

It is the intent of this author to begin discussions on this subject. Research was conducted to investigate the use and utility of a PID and an FID meter for the field headspace analysis of soil and water samples contaminated with unleaded gasoline. A series of laboratory and field experiments were conducted in order to:

(1) evaluate instrument operation and response characteristics
(2) evaluate the rate, extent, and chemistry of jar headspace development
(3) evaluate the effects of various physical and environmental factors on headspace development and instrument detection and response, and
(4) develop a preliminary correlation of "total organic vapor" headspace measurements with headspace and media concentrations of targeted gasoline constituents of concern, including benzene, toluene, ethylbenzene, and xylenes (BTEX).

BACKGROUND

In order to correlate or interpret TOV headspace screening data for a complex mixture such as gasoline, it is first necessary to gain an understanding of: (1) the chemistry and environmental fate of gasoline, (2) the volatilization process, and (3) the operational mechanics and selectivity of portable PID and FID instrumentation. While a detailed discussion of these three items is beyond the scope of this chapter, pertinent elements are briefly outlined in the following paragraphs.

The Chemistry and Fate of Gasoline

Industry specifications for gasoline products are based upon physical and performance-orientated criteria, and not upon a designated chemical formulation.[1] In addition to a variety of additives, refined gasolines may contain any combination of petroleum hydrocarbons from C_2 through C_{13}. While formulations are variable, Domask has reported, excluding additives, 42 specific constituents generally comprise about 75% of unleaded blends.[2]

With the notable exception of certain water-soluble and/or toxic additives, the monoaromatic compounds benzene, toluene, ethylbenzene, and xylenes (BTEX) are generally considered the most environmentally significant components of gasoline. The BTEX fraction typically constitutes approximately 15% of unleaded gasolines.[2] Because of its relatively high water-solubility, volatility, and toxicity, benzene, which generally comprises between 1% and 3% of gasoline, is normally targeted as the individual gasoline constituent of greatest concern.

In addition to the BTEX compounds, however, dozens of other volatile gasoline constituents will elicit a response on typical PID and FID field units. In this regard, three categories or component groupings may be designated within a gasoline formulation, as graphically illustrated in Figure 1, which depicts a (packed-column) chromatogram of an unleaded blend used during experimentation.

Peak elutions within Figure 1 are generally reflective of the vapor pressures of the individual (or co-eluting) gasoline components. The "first third" grouping represents a number of relatively low-boiling-point hydrocarbons, predominated by normal paraffins, isoparaffins, cycloparaffins, olefins, and cyclo-olefins. The "second third" grouping is composed primarily of the BTEX compounds. Finally, the "last third" components are comprised of heavier molecular weight (greater than 110), high-boiling-point hydrocarbons, and are predominately alkyl-aromatic compounds.

Once released to the environment, the chemical composition of gasoline residuals will be altered by environmental fate processes. In general, within the context of Figure 1, fate processes tend to result in a shift of relative component groupings from left to right: initial volatilization of "first third" hydrocarbons, with concurrent and subsequent leaching, volatilization, and biodegradation of the "second third" (BTEX) compounds. The specific extent and combination of fate processes, temporally and spatially, will dictate the media concentration and distribution of gasoline constituents, which in turn will dictate the extent and chemistry of headspace development and portable PID or FID unit response.

Volatilization and Headspace Development

While volatilization on a macro-environmental scale can significantly influence gasoline composition and sample chemistry, short-term volatilization on a micro-environmental scale is of principal concern in the analytical headspace screening of sample media.

Theoretical and empirical relationships used to describe and model volatilization are based upon mass transfer principles. Graphical depictions of the volatilization process for aqueous and soil systems are presented in Figures 2 and 3, respectively.

For aqueous systems, the "two-layer film" theory of mass transfer resistance is generally utilized to describe and model volatilization. "Henry's Law," an empirical relationship concerning the partitioning of volatile chemicals between the aqueous and vapor phases, is widely used to predict the extent of compound volatilization. According to this relationship, at equilibrium, the concentration of a gas dissolved in a liquid is directly proportional to the partial pressure of that gas above the liquid; the constant of proportionality being the Henry's Law constant.[3]

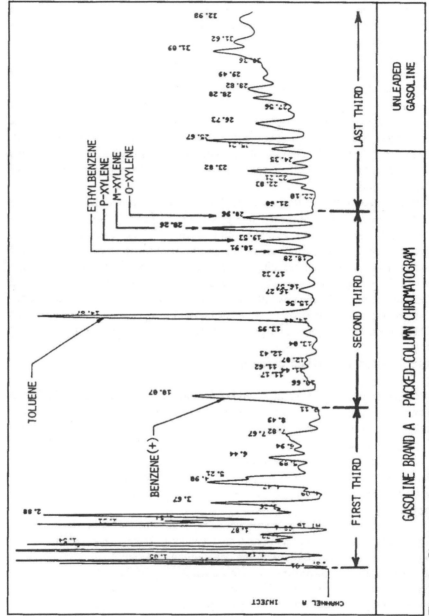

Figure 1.   Chromatogram of an unleaded blend gasoline.

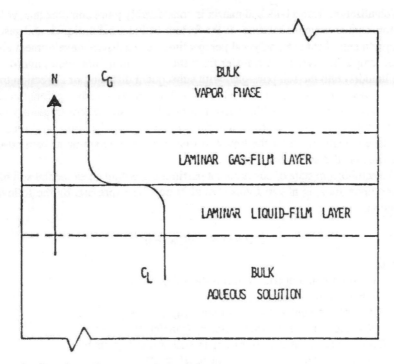

Figure 2. Two-layer film theory of mass transfer.

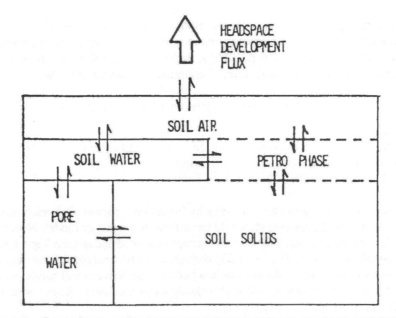

Figure 3. Contaminant partitioning and volatilization within an elemental soil block.

Volatilization from a bulk soil matrix is considerably more complex than volatilization from an aqueous solution, as the former includes solid, liquid, and gaseous compartments. From an analytical perspective, jar headspace development above a soil sample involves mass-transfer from the solid, immobile liquid, and/or mobile liquid(s) into the soil pore-gas, with subsequent diffusion or advection to the soil surface, and final diffusion/advection through the overlying containerized headspace. In moist soils without free (immiscible) gasoline, constituents volatilize from the soil/water interface into the soil pore-gas. The concentration at the interface is maintained by the liquid diffusion of the contaminant from adsorption sites on the soil solids.[3]

The equilibrium state of contaminant partitioning within an elemental soil block (based upon Jury, et al., in Lyman, et al.[3]) may be expressed by the following equation:

$$C(t) = \gamma Cs + \theta Cw + nCa \tag{1}$$

where:
   $C(t)$ = total contaminant concentration ($\mu g/cm^3$)
   γ = soil bulk density (g/cm^3)
   Cs = contaminant sorbed onto soil ($\mu g/g$)
   θ = volumetric soil water content (cm^3/cm^3)
   Cw = contaminant concentration in soil water ($\mu g/cm^3$)
   n = volumetric soil air content (cm^3/cm^3)
   Ca = contaminant concentration in soil air ($\mu g/cm^3$)

Substituting a Freundlich isotherm relationship for Cs, assuming adsorption is controlled by the organic carbon content of the soil, and substituting a Henry's Law relationship for Ca, it is possible to estimate the equilibrium state of a contaminant in terms of the soil water contaminant concentration, Cw:

$$C(t) = \gamma(Koc)(\%Soc)(Cw)/100 + \theta Cw + (n)(H')(Cw) \tag{2}$$

where:
   Koc = soil (organic carbon) partition coefficient ($\mu g/g$) / ($\mu g/cm^3$)
   $\%Soc$ = percentage of soil organic carbon (%)
   H' = dimensionless Henry's Law constant

Since gasoline is a mixture of many hydrocarbons, partitioning relationships between individual compounds would be difficult to predict or model. Nevertheless, for perspective, applying the relationships described in Equation 2 to a typical soil sample ($n=0.2$, $\theta=0.1$, $\gamma=1.58$) containing (only) benzene at a contaminant concentration of 1 $\mu g/g$, the equilibrium partitioning between soil solids, water, and air at a low, medium, and high organic carbon content ($\%Soc$) would be as shown in Table 1.

Significantly, mass percentage distribution of benzene (and the remainder of the "second third" BTEX compounds) favors the solid and water compartments. While this is even more true of the "last third" constituents, the high vapor pressures and Henry's Law constant of "first third" compounds result in substantially higher gaseous-phase partitioning. Though the relative mass partitioning of benzene soil gas may seem minor, it is important to note that even at a soil organic carbon content of 5%, the predicted equilibrium gaseous phase concentration of 0.02 $\mu g/cm^3$ represents a volumetric gaseous concentration of 6.2 ppm (20 °C and 1 atm), well within the detection limits of portable field PID and FID units.

Table 1. Equilibrium Partitioning in a Soil Sample.

Soc %	Cs $\mu g/cm^3$	Cs %mass	Cw $\mu g/cm^3$	Cw %mass	Ca $\mu g/cm^3$	Ca %mass
0.2	0.693	43.9%	0.624	39.5%	0.26	16.6%
2.0	1.400	84.5%	0.127	8.0%	0.05	3.2%
5.0	1.504	93.2%	0.005	3.4%	0.02	1.3%

Portable Field Instrumentation

In order to produce meaningful screening data, it is essential that the operating principles and mechanics of field PID and FID units be understood and that the manufacturers' recommendations on operation, maintenance, and calibration be instituted. In this regard, several items are worth noting:

(1) The difference in detector response and selectivity between PID and FID systems is significant. While FID response is relatively uniform for most volatile gasoline hydrocarbons, PID response increases with degree of compound unsaturation, resulting in an approximately 5 times greater response to benzene than hexane using a 10.2 eV lamp. Selectivity of this nature may be useful in focusing on the BTEX monoaromatic compounds within a media sample.

(2) Instrument flow rate and response characteristics vary. These parameters are especially relevant when attempting to analyze headspaces in small sampling containers.

(3) Normally, maximum jar headspace response is obtained 2 to 5 seconds following sampling initiation. Meter response and maximum deflection is difficult to discern on a unit with digital readout, unless a "maximum hold" feature is available.

(4) All field units are susceptible to environmental and temperature effects. Moreover, PID response may be significantly reduced under conditions of elevated humidity.

(5) Unlike PID units, FID systems will respond to methane, which may cause significantly higher TOV headspace concentrations in certain media samples and/or analytical settings.

EXPERIMENTAL PROCEDURES

Experimentation was conducted in three areas: (1) evaluation of portable unit operation, calibration, and jar headspace response characteristics, (2) aqueous headspace development and analysis, and (3) soil headspace development and analysis. Media experiments in areas (2) and (3) centered on the examination of a field headspace procedure comprised of the following:

(a) placement of a soil or water sample in a glass container (generally to one-half jar capacity)
(b) application of a sheet of aluminum foil to seal the glass jar
(c) allowance of a period of time for static or dynamic headspace development
(d) subsequent analysis of the jar headspace via the insertion of a PID or FID unit sampling probe through the aluminum foil seal, recording the maximum meter response as the "total organic vapor" headspace determination.

For increased relevance to "field" objectives and limitations, experimental evaluation focused on short-term development techniques under conditions and restrictions which may apply at a field site.

In all areas of experimentation, correlative headspace data was obtained by comparison of "total organic vapor" meter responses with the chromatographic analysis of a 5 mL split-sample. For comparative purposes, both PID and FID field units were calibrated to respond "as benzene" (i.e. assume a benzene response factor).

Media experiments were performed on water and soil samples contained in 5 oz (140 mL) through 16 oz (480 mL) glass jars in which basic headspace development parameters were varied, including:

- shape/volume of jar
- depth of sample
- headspace development temperature
- headspace development time
- static vs dynamic headspace development
- gasoline concentration

Aqueous samples were formulated by adding measured quantities of a well-characterized ("fresh" and evaporatively weathered) unleaded gasoline stock to distilled, organic-free water. Soil headspace analyses were conducted on silica sand samples contaminated with unleaded gasoline stock and subjected to various degrees of leaching and passive aeration. Limited analyses of actual field samples were also conducted.

In total, 183 aqueous samples and 128 soil samples were analyzed by an H-Nu Systems and/or Century Systems unit, with split-sample chromatographic analyses performed on 16 of the aqueous samples and 24 of the soil samples.

Physical and environmental variations within discrete experimental setups allowed a relative evaluation of the "mechanical" aspect of headspace development (i.e. sample size, agitation, temperature, etc.). Moreover, viewed externally with the results of other compatible setups, data obtained from individual and collective experimentation allowed a limited evaluation of the overall extent and kinetics of headspace development, headspace chemistry, and instrument response.

Two field units were evaluated during the course of this research project. The PID unit was an H-Nu Systems Portable Photoionization Analyzer, Model PI-101, with a 10.2 eV lamp probe. The FID unit was a Century Systems Portable Organic Vapor Analyzer, Model OVA-128, utilized on a direct FID response mode.

Chromatographic identification and quantification of gasoline and headspace vapor samples was accomplished on a Hewlett-Packard (HP) Model 5790 gas chromatograph equipped with either a packed or capillary column and a flame ionization detector. The HP gas chromatograph was interfaced with a Spectra-Physics Model SP 4270 Computing Integrator for FID signal processing. Chromatographic operating procedures were based upon a modification of a methodology developed by Bruell and Hoag at the University of Connecticut.[4]

EXPERIMENTAL RESULTS AND CONCLUSIONS

Experimental observations, results, and conclusions are summarized below:

Portable Instrumentation

(1) Suppression of H-Nu PI-101 (PID) meter response was observed under conditions of elevated atmospheric humidity. For benzene, response reduction was as high as 40% under certain conditions; suppression of other BTEX compounds may not be as pronounced.

This finding is consistent with information published by Barsky et al.[5] and Chelton et al.[6] Although not ionizable with a 10.2 eV lamp source, water vapor molecules apparently inhibit PID response (a) by absorbing emitted ultraviolet radiation, thereby decreasing photon intensity to ionizable sample compounds, and/or (b) by colliding and reacting with photoionized species.

Water vapor effects were not noted on the Century Systems OVA (FID) unit.

(2) H-Nu PI-101 response suppression was also noted on gasoline-vapor concentrations above 150 ppm (v/v) Total Organic Vapors (TOV) as benzene. Although apparently independent of the water vapor

condition, this phenomenon may be attributable to similar mechanisms of ultraviolet absorption and/or collisional deactivation.

(3) With both units calibrated to respond "as benzene," the H-Nu PI-101 to Century OVA (or PID/FID) response ratio ranged from 0.56 for "fresh" gasoline vapor to 1.35 for (evaporatively) weathered gasoline stock vapor. This finding is graphically displayed in Figures 4 and 5 and is consistent with detector selectivities and gasoline constituent distribution.

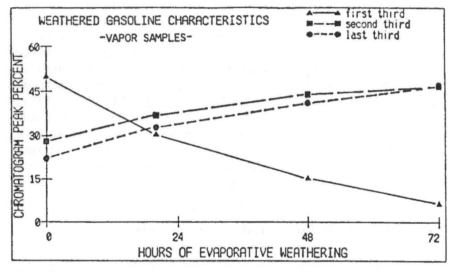

Figure 4. Weathered gasoline characteristics—vapor samples.

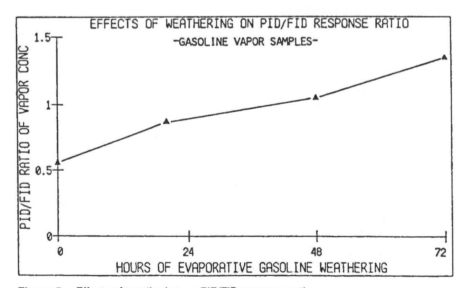

Figure 5. Effects of weathering on PID/FID response ratio.

Jar Headspace Analytical Screening Technique

(1) Using the "probe insertion" jar headspace sampling technique previously described, maximum H-Nu and Century OVA instrument response was generally obtained 2 to 5 seconds following probe insertion, as illustrated in Figures 6 and 7. This response value was recorded as the jar headspace sample concentration. Headspace concentrations above 150 ppm (v/v) TOV as benzene would on occasion produce erratic or significantly nonlinear meter responses. Accordingly, all data comparisons were limited to samples yielding less than this value.

(2) Using the "probe insertion" method, H-Nu meter response (accuracy) was observed to be approximately 50% of the true headspace vapor concentration in one-half filled 9 oz and 16 oz sample jars. OVA Model 128 response (accuracy) was approximately 65% for 16 oz jars, and 55% for 9 oz jars. Significantly lower response levels were noted in 5 oz sample jars, indicating insufficient sample headspace volume.

(3) In a comparison of 16 oz and 9 oz sampling jars subjected to the "probe insertion" technique, overall analytical screening reproducibility (precision) was superior in 16 oz jars. Using the H-Nu unit, the Relative Standard Deviation (RSD) of replicate sample headspace analyses was approximately 5% for 16 oz jars and 9% for 9 oz jars. For the OVA unit, based on a limited database, the RSD for 16 oz jars was approximately 5% to 10%, and approximately 10% to 15% for 9 oz jars. For comparative purposes, all sample jars were filled to one-half capacity.

(4) Jar agitation ("dynamic development") was generally observed to significantly increase short-term headspace development, especially in aqueous samples.

(5) The temperature of soil and water samples and the ambient temperature of headspace development did not significantly affect short-term dynamic headspace development over the range of temperatures examined. A majority of the media samples evaluated were cooled to about 12°C (to simulate ground/groundwater temperatures in Massachusetts) and developed at ambient temperatures which ranged from 20 to 25°C. Selected samples, however, were cooled to as low as 4°C and/or developed in ambient temperatures as low as −12°C. While effects on headspace development were minor, it is stressed that ambient temperature effects on the operation and response of field units may have a significantly more pronounced impact.

(6) The shape of sample jars did not significantly affect analytical results (headspace development and subsequent instrument response) of dynamically developed media samples.

(7) While a majority of experimental setups consisted of jars filled with soil or water samples to one-half capacity, a limited database obtained from selected samples indicated that short-term dynamic development results were lower (particularly in aqueous samples) in 1/4 filled jars and somewhat higher in jars filled to 3/4 depth. Due to decreased headspace volumes, however, the reproducibility (precision) of 3/4 filled jar samples were lower than in half-filled setups.

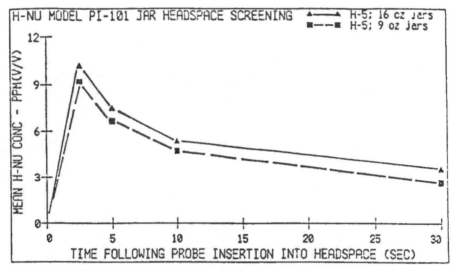

Figure 6. H-Nu Model PI-101 jar headspace screening.

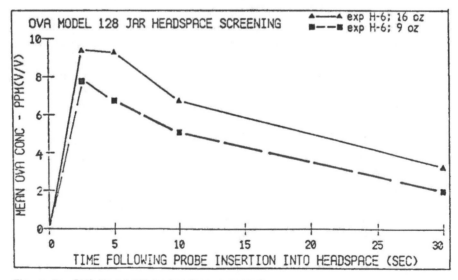

Figure 7. OVA Model 128 jar headspace screening.

Aqueous Headspace Evaluation

(1) In short-term evaluations (headspace development less than 30 min), headspace partitioning in agitated jars was one order of magnitude greater than in statically developed jars. Generally, over 90% of ultimate headspace development was achieved following 30 sec of vigorous agitation over a 5 to 10 min development period (see Figures 8 and 9).

(2) Aqueous headspace development was not significantly different between 9 oz and 16 oz sample jars.

(3) In "fresh" gasoline aqueous samples, during short-term dynamic development, benzene partitioning into the jar headspace ranged from 53% to 95% of the predicted Henry's Law (single component) equilibrium condition, with a mean value of 70% and a RSD of 18%. Total BTEX partitioning ranged from 25% to 56%, with a mean value of 40% and RSD of 24%. In (evaporatively) "weathered" aqueous stock samples, BTEX partitioning increased to 50% of the predicted equilibrium condition.

(4) Using the "probe insertion" technique, the BTEX headspace concentration comprised about 33% of the H-Nu meter TOV headspace concentration in "fresh" gasoline stock and about 50% in the "weathered" gasoline aqueous stock. The BTEX headspace concentration comprised about 20% of the Century OVA FID response in "fresh" stock and about 65% in "weathered" aqueous gasoline stock. This observation is consistent with the increased relative concentration of unsaturated ("second and last third") compounds in the more weathered samples and intrinsic PID/FID detector selectivity.

Soil Headspace Evaluation

(1) Experimental observation confirmed that headspace development and chemistry above gasoline-contaminated soils is subject to substantial variability, influenced by physical/chemical and moisture conditions of the soil and by the mechanism(s) of gasoline contamination and weathering. Due to this parameter variability and (relatively) limited scope and extent of experimentation, only preliminary conclusions may be drawn from examined data.

(2) Jar agitation significantly increased short-term headspace development over static conditions, with a 10% to 25% increase observed between 10 ppm (v/v) and 100 ppm (v/v) TOV and a 20% to 100% increase observed at concentrations less than 10 ppm (v/v) TOV.

(3) Contrary to aqueous experiments, headspace development values were consistently higher in 16 oz jars compared to 9 oz jars. Overall,

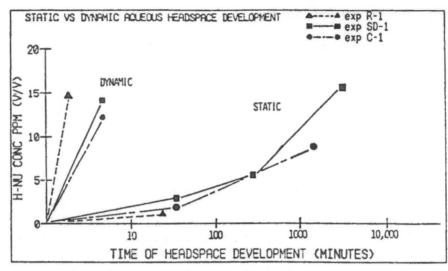

Figure 8. Static vs dynamic aqueous headspace development.

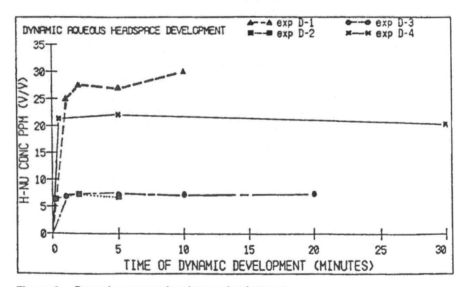

Figure 9. Dynamic aqueous headspace development.

the mean difference was approximately 14%, although differences were more pronounced at lower headspace concentrations (less than 10 ppm v/v TOV) where 16 oz jars were 20% to 50% higher than 9 oz jars.

(4) The observed relationship between "total organic vapor" PID and FID jar headspace measurements and volumetric headspace vapor

concentrations of benzene and BTEX constituents is graphically depicted in Figures 10 through 13.

(5) Based upon a limited database, short-term dynamic development of (moist) sandy soils yielded between 50% and 100% of the ultimate jar headspace development, and produced headspace benzene concentrations of 25% to 50% of the predicted equilibrium benzene condition. The total organic vapor (TOV) concentration in ppm (v/v) as benzene via the jar headspace procedure was seen to be 2 to 3 orders of magnitude greater than the soil benzene concentration in μg/g. As a preliminary conclusion, a (moist) sandy gasoline-contaminated soil sample producing a jar headspace concentration of 100 ppm (v/v) TOV as benzene would be expected to contain benzene in the bulk soil matrix in the range of 0.1 to 1.0 μg/g. Additional data is needed, however, to better define this relationship.

(6) Data analysis suggest the PID/FID response ratio may be a useful indicator of headspace and soil chemistry. In seven samples evaluated, the PID/FID response ratio "as benzene" varied from 0.42 to 2.9, the lower value representing less "weathered" gasoline contamination.

(7) Empirical observations and mathematical predictions indicate that the presence of even minute quantities of immiscible-phase gasoline within a soil sample will produce a TOV headspace concentration above 150 ppm (v/v), which is believed to be the upper limit of (relatively) linear PID response in contaminated soil samples.

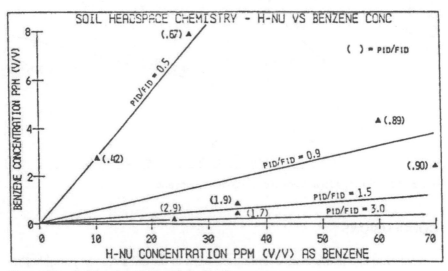

Figure 10. Soil headspace chemistry—H-Nu vs benzene.

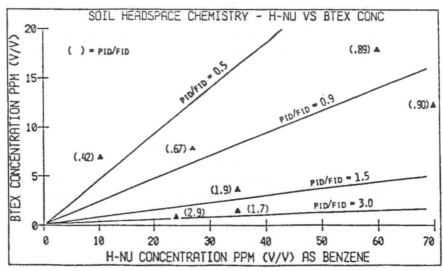

Figure 11.   Soil headspace chemistry—H-Nu vs BTEX.

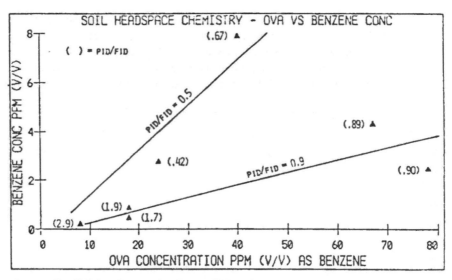

Figure 12.   Soil headspace chemistry—OVA vs benzene.

CONCLUSIONS

Analytical and interpretative limitations exist in the utilization of portable PID and FID response meters. Historically, improper and/or inconsistent use of these investigative tools has perhaps limited their acceptance and field utility. Nevertheless, certain, defined applications appear possible and desirable, particularly in

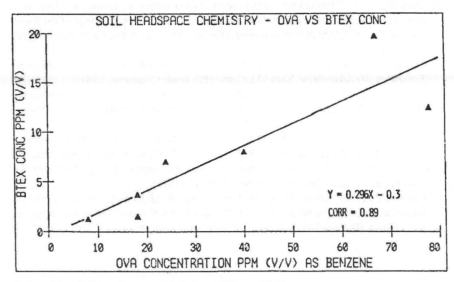

Figure 13. Soil headspace chemistry—OVA vs BTEX.

the areas of remedial response field evaluation and documentation.

Preliminary data has been obtained to aid in the interpretation of analytical headspace screening data. Additional investigation is necessary.

Based upon an evaluation of field headspace development and "probe insertion" sampling parameters, the following procedural elements are recommended:

(1) Utilize glass jars between 9 oz and 16 oz in total capacity; 16 oz jars are recommended, particularly for field instrumentation with sampling flowrates greater than 500 mL/min.
(2) Half-fill sample jars. Seal with a clean sheet of aluminum foil.
(3) Vigorously agitate sample jar for at least 30 sec over a 5 to 10 min headspace development period.
(4) Quickly insert the PID or FID sampling probe through the aluminum foil seal and record the maximum meter response as the "total organic vapor" headspace concentration. Maximum response should occur between 2 and 5 sec when using a 9 oz or 16 oz jar. Erratic meter responses should be discounted. PID measurements above 150 ppm (v/v) TOV as benzene may be significantly nonlinear.

REFERENCES

1. Price, D., and J. Chen. "Chemical-Physical Parameters and Processes Effecting Petroleum Fuel Migration," Draft Report, General Software Corporation, U.S. EPA Contract No. 68-02-3970, September 25, 1985.

2. Domask, W. G., "Introduction to Petroleum Hydrocarbons, Chemistry and Composition in Relation to Petroleum-Derived Fuels and Solvents," *Proceedings of the Workshop on Kidney Effects of Hydrocarbons*, (Boston, MA: American Petroleum Institute, 1983), pp. 5–33.
3. Lyman, W. J., W. F. Reehl, and D. H. Rosenblatt. *Handbook of Chemical Property Estimation Methods* (New York: McGraw-Hill Book Company, 1982).
4. Bruell, C. J., and G. E. Hoag. "Capillary and Packed-Column Gas Chromatography of Gasoline Hydrocarbons," in *Proceedings of Petroleum Hydrocarbons and Organic Chemicals in Groundwater*, National Water Well Association, 1984.
5. Barksy, J. B., S. S. Que Hee, and C. S. Clark. "An Evaluation of the Response of Some Portable, Direct-Reading 10.2 eV and 11.2 eV Photoionization Detectors, and a Flame Ionization Gas Chromatograph for Organic Vapors in High Humidity Atmospheres," *J. Am. Ind. Hyg. Assoc.*, 46:9–14 (1985).
6. Chelton, C. F., N. Zakraysek, G. M. Lautner, and R. G. Confer. "Evaluation of the Performance of the Bacharach TLV Sniffer and H-Nu Photoionization Gas Analyzer to Common Hydrocarbon Solvents," *J. Am. Ind. Hyg. Assoc.* 44:710–715.

CHAPTER 13

Problems Associated with Analysis of Petroleum-Derived Materials in the Environment

Pamela A. Kostle, Dennis R. Seeger, and James Knapp

As the state of Iowa's Public Health and Environmental Laboratory, the University Hygienic Laboratory (UHL) annually analyzes several hundred environmental soil and water samples for petroleum-derived fuels and solvents in samples submitted by state and local agencies. UHL has historically provided chromatographic identification and quantitation of petroleum-derived products introduced into the environment as a result of waste discharges, leaks, or spills. Recently, the requirement for identification and quantitation of these materials has become more intensive due to new state regulations and pending federal regulations concerning underground storage tanks.

The 1985 Iowa Groundwater Protection Act[1] specifies that all commercial underground storage tanks with a capacity of greater than 1100 gallons are required to be registered with the state. The Act further specifies that the condition of existing tanks must be checked via the installation of groundwater monitoring or sniffer wells. The initial deadline of May, 1988, for meeting these requirements was extended to January 1, 1989.[1]

To avoid the requirements of registration and installation of monitoring wells, many owners are opting to simply remove the underground storage tanks. The Iowa Department of Natural Resources (IDNR) is currently requiring that soil or groundwater samples from such excavation sites be analyzed to determine if contamination has occurred. The IDNR requests the determination of different analytes depending on the prior contents of the underground storage tank being excavated. The analyses requested for sites with tanks containing gasoline are for benzene, toluene, xylenes, and gasoline. For sites with tanks containing diesel fuel, motor oil, or other heavier hydrocarbons, analyses for total extractable hydrocarbons, benzene, toluene, and xylenes are most often requested.

The purpose of this chapter is to discuss some of the difficulties that were encountered while analyzing soil and water samples for gasoline, benzene, toluene, xylenes, and total extractable hydrocarbons. Different extraction techniques, concentration techniques, chromatographic columns and detectors were explored. A comparison of these techniques and recovery data are summarized.

EXPERIMENTAL

Reagents. Methylene chloride and methanol; American Burdick and Jackson. Mineral spirits, benzene, toluene, and xylenes; Chem Service, Inc. Organic free water; laboratory distilled water filtered through activated charcoal cartridges Model D0813; Barnstead. Gasoline and diesel fuel for standards were obtained from the University of Iowa motor pool. Kerosene and motor oil were obtained from a local service station.

Standards Preparation. Stock standards for gasoline, benzene, toluene, and xylenes were prepared in methanol. The daily working standards for gasoline, benzene, toluene, and xylenes were diluted from the stock standards into organic free water. The stock standards for diesel fuel, motor oil, kerosene, and mineral spirits were prepared in methylene chloride for direct injection.

Apparatus. Mixxor micro extractor; Xydex Corporation. Sonicator Model W-375; Heat Systems-Ultrasonics, Inc.

Instrumentation. Instrumentation and operating conditions are listed in Table 1.

Data System. Perkin-Elmer LIMS 2000/CLAS.

Table 1. Operating Conditions.

Section A	
Analysis	PID analysis of Gasoline, Benzene, Toluene, Xylenes
Instrument	Varian 3700
Detector	Photoionization detector (HNU)
Column	J & W Scientific DB-1 Megabore column 30 m x 0.53 mm ID, 5.0 micron film thickness
Flow Rates	Helium 10 mL/min.
Temperature Program	50° for 4 minutes, 8°/minute to 280 hold for 18 minutes
Analysis	Purge and trap concentration
Instrument	Tekmar LSC-2
Purge flow	40 mL/min

Table 1. (Continued)

Purge time	12 minutes
Desorbtion	40 mL/min Helium at 180°
Trap	OV-1/Tenax/Silica gel/Charcoal

Section B

Analysis	Gasoline, Benzene, Toluene, Xylenes
Instrument	Varian 2700
Detector	Flame ionization detector
Column	6' Stainless steel × ⅛" packed with 1% SP-1000 on Carbopak B
Flow Rates	Helium 30 mL/min
	Hydrogen 30 mL/min
	Air 300 mL/min
Temperature Program	50° for 3 minutes, 10°/minute to 220°, hold for 20 minutes
Analysis	Purge and trap concentration
Instrument	Tekmar LSC-1
Purge flow	45 mL/min
Purge time	12 minutes
Desorbtion	45 mL/min Helium at 180°
Trap	OV-1/Tenax/Silica gel/Charcoal

Section C

Analysis	Total Extractable Hydrocarbons
Instrument	Varian 3700
Detector	Flame ionization detector
Column	30 m × 0.32 mm ID Supelco SPB5
Flow Rates	Helium 20 mL/min, Hydrogen 30 mL/min, Air 300 mL/min
Temperature Program	50° for 4 minutes, 8°/minute to 280°, hold for 17 minutes
Injection volume	2 μL
Injection mode	Splitless for 0.5 minutes after injection

RESULTS AND DISCUSSION

Current UHL Procedures

Sample preparation methods employed for the determination of hydrocarbons are dependent on the sample matrix and the analytes requested. Analyses for volatile hydrocarbons (including gasoline, benzene, toluene, and xylenes) require purge and trap concentration of the sample using EPA method 5030[2] and detection using a modification of EPA method 602.[3] The method modifications include the use of a flame ionization detector. A gasoline standard chromatogram obtained under conditions listed in Table 1, Section B, is shown in Figure 1.

The sample preparation method employed for the determination of total extractable hydrocarbons (which would include diesel fuel, motor oils, kerosene, and mineral spirits) in an aqueous matrix is EPA method 3510[2], a methylene chloride liquid-liquid extraction procedure. For solid matrices EPA method 3550[2], a sonication procedure, is employed. Identification of sample contaminants is

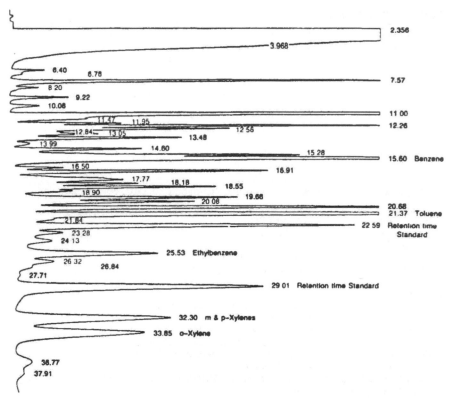

Figure 1. FID chromatogram of gasoline-7.8 μg: Conditions as in Table 1, Section B.

accomplished by comparison of the flame ionization chromatogram of the sample with those of dilutions of the commercial products. This method is loosely based on EPA method 8060[2] modified for the use of a capillary column with multipeak summations for quantitation of the multicomponent hydrocarbon products. For products producing chromatograms with reasonably well resolved peaks, the major peaks and their approximate relative intensities are compared to those of known products for identification. For products with poorly resolved components, the relative volatility of the unknown product, as evidenced by its elution time, is compared to that of known products. Chromatograms of this laboratory's standards for kerosene, diesel fuel, and motor oil are shown in Figures 2, 3, and 4 (the conditions for these chromatograms are in Table 1, Section C). The process of comparing the samples and standards is facilitated by overlaying the chromatograms on the laboratory's LIMS/CLAS graphic display terminals. This comparison is made based on the judgments of the analyst and supervisor; no formal pattern-matching algorithm is used. Diesel fuel spike recoveries of 68% from water and 61% from soil are routinely observed using this method. Recovery data for the purge and trap method are meaningless except for specific examples of extreme matrix effects because the standards are also analyzed by purge and

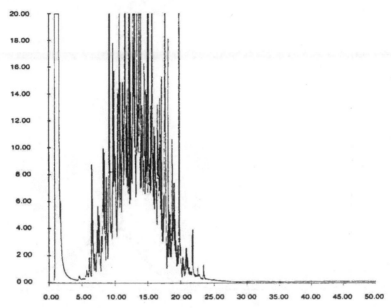

Figure 2. FID chromatogram of kerosene-26 μg: Conditions as in Table 1, Section C.

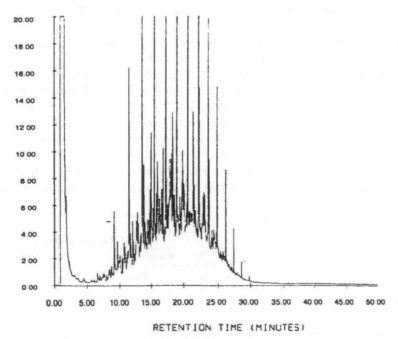

RETENTION TIME (MINUTES)

Figure 3. FID chromatogram of #2 diesel fuel-15 μg: Conditions as in Table 1, Section C.

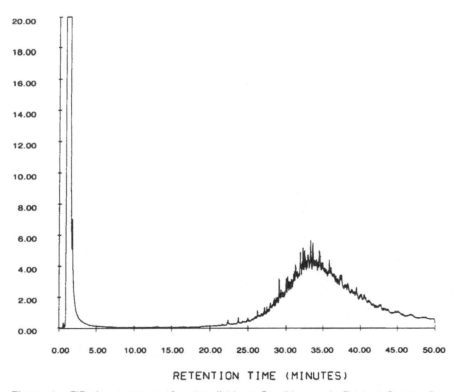

Figure 4. FID chromatogram of motor oil-11 μg: Conditions as in Table 1, Section C.

trap from aqueous solutions. The method quantitation limits summarized in Table 2 have been achieved utilizing the preceding methods.

A major difficulty encountered when utilizing the volatiles analysis described above occurs when a sample requiring the determination of benzene, toluene, or xylenes is contaminated with diesel fuel, fuel oil, motor oil, or other relatively heavy hydrocarbons. Within the limited temperature range of the column's liquid phase, the persistence of the heavier hydrocarbons on the packed column may cause an elevated baseline or excessive time for the heavier hydrocarbons to elute from the system. Due to the extensive problems caused by such determinations, alternative procedures were explored which would retain the relatively

Table 2.   Quantitation Limits.

| | Sample Matrix | |
Analyte	Water (ppb)	Soil (ppb)
Gasoline	20	100
Benzene	1	5
Toluene	1	5
Xylenes	1	5
Total Extractable Hydrocarbons	100	3000

low quantitation limits for the volatile hydrocarbons and yet allow identification and quantitation of the extractable hydrocarbons products. Retention of good recovery and a low detection limit for benzene was considered essential because it is a major component of petroleum products that may affect human health at low concentrations.[4]

Mixxor Micro Extraction

One procedure that was investigated utilized a Mixxor micro extraction apparatus. The recoveries from this procedure, which involved extracting 40 mL of sample with methylene chloride or pentane, are summarized in Table 3. The recoveries for the methylene chloride extraction were acceptable, but the design of the Mixxor apparatus precluded the use of much smaller volumes of heavier than water solvents. The quantitation limits for the 10 mL extraction volume ranged from 2 to 4 ppm for gasoline, benzene, toluene, and xylenes and 130 ppm diesel fuel. A final extract volume of less than 0.5 mL would be required to attain quantitation limits in the 0.1 ppm range for the volatile aromatics. A 5 mL extraction followed by a 5 mL solvent rinse of the apparatus indicated that volatile analyte recoveries may have been affected by poor solvent recovery while the extraction efficiency of diesel fuel decreased with reduced extraction solvent volume. The recoveries using two serial 5 mL extractions of pentane also yielded excellent recoveries; however, a single 5 mL pentane extraction resulted in significantly

Table 3.   Mixxor % Recoveries.

10 mL Methylene chloride	
Benzene	73
Toluene	86
m & p-Xylenes	93
o-Xylenes	76
Diesel fuel	67
5 mL Methylene chloride/5 mL Methylene chloride rinse	
Benzene	111
Toluene	121
m & p-Xylenes	99
o-Xylenes	101
Diesel fuel	45
Extracted twice with 5 mL Pentane	
Benzene	102
Toluene	86
m & p-Xylenes	91
o-Xylenes	92
5 mL Pentane	
Benzene	77
Toluene	67
m & p-Xylenes	64
o-Xylenes	65

decreased recoveries. Thus, reduction of extract to the volume that would be required to meet target quantitation limits was considered impractical.

Extract Concentration

Two extract concentration techniques, solvent evaporation by nitrogen blow-down and solvent distillation using a micro Snyder column apparatus, were investigated to determine which, if either, procedure could be used for quantitation limit reduction while retaining efficient volatile analyte recovery. Analyte recoveries for these concentration techniques with multiple solvents are summarized in Table 4. As anticipated, the lowest recovery (14% to 58%) was observed for benzene, the most volatile analyte.

Table 4.   Concentration Recoveries.

	Nitrogen Blowdown	Micro Snyder Column
	Methylene chloride	
Benzene	44	29
Toluene	71	60
m & p-Xylenes	72	64
o-Xylene	71	65
	Pentane	
Benzene	25	14
Toluene	51	37
m & p-Xylenes	55	47
o-Xylene	56	47
	Ethyl ether	
Benzene	38	25
Toluene	45	45
m & p-Xylenes	50	52
o-Xylene	50	53

Photoionization Detection

An option that was investigated to enhance current purge and trap chromatographic procedures included use of a megabore capillary column and a photoionization detector (PID). Use of a megabore capillary column decreased the time needed to remove heavier hydrocarbons from the system. Also, the PID in series with a flame ionization detector (FID) allowed for quantitation of aromatic hydrocarbons while retaining the FID identification characteristics. Preliminary analyses of aqueous solutions of diesel fuel, gasoline, benzene, toluene, and xylenes by purge and trap sample preparation, and gas chromatography using a megabore capillary column with a PID showed great promise. The chromatograms shown in Figures 5 thru 9 resulted from analyses using the conditions in Table 1, Section A. These demonstrate that the megabore capillary column effectively separated the benzene, toluene, and xylenes from other interferences, and the

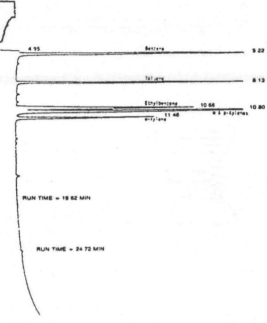

Figure 5. PID chromatogram of benzene-0.8 µg, toluene-0.2 µg, m-xylene-0.5 µg, o-xylene-0.4 µg, p-xylene-0.5 µg, ethylbenzene-0.5 µg: Conditions as in Table 1, Section A.

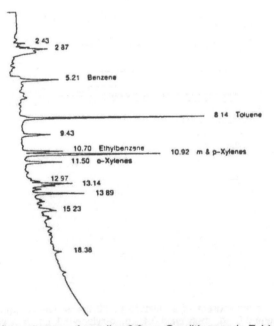

Figure 6. PID chromatogram of gasoline-3.3 µg: Conditions as in Table 1, Section A.

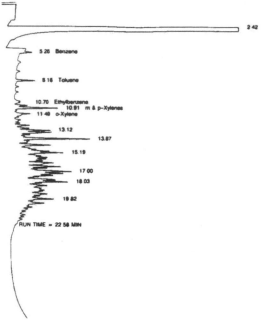

Figure 7.   PID chromatogram of #2 diesel fuel-43 μg: Conditions as in Table 1, Section A.

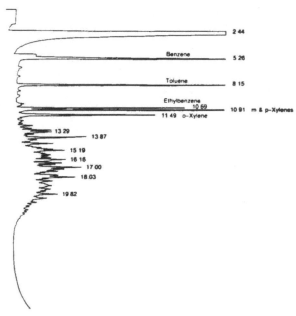

Figure 8.   PID chromatogram of a mixture of #2 diesel fuel-43 μg, benzene-0.5 μg, toluene-0.2 μg, m-xylenes-0.4 μg, o-xylene-0.3 μg, p-xylene-0.3 μg, ethylbenzene-0.3 μg, Conditions as in Table 1, Section A.

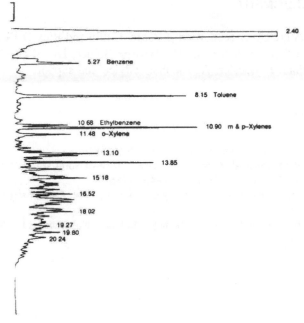

Figure 9. PID chromatogram of a mixture of gasoline-3.3 µg and #2 diesel fuel-43 µg: Conditions as in Table 1, Section A.

quantitation limits for the gasoline, benzene, toluene, and xylenes using the PID response remained the same while late eluting components were eluted in reasonable time. However, because of the poor purging efficiency of the heavier hydrocarbons, identification and quantitation of the full range of petroleum products is not possible using this procedure alone.

CONCLUSIONS

Because of the Mixxor micro extractor's unsuitability for solid matrices, high quantitation limits for direct extract analysis, and poor volatile aromatic recoveries for solvent concentration techniques, efforts to use solvent extraction procedures for hydrocarbon analyses were not continued. Current plans in this laboratory are to implement the use of purge and trap sample concentration and megabore capillary column gas chromatography with a PID/FID in series to provide acceptable volatile aromatic quantitation limits while minimizing interferences and analytical delays caused by the presence of heavier hydrocarbons from products such as diesel fuel and motor oil. Identification and quantitation of heavier hydrocarbon contamination will continue to require solvent extraction and capillary column gas chromatography of an aliquot of the concentrated extract.

ACKNOWLEDGMENT

The authors gratefully acknowledge Penny Catlin and Rick O'Donnell of the UHL staff for their assistance and Dr. George Breuer, Dr. Michael Wichman, and Dr. William J. Hausler, Jr. for their advice and support.

REFERENCES

1. Iowa Administrative Code, Chapter 567-135 (455)B.
2. "Test Methods for Evaluating Solid Waste, Office of Solid Waste," 3rd ed., U.S. Environmental Protection Agency SW846, 1986.
3. U.S. Environmental Protection Agency October 26, 1984, *Federal Register*, Vol. 49, No. 209.
4. U.S. Environmental Protection Agency Methods, November 13, 1985, *Federal Register*, Vol. 50, No. 219, p. 4688b.

CHAPTER 14

Three Common Misconceptions Concerning the Fate and Cleanup of Petroleum Products in Soil and Groundwater

James Dragun and John Barkach

INTRODUCTION

Petroleum and petroleum products have become the most widely utilized chemicals in history. For example, over 100 billion gallons of gasoline are consumed in the United States annually.

Extensive amounts of petroleum and its products are stored or transported underground. The release of large quantities of these chemicals to our environment has occurred from accidental spills, underground storage tanks and pipelines, transportation mishaps, land disposal facilities, and illegal dumping.

In the last 20 years, a considerable amount of research has occurred on the subject of environmental fate of petroleum and its products. Most of this work has focused on the fate of marine crude oil spills. Recently, due to the influence of federal legislation, more attention has been focused on petroleum in soils.

The published literature contains a few textbooks and conference proceedings addressing the fate of petroleum and its products in soils.[1,2,3] This literature addresses a wide spectrum of physical and chemical reactions that occur between soil and petroleum and its products, such as the migration of bulk hydrocarbons in soil, the migration and degradation of hydrocarbons dissolved in groundwater, the migration of petroleum and its products in soil air, and the biodegradation of petroleum and its products in soil and groundwater.

This chapter features a discussion of three aspects of the fate of petroleum and its products in soil that give environmental scientists and engineers problems during data interpretation.

THE INFLUENCE OF SOIL MACROPORES
ON THE MIGRATION OF BULK HYDROCARBONS IN SOIL

In general, water is the primary solvent in soil systems. When a relatively large volume of petroleum and its products (i.e., a bulk hydrocarbon) is released via spills and leaks, these chemicals can replace water as the primary solvent in localized areas.

As this relatively large volume of bulk hydrocarbon migrates through a unit mass of soil, a small amount of the total hydrocarbon volume will become attached to the soil particles due to capillary forces. The bulk hydrocarbon that is retained by soil particles is known as immobile or "residual saturation." If the migrating volume of bulk hydrocarbon is small relative to the soil surface area, the bulk hydrocarbon will eventually become exhausted as it is converted into residual saturation. When conversion is complete, downward migration ceases.

It is important to recognize that soil macropores can cause bulk hydrocarbons to migrate quickly downward toward the water table. Soil macropores are voids residing in soils that are greater than 100 microns in diameter. Macropores include voids that are present between soil peds, soil cracks, and fissures caused by soil shrinkage and cultivation techniques, and tubular pores formed by soil fauna and flora.

The ability of macropores to quickly transfer bulk liquids downward through the unsaturated zone has been known for several years. This phenomenon has been shown to be responsible for the rapid transport of surface-applied water.[4,5]

Some organic solvents can react with soil and cause soil shrinkage to occur. This results in a significant increase in soil permeability and sometimes causes the formation of soil macropores. Table 1 lists several organic solvents that can significantly increase the permeability of soil and clay. As a result, some bulk

Table 1.   Organic Solvents That Can Cause Significant Increases in Soil Permeability.[a]

acetone	aniine
carbon tetrachloride	diesel fuel
ethanol	ethanol/methanol
ethylene glycol	gasoline
heptane	isopropyl alcohol
kerosene	methanol
motor oil	naphtha
paraffin oil	phenol
trichloroethylene	xylene
xylene waste	

[a]Compiled from data presented in Dragun.[3]

organic solvents can be transported to great depths without leaving significant amounts of residual saturation in the unsaturated zone. Therefore, field investigations to determine the migration of petroleum and petroleum products in soil should always be planned in a manner that accounts for the effects of naturally occurring or induced soil macropores.

NATURALLY OCCURRING SOIL ORGANIC CHEMICALS THAT ARE FOUND IN PETROLEUM AND ITS PRODUCTS

Soils as well as petroleum and its products are comprised of hundreds of individual organic chemicals.[3] Table 2 lists those organic chemicals which have been identified in soils and in petroleum and petroleum products. A careful analysis of the chemicals listed in Table 2 indicates that the chemicals most commonly identified as "markers" for the presence of petroleum and petroleum products in soils and groundwater—benzene, toluene, xylenes, ethylbenzene (BTXE)—are also naturally occurring organic chemicals. These chemicals were first identified in 1966 as being naturally occurring in soils at concentrations of about one to five ppm.[16]

Therefore, the presence of low concentrations of BTXE in soil or groundwater should not be immediately and unequivocally interpreted as irrefutable evidence

Table 2.   Organic Chemicals Present in Various Petroleum Hydrocarbons That Are Also Naturally-Occurring Organic Chemicals Found in Soils.[a]

acetic acid	alkanes
benzene	1,2-benzofluorene
benzoic acid	butanoic acid
carbazole	decanoic acid
2,6-dimethylundecane	n-docosane
n-dotriacontane	n-eicosane
eicosanoic acid	ethanol
ethylbenzene	formic acid
n-heneicosane	n-hentriacontane
heptacosane	n-heptadecane
heptanoic acid	hexacosane
n-hexadecane	hexadecanoic acid
methane	methanethiol
methanol	naphthalene
n-nonacosane	n-nonadecane
nonanoic acid	n-octacosane
n-octadecane	octanoic acid
pentacosane	n-pentadecane
pentanoic acid	perylene
phenanthrene	propanoic acid
n-tetracosane	n-tetradecane
tetradecanoic acid	toluene
n-triacontane	n-tricosane
m-xylene	o-xylene
p-xylene	

[a]Compiled from data presented in Dragun.[3]

of a discharge of petroleum or its products to the environment. The natural occurrence of these chemicals should be evaluated before one reaches the definitive conclusion that a spill or leak has occurred.

WHY SOIL GAS SURVEYS MAY CAUSE MISLEADING CONCLUSIONS ON THE PRESENCE OR ABSENCE OF PETROLEUM AND ITS PRODUCTS

Petroleum and petroleum products are comprised of many organic chemicals that possess relatively high vapor pressures. In general, the higher the vapor pressure of a chemical, the more likely the chemical prefers to exist in the gaseous state as opposed to the liquid or solid state.

Organic chemicals with relatively high vapor pressures can be released from residual saturation or from the pancake layer. Because soils contain air-filled pores, these organic chemicals will migrate into these pores. Organic chemicals which have vaporized from residual saturation or the pancake layer are known as vapor-phase or gas-phase petroleum and products.

Traditional approaches to detecting the presence of petroleum and petroleum products in soil and groundwater involve soil and groundwater sampling and chemical analysis. Recently, the analysis of soil gas has been extensively used to determine the presence of petroleum and petroleum products in subsurface and groundwater. This approach involves the collection of a known volume of soil drawn from soil pores within a few feet of the soil surface. The soil gas is analyzed either in the field by a portable gas chromatograph or in an analytical laboratory.

Soil gas surveys can provide useful information regarding the presence of petroleum and petroleum products in soil and groundwater in some, but not all, soil systems. In general, any soil condition that inhibits or restricts the migration of an organic chemical in soil air can lead to the generation of unreliable data. One possible condition is the downward migration of a plume into an aquifer, which creates a layer of clean groundwater above the plume. Another possible condition is the presence of fine clay horizons in a sandy soil; clay can adsorb substantial amounts of water that can completely fill soil pores, which restrict the movement of organic chemicals in soil air. Another possible condition is the masking of a petroleum pancake by water that has accumulated on top of the pancake over several years.[3]

Information compiled from successful soil gas surveys published in the scientific literature are presented in Table 3. Because scientists and engineers generally do not publish the results of unsuccessful studies, one must first analyze the conditions under which successful studies and experiments were reported; then, one can use deductive logic to identify the conditions under which success will not be encountered.

Table 3. Summary of Hydrogeological and Sampling Data from
Successful Soil Gas Surveys.

Chemical(s) Present	Soil Type	Depth (ft) to Groundwater	Comments	Ref.
Chloroform	Gravelly Sandy Loam	9–12	Sampling probe placed 4 ft below grade; 75 cc soil gas samples withdrawn; sampling date unknown. Analysis via lab GC.	Kerfoot[6]
Chloroform	Gravelly Sandy Loam with Clay	6–12	Static samplers were placed in holes about 1.2 ft and sealed with a 14 day exposure time. Analysis via GC/MS.	Kerfoot et al.[7]
Butane Isooctane	unspecified	24 at most shallow point	Dynamic samplers were collected from depths of 2–7 ft below the ground surface. Analysis via lab GC.	Marrin & Kerfoot[8]
PCE 1,2-DCE	Sand & Gravel	15–20	250 cc sample withdrawn from 1–3 ft below grade in July, 1984. *Method useful only in source area.* Analysis via Photovac 10A10 field GC.	Wittmann et al.[9]
TCA TCE	Gravelly, Clayey Sand	5–20	Static trapping devices placed 0.5 ft below grade for 7–30 days; analysis via mass spectrometry; sampling date unknown.	Malley et al.[10]
TCA	Sand and Gravelly Sand	90	Dynamic soil vapor collectors used. Samples were taken at 4.5 ft below ground surface. Analysis via lab GC.	Marrin & Kerfoot[8]
PCE TCE	Sands & Gravels	100	Static soil vapor collectors used at an exposure time of 40 days. Analysis via GC/MS.	Mills[11]
1,1,2-TCE	Sandy Silt	24–27	Dynamic soil vapor collectors used. Probes were driven to a depth of 6–7 ft Analysis via lab GC.	Thompson & Marrin[12]
Toluene 2,3,3-TMP	Yellow-Brown Clay & Silt	7–10	Static soil vapor collected from 6.5–11.5 ft below the clay layer on three different sampling days over a four month period. Analysis via GC/MS.	Wallingford et al.[13]
1,1,1-TCE Toluene	Cobble & Crushed Limestone	1–2	Dynamic sample technique. Analysis via field GC.	Spittler et al.[14]
Hydrocarbons PCE	Silt & Fine to Medium-Grained Sand, Overlying Shale Bedrock	15–25	Static trapping samples were placed at an unspecified depth and retrieved after several days. Analysis via lab GC.	Markley[15]

An analysis of the information provided in Table 3 indicates the conditions which favor successful soil gas surveys: sandy soil textures, a shallow water table, static sampling methods for deeper water tables and for fine soil types, laboratory analytical instrumentation, and the presence of organic chemicals with very low molecular weights.

SUMMARY AND CONCLUSIONS

Three common misconceptions concerning the fate and cleanup of petroleum and its products in soil and groundwater were discussed.

First, soil macropores can cause bulk hydrocarbons to migrate quickly downward toward the water table without leaving significant amounts of residual saturation on unsaturated zone soil. In addition, some organic solvents can react with soil and cause soil shrinkage resulting in a substantial increase in soil permeability.

Second, soils as well as petroleum and its products are comprised of hundreds of individual organic chemicals, including benzene, xylene, toluene, and ethylbenzene. The presence of low levels of these organic chemicals in soil or groundwater should not be immediately and unequivocally interpreted as irrefutable evidence of a discharge of petroleum or its products. The natural occurrence of these chemicals should be evaluated before one reaches the definitive conclusion that a spill or leak has occurred.

Third, soil gas surveys can provide useful information regarding the presence of petroleum and petroleum products in soil and groundwater in some, but not all, soil systems. In general, any soil condition that inhibits or restricts the migration of a chemical in soil air can lead to the generation of unreliable data. An analysis of published scientific papers on successful soil gas surveys shows that the following conditions favor successful soil gas surveys: sandy and gravelly soil textures, a shallow water table, static sampling methods for deeper water tables and for finer soil types, laboratory instrumentation, and the presence of low molecular weight organic chemicals.

REFERENCES

1. Calabrese, E. J., and P. T. Kostecki, eds. *Petroleum Contaminated Soils, Volume 1.* (Chelsea, MI: Lewis Publishers, Inc., 1988a).
2. Calabrese, E. J., and P. T. Kostecki, eds. *Soils Contaminated by Petroleum: Environmental & Public Health Effects.* (New York: John Wiley & Sons, 1988b).
3. Dragun, J. *The Soil Chemistry of Hazardous Materials* (Silver Spring, MD: Hazardous Materials Control Research Institute, 1988).
4. Shuford, J. W., D. D. Fritton, and D. E. Baker. "Nitrate-Nitrogen and Chloride Movement Through Undisturbed Field Soil," *J. Environ. Qual.* 6:255–259 (1977).
5. Thomas, G. W., and R. E. Phillips. "Consequences of Water Movement in Macropores," *J. Environ. Qual.* 8:149–152 (1979).

6. Kerfoot, H. B. "Soil-Gas Measurement for Detection of Groundwater Contamination by Volatile Organic Compounds," *Environ. Sci. Technol.* 21(10):1022–1024 (1987).
7. Kerfoot, H. B., C. L. Mayer, and M. J. Miah. "Evaluation of a Passive Sampling Soil-Gas Survey Technique for Detecting Volatile Organic Compounds," *Proceedings of the NWWA/API Conference on Petroleum Hydrocarbons and Organic Chemicals in Ground Water—Prevention, Detection and Restoration*, Houston, TX, November 12–14, 1986. Dublin, OH: National Water Well Association, 1986.
8. Marrin, D. L., and H. B. Kerfoot. "Soil-gas Surveying Techniques: A New Way to Detect Volatile Organic Contaminants in the Subsurface," *Environ. Sci. Technol.* 22(7):740–745 (1988).
9. Wittmann, S. G., K. J. Quinn, and R. D. Lee. "Use of Soil Gas Sampling Techniques for Assessment of Groundwater Contamination," *Proceedings of the NWWA/API Conference on Petroleum Hydrocarbons and Organic Chemicals in Groundwater*, Houston, TX, November 13–15, 1985. Dublin, OH: National Water Well Association, 1985.
10. Malley, M. J., W. W. Bath, and L. H. Bongers. "A Case History: Surface Static Collection and Analysis of Chlorinated Hydrocarbons from Contaminated Groundwater," *Proceedings of the NWWA/API Conference on Petroleum Hydrocarbons and Organic Chemicals in Groundwater*, Houston, TX, November 13–15, 1985. Dublin, OH: National Water Well Association, 1985.
11. Mills, W. R., Jr. "Use of Static Soil Vapor Collectors to Identify Subsurface Contamination in Southern California Ground Water Basins," *Proceedings of the Sixth National Symposium and Exposition on Aquifer Restoration and Ground Water Monitoring*, Columbus, OH, May 19–22, 1986. Dublin, OH: National Water Well Association, 1986.
12. Thompson, G. M., and D. L. Marrin. "Soil Gas Contaminant Investigations: A Dynamic Approach," *Ground Water Monitoring Review* 7:88–93 (1987).
13. Wallingford, E. D., F. A. DiGiano, and C. T. Miller. "Evaluation of a Carbon Adsorption Method for Sampling Gasoline Vapors in the Subsurface," *Ground Water Monitoring Review* 8:85–92 (1988).
14. Spittler, T. M., W. S. Clifford, and L. G. Fitch. "A New Method for Detection of Organic Vapors in the Vadose Zone," in E. J. Calabrese and P. T. Kostecki, eds., *Soils Contaminated by Petroleum* (New York: John Wiley & Sons, 1988).
15. Markley, D. E. "A Comparative Evaluation of Soil Gas, Groundwater, and Soil Sampling Results," Presented at Haztech International Conference, St. Louis, MO, August 26–28, 1987.
16. Simonart, P., and Batistic, L. "Aromatic Hydrocarbons in Soil," *Nature* 212:1461–1462 (1966).

CHAPTER 15

The Effect of Water-Soluble Organic Material on the Transport of Phenanthrene in Soil

Brian R. Magee, Ann T. Lemley, and Leonard W. Lion

INTRODUCTION

Environmental fate modeling involves the difficult task of quantitatively describing the natural mechanisms that affect the mobility and concentration of a pollutant in the environment. Of primary importance for modeling the subsurface transport of many contaminants are dissolution in groundwater, which allows transport by advection, diffusion, and dispersion; and sorption to stationary soil particles, which retards transport. Many environmental contaminants, such as polynuclear aromatic hydrocarbons (PAHs), polychlorinated biphenyls, and dioxins, are hydrophobic and have low aqueous solubilities. Consequently, these compounds typically sorb strongly to soil and the risk to potential receptors distant from the waste site is assumed to be low. However, if dissolved macromolecules or colloidal particles are dispersed in the groundwater, they may sorb the pollutants, allowing them to be transported with these mobile carriers through the subsurface environment. The processes resulting in such "facilitated transport" are generally not incorporated into environmental fate models.

Reviews of the environmental fate of low-solubility compounds indicate that the simple models which describe the transport of the more soluble trace organics do not hold for the least soluble compounds, even at low water velocities.[1,2] Some investigators have found poor or no correlation between sorptive partition coefficients or calculated retardation values and observed transport of trace organics during land application of wastewater.[3,4] Sorption by mobile carriers may also explain the existence of groundwater PAH concentrations in excess of aqueous solubilities at coal tar sites.[5] Recent evidence that hydrophobic chemicals are being transported farther and at greater concentrations in the environment than mathematical models would predict suggests that a mechanism such as facilitated transport may be at work.[6-8] Pollutants sorbed to carriers could thus jeopardize water supplies thought to be a safe distance from a waste site.

Binding of hydrophobic compounds to colloidal particles and/or dissolved macromolecules has implications beyond enhanced transport. Sorption to dissolved macromolecules has been observed to reduce the rates of biodegradation, volatilization, and hydrolysis of pollutants, as well as to affect the rates of photolysis and bioaccumulation.[9-11] Furthermore, characterization of aquifers with respect to their carrier concentrations may aid in prioritization of waste sites for cleanup efforts, since sites with conditions which favor pollutant mobilization pose a greater risk. With respect to the cleanup of waste sites, flushing of contaminated soil in situ or soil washing aboveground can be enhanced if the addition of carriers brings strongly-sorbed pollutants into the mobile phase, thus reducing the extensive time and financial requirements such cleanup techniques involve.

Although many investigators have examined the partitioning of hydrophobic compounds into dissolved organic or colloidal materials (Table 1), less research has been conducted to assess the capability of such materials to serve as mobile carriers of sorbed pollutants, especially in groundwater. Of the few investigations of facilitated transport through soil, most have attempted to develop merely a qualitative understanding of carrier effects on retardation. Enhanced transport of hydrophobic pollutants has been observed in laboratory soil columns in the presence of suspended solids from sewage effluent,[32] humic material,[19] and well-defined surrogates for dissolved and colloidal organics.[29]

Table 1.   Investigations of Non-Settling Particles as Sorbents.

Sorbent	Sorbate	Variables	Researchers
Fulvic acid	Dialkyl phthalates	pH	Matsuda & Schnitzer[12]
Dissolved organic matter from sea water	Hydrocarbons	pH, salt content	Boehm & Quinn[13]
Suspended particulate matter from sea water	pp'-DDT	Compared to clay and sediment	Pierce et al.[8]
Dissolved organic matter from river, sewage particles	Cholesterol and a PCB	Compared to river and sewage particles	Hassett & Anderson[14]
Humic acid	DDT	pH, Ca, and ionic strength	Carter & Suffet[9]
Estuarine colloids	Atrazine and linuron	pH and salinity	Means & Wijayaratne[15]

Table 1.   (Continued)

Sorbent	Sorbate	Variables	Researchers
Estuarine colloids	Benzene, naphthalene, anthracene	pH	Wijayaratne & Means[16]
Humic acid	Hydrazine hydrate	Exchanged ions on sorbent, pH	Isaacson & Hayes[17]
Estuarine colloids	Atrazine	Oxidation	Wijayaratne & Means[18]
Humic materials	Hexachlorbenzene and DDT	–	West[19]
Humic, fulvic, tannic acids, lignin, zein, and cellulose	Trichloroethylene and toluene	Elemental content of sorbent	Garbarini & Lion[20]
NSPs from sediment	PCBs	Prewashing of sediment	Gschwend & Wu[21]
Fulvic acid	Atrazine	pH, metal ion concentration	Haniff et al.[22]
Humic material	Naphthalene and benzo(a)pyrene	Bioavailability	McCarthy & Jimenez[10]
Humic acid	DDT and lindane	Distribution	Caron et al.[23]
Humic acid	Benzo(a)pyrene	Reversibility of sorption	McCarthy & Jimenez[24]
Water-soluble soil organic material	Herbicides (6)	Sorbate's ionization, size, and structure	Madhun et al.[25]
Humic and fulvic acids, organic polymers	DDT, PCBs, TCB, and lindane	Sorbent's polarity, structure, size, and source	Chiou et at.[26]
Fulvic acid	Atrazine	–	Gamble et al.[27]
Colloids	PCBs	–	Baker et al.[7]
Humic and fulvic acids	Phenanthrene, pyrene, anthracene	–	Gauthier et al.[28]
Bovine serum albumin, Triton X-100 micelles	Naphthalene and phenanthrene	–	Kan & Tomson[29]
Humic and fulvic acids	Pyrene	Structure and composition of sorbents	Gauthier et al.[30]
Humic materials	DDT and PCBs	Aquatic vs. commercial humics	Chiou et al.[31]

The objectives of the research described in this chapter were to evaluate several organic carriers which might facilitate the transport of a hydrophobic pollutant (phenanthrene), to quantify the extent to which transport through soil may be enhanced by a selected carrier, and to model the facilitated transport using distribution coefficients derived from independent column and batch experiments.

THEORETICAL DEVELOPMENT

Models used to describe the movement of a chemical solute through a porous medium (i.e., soil) are often derived from the advection-dispersion equation which, assuming steady-state flow, a linear sorption isotherm, and local equilibrium, has the form

$$\frac{\delta C}{\delta t}(1 + \frac{\rho_b K_d}{n}) = D\frac{\delta^2 C}{\delta z^2} - v\frac{\delta C}{\delta z} \qquad (1)$$

where C=dissolved chemical concentration; t=time; ρ_b=the bulk density of the porous medium, K_d=the equilibrium distribution coefficient of the pollutant between the sorbed and dissolved phases, n=the effective porosity (water content); v=interstitial flow velocity (distance per time), D=apparent dispersion coefficient, a function of n and v; z=depth; and δ indicates a partial derivative. The model incorporates diffusion, mechanical dispersion, and mass flow (advection) as transport mechanisms. Detailed derivation of this relationship can be found elsewhere.[33,34]

In Equation 1, the term $(1 + \rho_b K_d/n)$ is defined as the retardation factor, R. It relates the transport of a sorbing compound (R > 1) to that of a nonsorbing one (R = 1). For example, a compound with a given K_d value that results in an R = 10 will travel at one-tenth the average velocity of a nonsorbing tracer, such as ^3H_2O. R can also be less than 1 when most of the solute elutes in less than one pore volume. This can occur when an ionic molecule is repelled from soil surfaces with the same charge, i.e., "negative sorption"[35]; or when macromolecules are excluded from small pores due to their size. This latter effect should properly be accounted for in the velocity and dispersion terms in the advection-dispersion equation rather than in R, unless it is envisioned that the pore exclusion significantly alters the effective porosity, n. R can also be written as the ratio of the average velocities of the nonsorbing tracer (v_w) and sorbing compound (v_{sv}):

$$R = v_w/v_{sv} \qquad (2)$$

In the presence of potential carriers such as dissolved organic matter, West[19] and Hutchins et al.[4] have shown that R can be redefined to account for additional pollutant sorbed to the carrier:

$$R' = 1 + \frac{K_d'\rho_b/n}{1 + K_d^{om}DOM} \qquad (3)$$

where K_d' and K_d^{om} are the distribution coefficients for the pollutant into soil and mobile organic matter, respectively, and DOM is the aqueous concentration of mobile organic matter. This relationship describes transport of a compound in the presence of a nonadsorbing carrier. R' can also be defined on an organic carbon basis as

$$R' = 1 + \frac{K_{oc}{}^s \rho_{boc}/n}{1 + K_{oc}{}^{om} DOC} \tag{4}$$

where $K_{oc}{}^s$ and $K_{oc}{}^{om}$ have units of mL/g of organic carbon (OC) in soil and mobile organic matter, respectively; ρ_{boc} is the g OC in soil per cm^3 bulk volume; and DOC is the dissolved organic carbon concentration (g OC per mL solution).

A carrier that sorbs to soil will not facilitate transport to the extent represented by Equation 2, since the average velocity of the carrier, v_{om}, would not be the same as the water velocity. Therefore, one needs to modify the expression for R' to account for the sorption of the carrier. Incorporating the retardation of the carrier, the expression for the compound's retardation factor becomes[36]

$$R^* = \frac{1 + K_d{}^{om} DOM + K_d{}^s \rho_b/n}{1 + \dfrac{K_d{}^{om} DOM}{1 + (K_{dom}{}^s \rho_b/n)}} \tag{5}$$

where $K_{dom}{}^s$ is the soil-water distribution coefficient for DOM. On an organic carbon basis, $K_{dom}{}^s$ becomes $K_{ocom}{}^s$ and R^* is written as

$$R^* = \frac{1 + K_{oc}{}^{om} DOC + (K_{oc}{}^s \rho_{boc}/n)}{1 + \dfrac{K_{oc}{}^{om} DOC}{1 + (K_{ocom}{}^s \rho_{boc}/n)}} \tag{6}$$

The difference between Equations 3 and 6 is incorporated into the carrier retardation term, $1 + (K_{dom}{}^s \rho_b/n)$. When the carrier does not partition into the soil ($K_{dom}{}^s = 0$), this term equals 1 and Equation 6 becomes equivalent to Equation 3.

If the interactions illustrated in Figure 1 are sufficient to describe phenanthrene behavior in the presence of a carrier, then the individual distribution coefficients $K_d{}^{om}$, $K_d{}^s$, and $K_{dom}{}^s$ determined in independent experiments should predict the behavior of pollutant-carrier mixtures (R^*).

METHODS AND MATERIALS

A dark sand obtained from a quarry in Newfield, NY, was used in all the experiments, while a second soil, a clayey silty loam from eastern New York State (Rhinebeck) was used in some column experiments for comparison. The characteristics of the soils are provided in Table 2.

Radiolabeled phenanthrene-9-^{14}C with a specific activity of 10.9 mCi/mmol was obtained from Sigma Chemical Company, St. Louis, MO. Purity was tested by high-performance liquid chromatography (HPLC) on a C-18 column, and 99+% of the peak area and activity were found to occur at the elution time for phenanthrene. Radiolabeled phenanthrene stock solutions were prepared by dissolving the phenanthrene in a 50:50 mixture of HPLC-grade methanol and distilled-dionized water (Coring Mega-Pure system). Methanol fractions in all experiments

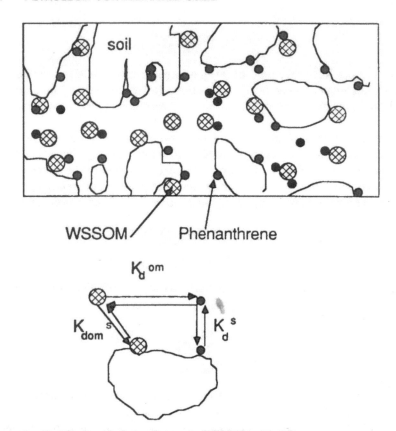

Figure 1. Partitioning of phenanthrene and WSSOM with soil.

Table 2. Soil Characteristics.

	Rhinebeck	Dark Sand
Organic matter, %	5.9	0.4
Organic carbon, %	3.13	0.11
Clay, %	35.5	ND
Silt, %	52.3	ND
Silt and clay, %	ND	1.2
Sand, %	12.2	98.7
Very fine (.1–.05 mm)	ND	3.7
Fine (.25–.1 mm)	ND	47.2
Medium (.5–.25 mm)	ND	47.6
Coarse (>.5 mm)	ND	0.2
pH, H_2O extract	6.7	8.6
Cation exchange capacity, meq/100g	27.0	3.0
Calcium	14.66	ND
Magnesium	2.451	ND
Sodium	0.14	ND
Potassium	0.204	ND
Acidity	9.0	0.9

ND = Not determined.

ranged from 0.0010 to 0.0015 and were deemed small enough to avoid significant solvophobic effects.[37] Nonradiolabeled zone-refined phenanthrene (99.5% pure) used to measure binding to WSSOM was obtained from Aldrich Chemical Co. Reagent-grade calcium sulfate was obtained from Mallinckrodt and sodium azide from Fisher Scientific.

Evaluation of Potential Carriers

Several types of organic macromolecules were tested for their elution rates from soil columns. Tannic acid (Fisher Scientific) at 12 mg/L and 97 mg/L was applied to soil columns of dark sand. Detection was by ultraviolet absorbance at 278 nm (Beckman 3600 UV-Visible Spectrophotometer). Following more than 12 pore volumes, no tannic acid was detected in the effluent. Since the pH of the tannic acid in the calcium sulfate solution was rather low (pH 4.45 for 97 mg/L), a sodium hydroxide solution was added to raise the pH to circumneutral (pH 7.5). Tannic acid still did not elute from the column. Early eluate fractions had a high pH and were viscous and blue, as was observed when the tannic acid solutions exceed pH 10. Thus it appears that the added base eluted faster and was chromatographically separated from the tannic acid. Cellulose (Sigma Chemical) was also tested as a carrier but did not form a stable suspension.

Solutions of water-soluble soil organic material (WSSOM) were prepared by extracting the soluble fraction of soil (see details below). One soil (Rhinebeck) was a clayey silty loam from east-central New York State. WSSOM from a second soil, an organic muck from Sapsucker Woods in Ithaca, NY, was also prepared. While WSSOM from both sources had a high DOC concentration, WSSOM from Sapsucker Woods did not pass through the soil column as readily; therefore, WSSOM from the Rhinebeck soil was selected for further study. It should be noted that while there may be many sources of DOC in groundwater or surface water, not all types are mobile, and only the mobile types are potential carriers.

Production of WSSOM

To produce WSSOM, Rhinebeck soil (50 g) and a solution (250 mL) of 0.01 N calcium sulfate and 1.0 g/L sodium azide were combined in 250 mL nalgene bottles. Calcium sulfate provided a background electrolyte and sodium azide inhibited microbial degradation. After approximately 18 hours on a wrist-action shaker, the bottles were centrifuged for 30 minutes at 5400 g. The supernatant was filtered through a Whatman 2V filter and refrigerated until use. A typical WSSOM sample contained approximately 80 ppm total organic carbon and had a pH in the range of 6.4 to 7.5.

Characterization of WSSOM

WSSOM was analyzed by infrared (IR) spectrophotometry to determine its prominent functional groups. WSSOM samples were freeze-dried and approximately 1 g of the lyophilized material was analyzed as potassium bromide pellets on a Perkin-Elmer Infrared Spectrophotometer (Model 683). WSSOM collected after passing through a soil column was also analyzed to determine whether only certain fractions of the soluble organic matter penetrated the column.

Since humic acids are insoluble at pH values below 2, by lowering the pH one could determine whether the WSSOM contained fulvic acids, which are soluble at all pH values, and/or humic acids, which would precipitate. To accomplish this, a WSSOM sample was first passed through a 0.45 μ filter membrane to remove any suspended organic matter small enough to pass the Whatman 2V filter, and the concentration of DOC was measured. Sulfuric acid was added to lower the pH to 1.0 and the sample was again passed through a 0.45 μ filter membrane. The sample was then brought to approximately pH 7.0 with the addition of sodium hydroxide, and analyzed for DOC.

Soil Column Experiments

The transport of the carrier or compound of interest was observed by continuous delivery of a solution containing the compound through a soil column at a constant pore water velocity. Experiments were run in the dark to inhibit photolysis of phenanthrene and in a temperature-controlled room maintained at the temperature of groundwater. Parameters and equipment for the column experiments are provided in Table 3.

A short (5.5 cm) column was used to decrease the time to breakthrough for phenanthrene. The 21.5-cm column was used for WSSOM breakthrough to allow for the collection of larger samples per pore volume for analysis of DOC, which required larger sample volumes.

Prior to the start of each column experiment, approximately 20 pore volumes of the calcium sulfate-sodium azide solution were applied in an upflow mode to equilibrate the column packing. If WSSOM was to be used as a carrier, it was also included in the equilibration solution.

A problem often encountered when studying the transport of hydrophobic compounds is the adsorption of the compound to experimental apparatus. Therefore, phenanthrene was pumped through a second pump until the concentration in the effluent from the pump was stable. Once equilibration of the column was complete, delivery was switched to a solution containing the substance of interest (phenanthrene or WSSOM), which was applied as a step input. The column was inverted so that the effluent would drip from the bottom of the column directly to test tubes in the fraction collector. After saturation, the column was weighed to determine the water content, as well as at the end of the experiment. The mass of the wetted soil column was determined to be approximately the same or slightly greater after the step input than before, indicating that the column remained saturated in the downflow mode.

The step input also contained a nonsorbing tracer, ^3H_2O, which has a theoretical retardation of 1.0. Phenanthrene or WSSOM breakthrough was compared to the breakthrough of the nonsorbing tracer.

Table 3. Initial Parameters and Analytical Equipment for Soil Column Experiments.

	Dark Sand	Rhinebeck
WSSOM Breakthrough		
WSSOM DOC (mg/L)	59.3	81.2
Column length (cm)	21.4	5.5
Bulk density (g/cm³)	1.94	1.12
Wetted pore volume (mL)	29.1	14.2
Volumetric flow (mL/hr)	13.6	25.8
Phenanthrene Breakthrough		
Phenanthrene (μg/L)	47.6, 52.9	62.1, 53.3
Column length (cm)	5.5	5.5
Bulk density (g/cm³)	1.65, 1.65	1.20, 1.16
Wetted pore volume (mL)	8.54, 9.26	12.8, 13.35
Volumetric flow (mL/hr)	14.7, 16.8	23.2, 24.3
Phenanthrene Breakthrough with WSSOM		
Phenanthrene (μg/L)	39.2, 56.8	49.7, 58.6
WSSOM DOC (mg/L)	99.0, 72.6	84.3, 81.2
Column length (cm)	5.5	5.5
Bulk density (g/cm³)	1.61, 1.58	1.10, 1.12
Wetted pore volume (mL)	10.2, 9.49	14.1, 14.2
Volumetric flow (mL/hr)	18.6, 17.3	25.6, 25.8

For all experiments:
Solution delivery: Sage syringe pump model 220
Pore water velocity: 10 cm/hr
Soil column: Beckman preparative column, 2.5 cm inner diameter
Temperature: 15 ± 0.3°C
Dissolved organic carbon: Persulfate oxidation to CO_2 and IR analysis (O.I. Corporation, TOC Analyzer Model 700)
Radiolabel analysis: Liquid scintillation counting on Beckman LS 9800; 1.5 mL sample in 10 mL ScintiVerse E (Fisher Scientific). Samples were counted 20 minutes or to 1% error, whichever occurred first. Quenched curves were produced for ^{14}C and ^{3}H, and were used to correct counts.

Three types of column experiments were conducted: breakthrough of WSSOM alone, breakthrough of phenanthrene alone, and breakthrough of phenanthrene with WSSOM. Each phenanthrene experiment was replicated, and initial parameters for each experiment are presented in Table 3. Retardation factors were determined by measuring the area above the breakthrough curve up to $C/C_o = 1.0$.[38]

Phenanthrene Binding to WSSOM

The fluorescence of phenanthrene has been shown to be quenched when bound to dissolved or colloidal material and can be used to measure the extent of binding.[24,28] Fluorescence was measured on a Perkin-Elmer MPF-44B fluorescence spectrophotometer with slit widths set for band widths of 2 nm on the excitation monochromator and 4 nm on the emission monochromator. Fluorescence was measured as a function of added humic material at 288 nm excitation wavelength and 364 nm emission wavelength. A procedure similar to that described by Gauthier et al.[28] was employed. The fluorescence of 2 mL of a phenanthrene solution in a quartz fluorescence cuvette was measured after equilibration with the cuvette walls. Aliquots (0.050 mL) of WSSOM (99.0 ppm) were added to the cuvette. After each addition, the solution was shaken for 30 seconds and allowed to stand quiescent for about 4 minutes before recording the fluorescence intensity.

While WSSOM was being added and during equilibration, the shutter of the spectrofluorometer was closed to protect the phenanthrene from photodegradation. In a trial run in which a phenanthrene solution sample was continually exposed to the UV excitation source, no decrease in intensity was observed for the time period required for the addition of 10 aliquots (about 1 hour).

To correct for possible fluorescence by added WSSOM, the fluorescence of a solution of WSSOM alone was measured with the same concentrations and instrumental conditions as for the measurements with WSSOM and phenanthrene. The fluorescence intensity measured for WSSOM was subtracted from the total fluorescence intensity measured for phenanthrene in the presence of WSSOM. The maximum fluorescence intensity of WSSOM was 2.6% of the total.

The binding of phenanthrene with WSSOM can be represented by

$$K_d^{om} = \frac{[\text{phen-WSSOM}]}{[\text{phen}][\text{WSSOM}]} \tag{7}$$

Performing a mass balance on phenanthrene and assuming that the fluorescence intensity is proportional to the concentration of free phenanthrene[phen],[28] then

$$F_o/F = 1 + K_d^{om}[\text{WSSOM}] \tag{8}$$

where F_o and F are the fluorescence intensities in the absence and presence of WSSOM, respectively. Since a significant excess of WSSOM over phenanthrene was present, [WSSOM] was taken as the amount of added WSSOM without correction for the fraction of humic that was associated with phenanthrene. K_d^{om} was then obtained from a plot of F_o/F versus WSSOM concentration.[28]

RESULTS AND DISCUSSION

Characterization of WSSOM

A freeze-dried sample of WSSOM produced the IR spectrum shown in Figure 2. The absorption bands indicated the presence of various functional groups, both hydrophilic and hydrophobic (Table 4). The absorption bands at 3400, 1620, 1440, and 1220 are also commonly detected in humic materials.[39,40] In spite of the widespread view that humic substances have a major carboxylic character, MacCarthy and Rice[39] noted that the OH absorption band at 3400 cm^{-1} occurs in humics

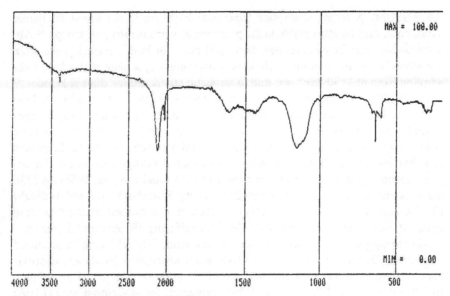

Figure 2. Infrared spectrum of WSSOM.

Table 4. Probable Prominent Functional Groups on WSSOM Based on Infrared Spectrum.

Wavenumber (cm⁻¹)	Probable Functional Groups[a]
3400	O-H stretch of intermolecularly-bonded alcohols and phenols; hydrogen bonds
2100	Nitrogen groups—aromatic isonitriles (RNO) and/or isocyanates (RNCO)
1620	Aromatic C = C bonds, unsaturated ketones
1440	Aliphatic (C-H) bending vibration
1220	Carboxyl (C-O stretching and O-H bending)
1130	Aliphatic ethers, dialkyl ketones

[a]Silverstein et al.[41]; MacCarthy and Rice.[39]

(as well as in Rhinebeck WSSOM) in a region closer to that characteristic of phenolic and alcoholic OH groups. The weak 1220 cm⁻¹ band, assigned to C-O stretching and OH bending vibrations also suggested a rather low concentration of carboxyl groups; this band disappeared on producing the salt form. The absence of a small absorption band at 1720 cm⁻¹ which is usually observed in humic samples provided further evidence that carboxyl groups were not prominent in the WSSOM. A strong absorption band near 2100 cm⁻¹, not found for humic substances, may be attributable to the presence of various nitrogen groups.[42] Absorption bands at frequencies less than 1200 cm⁻¹ for humic materials are often not identified in the literature. However, one might speculate that a broad absorption band at 1130 cm⁻¹ was due to aliphatic ethers and/or dialkyl ketones.[41]

The WSSOM IR spectrum differed somewhat from that obtained by Madhun et al.[25] for WSSOM extracted from Semiahmoo mucky peat soil in Oregon. Specifically, the authors observed a strong absorption band at 1720 cm⁻¹, but none at 2100 cm⁻¹, suggesting the Semiahmoo WSSOM was richer in carboxylic groups and, like most humic materials, without substantial nitrogen moieties. The organic carbon content of both the peat soil (27.8%) and the peat WSSOM (238 mg/L) were higher than that of the clayey silty Rhinebeck soil and WSSOM (3.13% and 66.3 mg/L, respectively). Therefore it is not surprising that these quite different soils produced WSSOM with differing IR absorption spectra.

After being passing through a column of dark sand, WSSOM samples produced a spectrum (Figure 3) with the same five major absorption bands and approximate relative strengths as shown in Figure 2. In addition, the pH of the influent (6.78) matched that of the effluent (6.84). However, the absorption band at 1620 cm⁻¹ was somewhat weaker, suggesting the loss of some of the hydrophobic aromatic groups via sorption to soil organic carbon. A slight decrease in relative band strength for 1440 cm⁻¹ (aliphatics) may be attributable to the same mechanism.

The DOC content of WSSOM changed little, from 66.8 to 63.7 ppm (approximately 4.7%) upon lowering of the pH and filtering. Since acidification did not

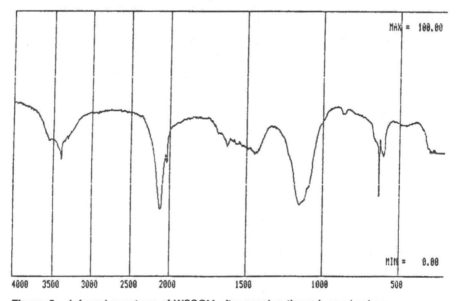

Figure 3.   Infrared spectrum of WSSOM after passing through sand column.

produce any visible precipitate nor a significant drop in DOC, WSSOM appeared to contain no significant amounts of humic acid. The humic materials contained in the WSSOM were, therefore, primarily the more-soluble fulvic acids.

Transport Through Dark Sand Columns

The breakthrough curve (BTC) of WSSOM from a dark sand column was compared to the BTC of ^3H_2O. The BTC of ^3H_2O alone is shown in Figure 4. By measuring the area above the BTC up to $C/C_o = 1$, R was estimated at 0.99, confirming that ^3H_2O behaved as a nonadsorbing tracer.

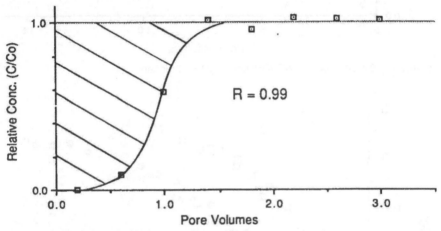

Figure 4.   Breakthrough of ^3H_2O in dark sand.

The BTC of WSSOM (measured as DOC), shown in Figure 5, indicated that the carrier was somewhat retarded compared to the tritiated water, with a measured R of 3.2. The curve is very asymmetric, with rapid breakthrough of approximately 75% of the WSSOM DOC, almost matching the breakthrough of ^3H_2O. However, there appeared to be a WSSOM fraction that eluted much more slowly, causing the BTC to level off. West found a similarly-shaped BTC for a humic material.[19] Since WSSOM is a heterogeneous mixture of molecular sizes and functional groups, it is possible that a fraction that is more hydrophobic or smaller (i.e., less subject to pore exclusion) takes longer to elute. The measured R can be considered an average for all WSSOM fractions. Skewed breakthrough curves of this type may also be produced when local equilibrium assumptions break down and kinetic or mass transfer limitations occur.[42]

Phenanthrene eluted more quickly in the presence of WSSOM than in its absence (Figure 6). The average R (from duplicate experiments) was reduced from 79 without the carrier to 44 with it; i.e, transport was 1.8 times faster. Although the dark sand is low in organic carbon, it has been shown that significant DOC

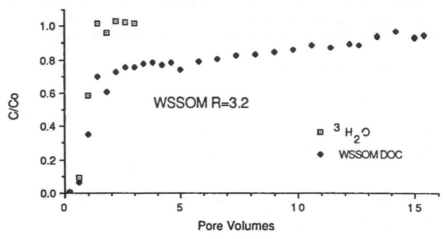

Figure 5.   Breakthrough of WSSOM and ^3H_2O in dark sand.

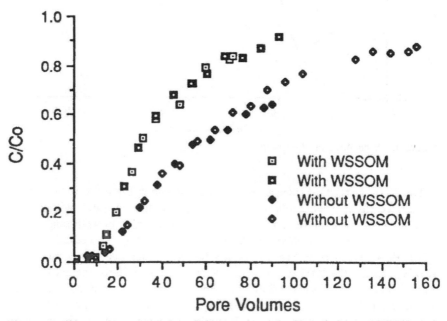

Figure 6.   Phenanthrene breakthrough from dark sand, with and without WSSOM, replicate experiments.

concentrations can be reached in low-carbon aquifer materials, since the source of the recharge water can control the DOC concentration.[43] The long slow breakthrough of phenanthrene suggested nonideal sorption, i.e., failure to conform to the assumption of local equilibrium.[42] It is likely that the sorption of phenanthrene involves a fast and a slow kinetic component. The slow component may involve slow phenanthrene diffusion to sorptive sites or hindered diffusion into organic matter within the soil.[42] Furthermore, sorption by the Teflon® retainers on either end of the soil column may have contributed to the slow approach to $C/C_o = 1$.

Transport Through Rhinebeck Soil Columns

Phenanthrene was applied to Rhinebeck soil with and without WSSOM to qualitatively determine whether facilitated transport occurred to a similar extent in a soil with very different characteristics than the dark sand. As expected, phenanthrene breakthrough was much slower due to the higher organic content of the Rhinebeck soil. However, phenanthrene eluted more quickly from the Rhinebeck soil without WSSOM than with it (Figure 7). It was initially postulated that the WSSOM, which was derived from Rhinebeck soil, was reincorporated to a certain extent into the stationary soil particles. Much of this WSSOM would have been carrying phenanthrene with it, thus slowing the breakthrough of phenanthrene. If this were the case, WSSOM should be significantly retarded by the Rhinebeck soil. However, the observed breakthrough of WSSOM (Figure 8) did not substantiate this hypothesis. In fact, R for WSSOM was approximately 1.4, less than that of WSSOM through dark sand. Size exclusion from smaller pores in the Rhinebeck may have reduced breakthrough time for the WSSOM relative to its behavior in the sand column. In any case, there was no indication that WSSOM alone was retarded to any great extent during flow through the Rhinebeck soil.

Another possible explanation for the slow breakthrough of phenanthrene in the presence of WSSOM is the interaction of the phenanthrene-WSSOM complex with the stationary soil phase. Removal of a WSSOM phenanthrene complex may be influenced by a change in conformation. Ghosh and Schnitzer[44] have shown

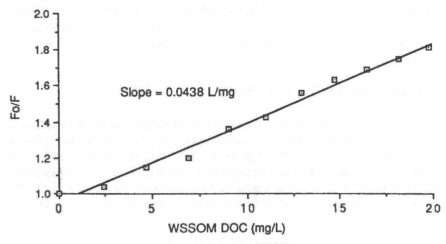

Figure 7. Isotherm for phenanthrene binding to WSSOM.

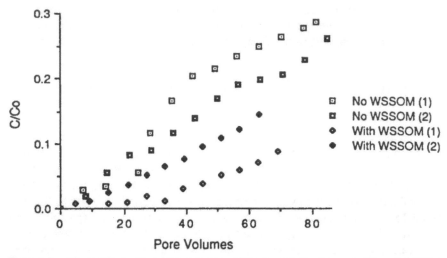

Figure 8. Phenanthrene breakthrough from Rhinebeck soil, with and without WSSOM.

that humic materials can exist in a variety of conformations depending on the solution conditions. Freeman and Cheung[45] have attributed slow desorption kinetics of lipophilic organics in soils to hindered diffusion from a swollen humic gel-like structure. If the binding of phenanthrene were to cause WSSOM molecules to change conformation, the frequency of WSSOM collisions with the porous media would be altered. If, in addition, the WSSOM-phenanthrene complex were more hydrophobic than WSSOM alone, the attachment frequency resulting from collisions would increase and retardation of the complex may, in such a case, be greater than retardation of either phenanthrene or WSSOM alone. This effect was not observed in the dark sand, possibly because of the larger pore size of the sand compared to the clayey silty loam. The clay in the Rhinebeck soil, which swells upon wetting, can yield a significantly smaller average pore size. Unfortunately, the mechanism by which phenanthrene might change the conformation of WSSOM molecules is not clear, nor is it clear that an interaction of this nature slowed phenanthrene breakthrough from the Rhinebeck soil.

Phenanthrene Binding to WSSOM

Measurements of fluorescence quenching of phenanthrene by WSSOM were used to determine K_d^{om}. The resulting isotherm ($r^2 = .989$, shown in Figure 9) has a slope (K_d^{om}) of 0.0438 L/mg, or 43,800 mL/g (log $K_d^{om} = 4.64$). Since WSSOM is measured in DOC, the distribution coefficient is on an organic carbon basis ($K_d^{om} = K_{oc}^{om}$). For comparison, Gauthier et al. measured $K_{oc} = 50,000$ mL/g for phenanthrene with a humic acid by the fluorescence quenching technique.[28]

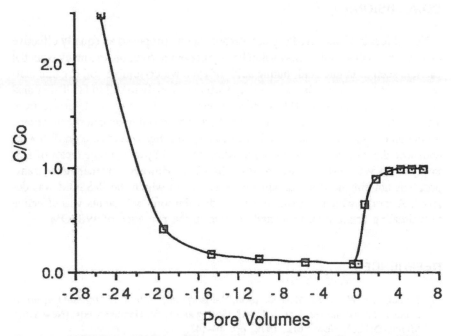

Figure 9. WSSOM (as DOC) breakthrough from Rhinebeck soil. Since some DOC was released from the soil, the DOC in the column effluent was allowed to stabilize prior to WSSOM input, which occurred at 0 pore volumes.

Modeling Transport with WSSOM from Three Distribution Coefficients

Phenanthrene and WSSOM partition coefficients with the dark sand may be calculated from their retardation in the column studies shown in Figures 5 and 6. It was determined that $K_{dom}{}'=0.60$ mL/g from the WSSOM BTC, $K_d{}'=16$ mL/g from the average of the phenanthrene BTCs, and $K_d{}^{om}=0.0438$ L/mg from the fluorescence batch experiment. Combining the results of these column and batch experiments, Eq. 5 can be used to predict R^* for phenanthrene in the presence of WSSOM. Therefore, using the average soil column characteristics ($\rho_b/n=4.38$ g/mL) and WSSOM concentration (85.8 mg/L) from the column experiments of phenanthrene with WSSOM, Eq. 5 predicts $R^*=37$, a reasonable estimate of the observed R^* of 44. The underestimation of the observed R^* may be attributable to the complex sorbing to soil differently than WSSOM alone. The soil-complex distribution coefficient, not modeled here, may be more significant during transport through the clayey, organic Rhinebeck soil. In general, however, the model appears to provide a reasonable estimate of retardation of a hydrophobic pollutant through a sand column in the presence of a carrier.

CONCLUSIONS

Not all forms of dissolved organic carbon can be proposed as equally effective carriers. Even though they may solubilize hydrophobic compounds, many potential carriers such as tannic acid, humic acid, or even water-soluble soil organic matter (WSSOM) from certain sources may sorb too strongly to soil to facilitate transport. In some cases, WSSOM can be an effective carrier. WSSOM derived from a clayey silty loam contained a variety of hydrophilic and hydrophobic functional groups and was relatively unaltered by its passage through a sand column. WSSOM enhanced the transport of phenanthrene through sand by an average factor of 1.8, even though WSSOM itself was retarded ($R=3.2$). However, phenanthrene transport was not enhanced in the Rhinebeck soil from which the WSSOM was derived. A model which incorporates three distribution coefficients was effective in estimating retardation in a sand column in the presence of WSSOM.

REFERENCES

1. Hamaker, J. W., and J. M. Thompson, "Adsorption," Ch. 2. in *Organic Chemicals in the Soil Environment*, Vol. 1, C.A.I. Goring and J. W. Hamaker, eds. (New York: Marcel-Dekker, New York, 1972, pp. 49-143.
2. Rao, P. S. C., and J. M. Davidson. "Adsorption and Movement of Selected Pesticides at High Concentrations in Soils, *Water Res.*, 13:375-380 (1979).
3. Tomson, M. B., J. Dauchy, S. R. Hutchins, C. Curran, C. J. Cook, and C. H. Ward. "Ground Water Contamination by Trace Level Organics from a Rapid Infiltration Site," *Water Res.* 15:1109-1116 (1981).
4. Hutchins, S. R., M. B. Tomson, P. B. Bedient, and C. H. Ward. "Fate of Trace Organics During Land Application of Municipal Wastewater," *CRC Crit. Rev. Env. Control*, 15:355-416 (1985).
5. Villaume, J. F. "Coal Tar Waste: Their Environmental Fate and Effects," in *Hazardous and Toxic Wastes Technology, Management and Health Effects*, K. Majumdar and E. W. Miller, eds., Pennsylvania Academy of Science, 1984, pp. 362-375.
6. Enfield, C. G., D. M. Walters, R. F. Carsell, and S. Z. Cohen. "Approximating Transport of Organic Pollutants to Groundwater," *Ground Water*, 20:711-722 (1982).
7. Baker, J. E., P. D. Capel, and S. J. Eisenreich. "Influence of Colloids on Sediment-Water Partition Coefficients of Polychlorobiphenyl Congeners in Natural Waters," *Environ. Sci. Technol.*, 20:1136-1143 (1986).
8. Pierce, R. H., Jr., C. E. Olney, and G. T. Felbeck, Jr. "pp'-DDT Adsorption to Suspended Particulated Matter in Sea Water," *Geochimica et Cosmochim. Acta*, 30:1061-1073 (1974).
9. Carter, C. W., and I. H. Suffet. "Binding of DDT to Dissolved Humic Materials," *Environ. Sci. Technol.*, 16:735-740 (1982).
10. McCarthy, J. F., and B. K. Jimenez. "Reduction in Bioavailability to Bluegills of Polycyclic Aromatic Hydrocarbons Bound to Dissolved Humic Material," *Environ. Tox. Chem.*, 4:511-521 (1985).

11. Guy, R. D., D. R. Narine, and S. DeSilva. "Organo-Cation Speciation. I. A Comparison of the Interactions of Methylene Blue and Paraquat with Bentonite and Humic Acid, *Can. J. Chem.*, 58:547–554 (1980).

12. Matsuda, K., and M. Schnitzer. "Reactions Between Fulvic Acid, a Soil Humic Material, and Dialkyl Phthalates," *Bull. Env. Contam. Tox.* 6:200–204 (1971).

13. Boehm, P. D., and J. G. Quinn. "Solubilization of Hydrocarbons by the Dissolved Organic Matter in Seawater," *Geochimica et Cosmochim. Acta* 37:2459–2477 (1973).

14. Hassett, J. P., and M. A. Anderson. "Effects of Dissolved Organic Matter on Adsorption of Hydrophobic Organic Compounds by River and Sewage-Borne Particles," *Water Res.* 16:681–686 (1982).

15. Means, J. C., and R. D. Wijayaratne. "Role of Natural Colloids in the Transport of Hydrophobic Pollutants," *Science* 215:968–970 (1980).

16. Wijayaratne, R. D., and J. C. Means. "Sorption of Polycyclic Aromatic Hydrocarbons by Natural Estuarine Colloids," *Mar. Environ. Res.* 11:77–89 (1984).

17. Isaacson, P. J., and M. H. Hayes. "The Interaction of Hydrazine Hydrate with Humic Acid Preparations at pH 4," *J. Soil Sci.* 35:79–92 (1984).

18. Wijayaratne, R. D., and Means, J. C. "Affinity of Hydrophobic Pollutants for Natural Estuarine Colloids in Aquatic Environments," *Environ. Sci. Technol.* 18:121–123 (1984).

19. West, C.C. "Dissolved Organic Carbon Facilitated Transport of Neutral Organic Compounds in Subsurface Systems," Ph.D. Thesis, Rice University, Houston, TX, 1984.

20. Garbarini, D. R., and L. W. Lion. "Influence of the Nature of Soil Organics on the Sorption of Toluene and Trichloroethylene," *Environ. Sci. Technol.* 20:1263–1269 (1986).

21. Gschwend, P. M., and S. Wu. "On the Constancy of Sediment-Water Partition Coefficients of Hydrophobic Organic Pollutants," *Environ. Sci. Technol.* 19:90–96 (1985).

22. Haniff, M. I., R. H. Zienius, C. H. Langford, and D. S. Gamble. "The Solution Phase Complexing of Atrazine by Fulvic Acid: Equilibria at 25°C.," *J. Environ. Sci. Health* B20(2): 215–262 (1985).

23. Caron, G., I. H. Suffet, and T. Belton. "Effect of Dissolved Organic Carbon on the Environmental Distribution of Nonpolar Organic Compounds," *Chemosphere* 14:993–1000 (1985).

24. McCarthy, J. F., and B. K. Jimenez. "Interactions Between Polycyclic Aromatic Hydrocarbons and Dissolved Humic Material: Binding and Dissociation," *Environ. Sci. Technol.* 19:1072–1076 (1985).

25. Madhun, Y. A., J. L. Young, and V. H. Freed. "Binding of Herbicides by Water-Soluble Organic Materials from Soil," *J. Environ. Qual.*, 15: 64–68 (1986)

26. Chiou, C. T., R. L. Malcolm, T. I. Brinton, and D. E. Kile. "Water Solubility Enhancement of Some Organic Pollutants and Pesticides by Dissolved Humic and Fulvic Acids," *Environ. Sci. Technol.* 20:502–508 (1986).

27. Gamble, D. S., M. I. Haniff, and R. H. Zienius. "Solution-Phase Complexing of Atrazine by Fulvic Acid: A Batch Ultrafiltration Technique," *Anal. Chem.* 58:727–731 (1986).

28. Gauthier, T. D., E. C. Shane, W. F. Guerin, W. R. Seitz, and C. L. Grant. "Fluorescence Quenching Method for Determining Equilibrium Constants for Polycyclic Aromatic Hydrocarbons Binding to Dissolved Humic Materials," *Environ. Sci. Technol.*, 20:1162–1166 (1986).

29. Kan, A. T., and M. B. Tomson. "Facilitated Transport of Naphthalene and Phenanthrene in a Sandy Soil Column with Dissolved Organic Matter—Macromolecules and Micelles," in *Proceedings of the Petroleum Hydrocarbons and Organic Chemicals in Groundwater Conference*, Houston, TX, 1986, pp. 93–105.

30. Gauthier, T. D., W. R. Seitz, and C. L. Grant. "Effects of Structural and Compositional Variations of Dissolved Humic Materials on Pyrene Koc Values," *Environ. Sci. Technol.* 21:243–252 (1987).

31. Chiou, C. T., D. E. Kile, T. I. Brinton, R. L. Malcolm, J. A. Leenheer, and P. MacCarthy. "A Comparison of Water Solubility Enhancements of Organic Solutes by Aquatic Humic Materials and Commercial Humic Acids," *Environ. Sci. Technol.* 21:1231–1234 (1987).

32. Vinten, A. J. A., B. Yaron, and P. H. Nye. "Vertical Transport of Pesticides into Soil When Adsorbed on Suspended Particles, *J. Agric. Food Chem.* 31:662–664 (1983).

33. Davidson, J. M., P. S. C. Rao, and P. Nkedi-Kizza. Physical Processes Influencing Water and Solute Transport in Soils," in *Chemical Mobility and Reactivity in Soil Systems*, D. W. Nelson, D. G. Elrick, and K. K. Tanji, eds., American Society of Agronomy, SSSA Special Publication No. 11, Madison, WI, 1983, pp. 35–47.

34. Rao, P. S. C. and R. E. Jessup. "Sorption and Movement of Pesticides and Other Toxic Organic Substances in Soils," in *Chemical Mobility and Reactivity in Soil Systems*, D. W. Nelson, D. G. Elrick, and K. K. Tanji, eds., American Society of Agronomy, SSSA Special Publication No. 11, Madison, WI, 1983, pp. 183–201.

35. Gamerdinger, A. P., R. J. Wagenet, and M. Th. van Genuchten. Two-Site/Two-Region Models for Pesticide Transport and Degradation 2. Experimental," *Soil Sci. Soc. Am. J.* (submitted).

36. Magee, B. R. "The Transport of Phananthrene in Soil in the Presence of Water-Soluble Organic Material." M.S. Thesis, Cornell University, Ithaca, NY, 1989.

37. Nkedi-Kizza, P., P. S. C. Rao, and A. G. Hornsby. "Influence of Organic Cosolvents on Sorption of Hydrophobic Organic Chemicals by Soils," *Environ. Sci. Technol.* 19:975–979 (1985).

38. Nkedi-Kizza, P., P. S. C. Rao, and A. G. Hornsby. "Influence of Organic Cosolvents on Leaching of Hydrophobic Organic Chemicals Through Soils," *Environ. Sci. Technol.*, 21:1107–1111 (1987).

39. MacCarthy, P., and J. A. Rice. "Spectroscopic Methods (Other Than NMR) for Determining Functionality in Humic Substances," in *Humic Substances in Soil, Sediment, and Water.*G. R. Aiken, D. M. McKnight, R. L. Wershaw, and P. MacCarthy, eds. (New York: John Wiley and Sons, 1985), pp. 527–560.

40. Stevenson, F. J., and K. M. Goh. "Infrared Spectra of Humic Acids and Related Substances," *Geochimica et Cosmochim. Acta* 35:471–483 (1971).

41. Silverstein, R. M., G. C. Bassler, and T. C. Morrill. *Spectrometic Identification of Organic Compounds*, 3rd ed. (New York: John Wiley and Sons, 1974), pp. 135–139.

42. Brusseau, M. L.; and P. S. C. Rao. "Sorption Nonideality During Contaminant Transport in Porous Media," *CRC Crit. Rev. Env. Control*, 19:33–99 (1989).

43. Thurman, E. M. "Humic Substances in Groundwater," in *Humic Substances in Soil, Sediment, and Water*, G. R. Aiken, D. M. McKnight, R. L. Wershaw, and P. MacCarthy, eds., (New York: John Wiley and Sons, 1985), pp. 87–104.

44. Ghosh, K. and M. Schnitzer. "Macro-Molecular Structure of Humic Substances," *Soil Sci.*, 129:266–276 (1980).

45. Freeman, D. H., and L. S. Cheung. "A Gel Partition Model for Organic Desorption from a Pond Sediment," *Science* 214:790–792 (1981).

CHAPTER 16

Stabilized Petroleum Waste Interaction with Silty Clay Subgrade

Joseph P. Martin, Francis J. Biehl, and W. Terry Robinson

INTRODUCTION

Concern for contamination of soil and groundwater by petroleum products is generally associated with spills or leaks from product storage and distribution facilities. However, accidental releases are also of concern at refineries which handle large quantities of hydrocarbons over extended periods. In addition, crude oil refining generates unusable by-products. This has decreased in recent years due to improved process efficiencies that have been spurred by both price increases and environmental concerns. However, near many refineries are tracts containing wastes from earlier years which remain in outdoor storage.

Destruction of oily wastes is often done by incineration or sludge farming, but degradation is not always feasible for the older wastes. The storage methods used in the past are often not acceptable by today's standards. The central problem in almost all waste disposal sites is that they do not have conventional design lives. Waste storage or disposal on land is implicitly "forever." Consequently, the risks of long-term contaminant releases must be considered should an older or present-day standard waste containment not work entirely as intended. Moreover, high levels of maintenance, monitoring, and ongoing treatment such as leachate handling cannot be sustained indefinitely.

Similar concerns exist with cleanup of liquid fuel spills, especially with regard to residuals handling and postremediation performance of the affected soil body.

For example, excavation of hydrocarbon-soaked soil and landfilling of the spoil only moves the problem from one point to another. The fate of the pollutants in the "final" depository must be addressed. Back at the spill site, some fraction of a product spill that has penetrated soil strata generally remains even after remedial methods such as pumping or volatilization are carried out as far as possible.

The dual goals of modern disposal site design are to provide a place to put unwanted material while preventing release of contaminants to the local environment. The latter goal dominates postremediation planning at a spill site.

Design of waste disposal facilities for inground residuals containment revolves around restricting one or more of the requirements for contaminant transport:

- The contaminant must be in a mobile form such as a gas, liquid, particulate or solute.
- The mobilized contaminant must have access to a continuous pathway to and through the local environment.
- A potential gradient (gravity, hydraulic, concentration, etc.) must exist to induce movement.

For a given spill or waste deposition site, there are three main time frames in terms of potential for control of contaminant movement:

- active operation, construction or remediation
- postremediation or postclosure site management (leachate collection, cap maintenance, etc.)
- "forever," the indefinite period thereafter

At some point, the control must be reduced for reasons of economy, and the site essentially abandoned except for monitoring. However, if contaminants are mobilized and transport commences, several media must be crossed before the site boundary or a water supply source is reached:

- the deposited or initially affected volume
- the boundary control system (landfill liners, recovery wells, etc.)
- the natural subgrade beyond the control system subsurface volume

In these locations, there may be some opportunities for contaminant retardation by adsorption, precipitation, biodegradation, etc.[1,2] While not favored as mechanisms to reduce impact, dispersion and dilution in groundwater also reduce concentrations. This may be especially important if products of biodegradation and other transformations are not fully mineralized, but are more mobile than the source material.[3]

Planned control can only be exerted for the active operation and maintenance periods, and generally, for only the main waste body and its boundary. It is thus advisable to not only provide the safest possible situation when active control ceases but also to assess the ability of the near-subsurface environment to mitigate unforeseen contaminant releases following abandonment.

The emphasis in petroleum spill cleanup has been on reducing contamination by removing as much of an inherently mobile material as possible from the environment: first, recovering as much free product as possible, and then, controlling and accelerating natural processes that would mobilize residue later on, e.g., volatilization and biodegradation.

In land disposal of waste, the potential contaminant is consciously put in the ground. At a minimum, mobility restriction starts by not depositing liquids or sludge in a landfill. The standard practice today is to also install and operate an external envelope that prevents fluid movement across the boundaries. Included in this envelope are liners and caps that are fluid barriers (pathway control), and leachate and gas fluid interceptors that provide gradient control. Waste stabilization before or at the time of deposition extends restriction of mobilization and transport into the landfill itself. However, this is difficult for organic wastes, especially bulk nonpolar hydrocarbons. Therefore, a backup analysis of mitigation of uncontrollable long-term releases is necessary.

SITE AND MATERIALS DESCRIPTION

The wastes on the site include acid hydrocarbon sludges that are the waste products of a variety of crude oil refining operations employing sulfuric acid. The sludge emulsion includes 2% to 8% sulfur, saturated long-chain aliphatics and complex aromatics, and has a pH in the range of 2.0 to 2.5. The ash content is quite variable, but is generally sufficient to produce a sludge density slightly above that of water. The most important property with respect to both groundwater impact and difficulty in remediation is the consistency. This varies with temperature, but ranges roughly between the viscosity of a semisolid and a liquid asphalt.

The lagoons are up to about 25' deep, and were constructed by both excavation and dike construction, as shown on Figure 1. Volatile emissions from undisturbed surfaces are very low, apparently due to several decades of exposure. Mitigation of emissions and removal of more soluble fractions is also assisted by detention of site runoff in some of the lagoons prior to staged treatment and discharge. Sludge impoundment has also produced lagoon stratification, such that a bottom layer of extremely hard "asphalt" functions as an impermeable barrier.

The other waste material is an oily spent clay (fuller's earth) that was used in lubricant processing. It was discarded and placed in high piles when the hydrocarbon content approached about 20% (by weight). After deposition, the highly plastic (attapulgite) material consolidated under increasing overburden and hydrated in the humid climate. These processes result in petroleum expulsion, such that the oil and grease content is now in the 4% to 8% range. Moisture content near the surface can be as high as 90% (by weight), producing a very sticky but weak material. Drying to about 60% moisture or treatment with lime facilitates handling of the spent clay. The clay has a medium cation exchange capacity, with values ranging from 25.1 to 38.8 milliequivalent per gram.

The sludge and spent clay deposits are on, or on the sides of, two ridges that bracket a local stream. The metamorphosed crystalline bedrock is relatively intact, and its mantle of weathered silty clay is up to about 25' thick on the elevated portions of the site. The soil is friable and of a loamy consistency near the surface. The saturated hydraulic conductivity measured on undisturbed samples is about 10^{-5} cm/sec in the horizontal direction, and 10^{-6} cm/sec vertically. However, the material can be easily remolded and compacted to yield much lower hydraulic conductivities, as discussed later.

Groundwater impact from the lagoons is very localized, but a pool of petroleum lies on the water table at the foot of one clay pile. A series of conventional recovery wells have been installed between the brook pile and the clay pile, as illustrated in Figure 1.

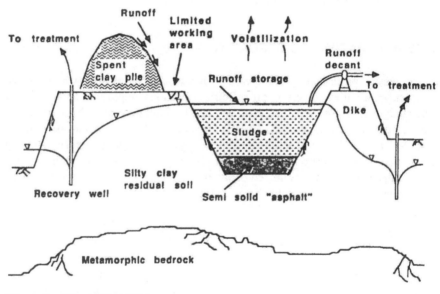

Figure 1. Site cross section.

REMEDIATION NEED

The water table is generally near the soil/rock interface, local recharge rates are low, and the well-populated area does not rely on local groundwater. The stream, conveying both natural flow and treated runoff, meets state standards and discharges into a tidally influenced river. The site is not abandoned, but rather, is regarded as being an inactive part of an otherwise busy refinery. With the carefully maintained facilities, no measurable volatiles release, and compliance with discharge permit requirements, the question arises as to why it is necessary to do more than to monitor and maintain the internal containments (dikes, wells, etc.).

The answer lies in the very large waste volume, with several hundred thousand cubic yards of the sludge alone. The failure of a lagoon dike could have disastrous impacts on the downstream environment. At the spent clay piles, hydrocarbon seepage from dispersed clay piles is being intercepted, not prevented. In this context, the state cannot sanction the maintenance of such a large private site "forever" by a corporate entity as a permanent solution. The owner is also concerned with liability. A permanent solution that entails less exposure was required.

Various alternative solutions have been investigated. The objective and public relations risks of transportation offsite for disposal or treatment are unacceptable. However, onsite solutions are restricted by the site congestion. Sludge farming or other surface-intensive treatments would be quite difficult, and the pretreatment and extended exposure required generate serious concerns about air quality deterioration. Onsite incineration is extremely expensive. An acceptable solution must include as little exposure of disturbed material as possible to minimize volatilization of the oil product in the clay and the unknown constituents in the deeper zones of the sludge lagoons.

Consequently, onsite redeposition in capped landfills is the primary option. Support for the caps and prevention of containment corrosion requires mechanical improvement and neutralization of the sludge. However, reducing the liquidity of the sludge and depositing it inside a secure landfill envelope is not in itself sufficient. The integrity of the external containment (cap and liner) and the internal fluid diversions would have to be maintained as long as the threat to the environment existed. Internally immobilizing the petroleum by-products was seen as a desirable short- and long-term goal.

STABILIZATION

Stabilization is difficult to define, as the term is often used interchangeably with other expressions such as solidification, immobilization, fixation, and microencapsulation. In general, stabilization takes a raw waste material with high constituent mobility and makes it reasonably safe to store, transport, or landfill with significantly reduced concern for accidental release of contaminants. The major point is internal, bulk modification of the waste, reducing reliance on external containments such as drums, tanks, and landfill liners.

One approach focuses on contaminant immobilization on the microscale.[4,5] The ideal result is incorporation of molecules into crystals or reversible reactions by reducing solubility with pH control. Stabilization is also used to describe physicochemical treatments that cement, bind, embed or otherwise incorporate a waste into a self-supporting matrix. When the waste is liquid or sludge, the most popular method is admixture of a hydraulic cement to solidify the waste. Often other waste by-products such as fly ash are used as components in pozzolanic cement reactions. The intent is often to form a continuous, hard monolith. The integrity of the mass is as important as the microscopic immobilization, as this reduces the surface area of contaminant exposure to the surrounding environment.

Stabilization is functionally described in terms of three properties or characteristics of the final product:

1. Mechanical stabilization entails reducing the bulk mobility or liquidity of the waste to make it self-supporting, durable, and dimensionally stable. The product may be continuous or granulated, and either incorporate the contaminants in a solid or have an artificial solid skeleton enclosing pores containing the waste. In either case, it is essential that the solid portion resist deterioration or excessive deformation by mechanical, biochemical, or climatic stresses in the final depository environment.
2. Immobilization by fixation (adsorption, precipitation, etc.) reduces the local mobility of individual contaminant molecules. The extent of fixation is indicated by leaching tests, especially those which indicate an assumed equilibrium between stabilized material and a potential aggressive environment. For example, the EP toxicity test exposes pulverized material to an acid solution.
3. Immobilization by isolation. Even if a contaminant is, or becomes, mobile, protection against transport to the boundary of the deposited mass is afforded by restricting internal transport pathways. Embedding contaminants with or without local fixation in a matrix with low permeability and diffusion coefficient values is one example.

All three characteristics are often sought with hydraulic cement stabilization.[6] The hardening precipitate binds and strengthens the mass, incorporates free moisture in the hydration, blocks channels between pores or coats waste particles, and incorporates some contaminant molecules in the siliceous solid.

SLUDGE/SPENT CLAY STABILIZATION

It was assumed that any solution must include neutralization and immediate improvement of the consistency of the sludge for construction feasibility. The first stabilization trials were with the simplest and least expensive technique, the addition of a lime/fly ash mixture. The lime provided pH control and the silt-sized coalburning powerplant fly ash contributed surface area. Lime/fly ash solidification is often used successfully with inorganic sludges. The desired result in this case was micro–encapsulating the sludge droplets inside a fine, stiff, porous structure.

However, the mixed texture could best be described as a very fine tar sand, and the soluble hydrocarbon fraction apparently interfered with the pozzolanic cementing reaction. Strength and stiffness improvement after compaction of the soil-like material was minimal, and the nonadsorbing, interconnected skeletal structure neither fixated nor isolated the hydrocarbons. Permeabilities were high (10^{-4} to 10^{-5} cm/sec) and the oil could be readily leached, either as a solute or as a separate liquid phase.

Portland cement was then added to improve the interparticle binding. With a high cement proportion and exposure to hydration moisture, mechanical properties improved greatly. Permeabilities were also reduced to the 10^{-7} range, depending upon sludge consistency and additive proportioning. The pore restrictions prevented displacement of hydrocarbons as a separate-phase even under high gradients.[7] However, permeation with water or weak acid yielded an effluent with very high dissolved hydrocarbon content. This was readily indicated by clarity as well as more definitive indicators such as TOC (Total Organic Carbon) or COD (Chemical Oxygen Demand). In a stabilized landfill configuration such as the one shown in Figure 2, it would have been necessary to maintain the leachate collection system indefinitely unless a very secure external containment was installed.

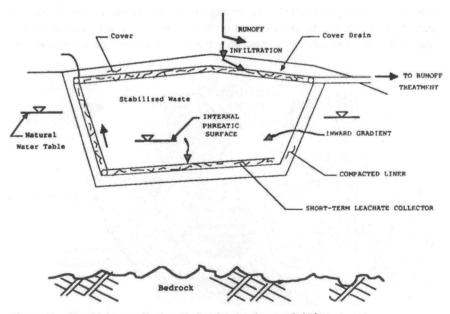

Figure 2. Short-term gradient control using barriers and drains.

A traditional earthwork technique was successful in stabilizing the oily spent clay. Lime admixture decreases the water affinity or plasticity, thus reducing the thickness of the "diffuse ionic double layer" of water surrounding clay particles. The practical effects intended were to improve compactibility and also to reduce the swell potential and its threat to the dimensional integrity of the cap. The improved compactibility from the lime treatment also reduced the permeability and entrapped oil globules in a more tortuous pore structure, as had also occurred with the cement stabilization of the sludge. However, the solubility of the hydrocarbons was also dramatically reduced, such that permeated effluent clarity improved significantly. The double-layer shrinkage and pH adjustment appeared to enhance the capacity of the clay to adsorb or otherwise fixate hydrocarbon molecules.

The success of the spent clay stabilization led to a combined program, a proportioning of sludge, spent clay, and lime/fly ash pozzolan additives. This provided monetary and volumetric economy compared to treating the two wastes separately, due to a lessened need for importation of materials from offsite.

The microscopic result is shown conceptually in Figure 3. The current process model postulates that the clay provides the porous solid skeleton, assisted to some extent by the fly ash. Neutralization causes the sludge emulsion to separate, allowing hydrocarbon droplets to become entrapped in the pore structure with mixing and compaction. Calcium and magnesium ions released in the neutralization reaction not only stabilize the clay fraction, but also produce alkaline conditions that etch and dissolve the glass portion of the fly ash. This initiates the pozzolanic reaction, and the precipitate binds the expanded clay structure. The results, depending upon material proportions and moisture content, include gains in mechanical properties, and further increase pore channel tortuosity. As in the clay-only

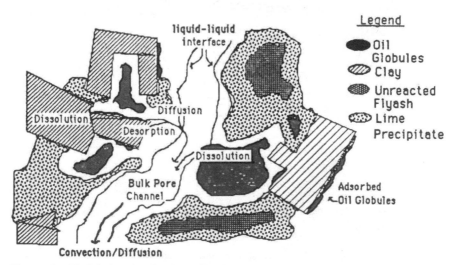

Figure 3. Model of contaminant mobilization.

situation, hydrocarbon solubility also decreased, as will be discussed in more detail later.

Figure 3 also implies that movement of a contaminant to the boundary of the stabilized mass involves two basic sets of activities. A contaminant must first be put in a dissolved, mobile form and then migrate across the "bound water" layer and the idealized liquid/liquid boundary with internal continuous pore channels. This stage is broadly called mobilization. Analysis of the individual microscale mechanisms involved appears to be impossible at this time.

Once in contact with continuous pore channels, transport through the stabilized material occurs under global hydraulic or concentration gradients. It is assumed that the entrapped separate-phase fraction of the sludge or oil is already in contact with the channels, but must still dissolve to mobilize. Empirical studies indicate that many factors affect contaminant mobilization and transport. Some can be controlled or optimized, such as material proportioning, raw sludge and clay consistency, pH, mixing effort, etc.

INTERNAL TRANSPORT MECHANISMS AND FILL LAYOUT

Figure 4 illustrates mechanisms that could induce transport of mobilized contaminants within the stabilized fill. They would eventually enter the local groundwater if not intercepted or attenuated by a leachate collection-liner system. Each mechanism has different time and space characteristics, as described further in Martin et al.[8,9]

The first is drainage of liquid in excess of the equilibrium capillary retention for the cemented but particulate and porous media. This is not unlike the movement of an oil spill from the vadose zone to the water table. The stabilized sludge/spent clay mixture appears to easily entrap separate-phase hydrocarbons, but water containing solutes could drain. Excess moisture is of most concern during cell construction, as some incident rainfall might not run off a rough working surface.

Quantification of the drainage contribution to leachate volume and rate is difficult, especially as the liquid must work its way down through underlying compacted and solidified material. Some moisture might be tied up in further pozzolanic cementation or clay adsorption. Despite the difficulty in estimating the flow rate, it is reasonable to assume that capillary drainage would cease some time after final cap placement.

Another leachate production mechanism is consolidation expulsion from the lower portions of the fill. Compression from overburden loading must be assumed to consist primarily of reduction in the void space. Hence, the fluid in the pores will be pressurized and expelled. Consolidation drainage would be expected to be primarily toward the upper and lower boundaries.

The contribution to leachate production from gross volume change can be estimated by traditional geotechnical methods. Some of the many parameters which

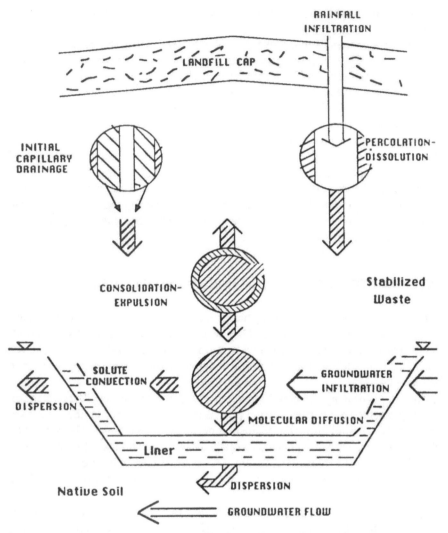

Figure 4. Mechanisms of internal mobilization and internal-external transport.

influence the leachate production can be controlled: stiffness, overburden loads, permeability, path length, etc. One uncertainty, however, is the extent of secondary consolidation after the primary consolidation relief of excess fluid pressures. Another is the potential for deterioration of the solid skeleton by long-term in situ conditions. Loss of stiffness would induce further consolidation.

The most commonly addressed mechanism of leachate generation is seepage of rainfall through the fill, with dissolution and convection of contaminants en route. This is of most concern during the exposed construction period discussed above. Not only is the fill uncovered, but permeability is highest with the fresh, uncompressed material. Following hardening, capping, and compression of lower zones, leachate production from rainfall seepage would be expected to decrease dramatically.

Unlike conventional loose-fill landfills, which rely entirely on the cap to suppress permeation, the whole stabilized waste mass is a hydraulic barrier with a permeability in the range of 10^{-8} cm/sec or less. The long-term gradient is at most unity, e.g., gravity. Consequently, even if the cap were to leak, rainfall seepage would be quite low. Cap performance would be expected to be much better than the equivalent arrangement at loose-fill landfills. A cap is a roof like any other; its performance depends as much on structural support and runoff-inducing slopes as on the basic imperviousness of the materials used. A conventional landfill is a poor roof support indeed, whereas the stabilized fill maintains dimensional stability.

While some of these first three mechanisms of leachate generation may not produce a measurable flow, it must be assumed that there will be a finite movement of leachate to the bottom of the stabilized fill. However, the production rate will be far lower than that at a conventional loose-fill landfill of equivalent area. Each drop of rainfall incident to such a facility that does not evaporate must be assumed to appear later as leachate. Consequently, only a low-volume leachate collection system will be necessary, underlain by a liner to prevent leachate entry into the native subgrade. This is shown in Figure 2.

For the period when detectable leachate production occurs, pumping from the collector will be required. In this interval, all three restrictions on contaminant transport described earlier (mobility, pathway, and gradient control) would be in effect. It is reasonable to assume that the leachate production rate will decrease markedly over the first few years, such that intermittent collection and possibly, cessation of pumping, might be justified thereafter.

The fourth potential mechanism of leachate transport is flooding with groundwater. It must be assumed that at some future date, the water table will rise above the fill base, and thus threaten to inundate the fill. A combination of gradient control, liner protection, and the low permeability of the stabilized waste monolith are proposed to minimize the potential impact. Figure 5a shows the idealized groundwater flow pattern prior to landfill installation. If the water table rises to the fill during the term of leachate pumping, the gradient will be into, not out of the fill, as shown on Figure 5b. In this case, the liner acts not so much to control loss of leachate from inside the facility as to minimize the rate of groundwater entry and excess volume to be treated.

Following cessation of leachate pumping, a restored groundwater flow condition will tend to bypass the fill even if the water table rose substantially. The stabilized waste will be a monolith with a permeability about two orders of magnitude below that of the natural soil. Hence, by taking the path of least resistance significant groundwater flow through the fill is highly unlikely. The problem, however is to prevent groundwater short-circuiting across the buried leachate collection system. The liner must remain intact indefinitely.

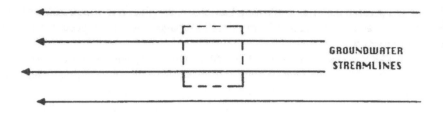

A.) Pre-Construction.

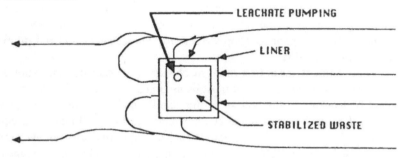

B.) Short-term: Pumping from enclosure.

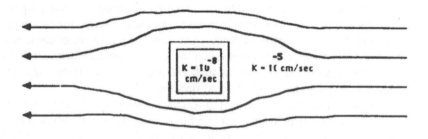

C.) Long-term: Monolith bypass.

Figure 5. Variation in groundwater flow with gradient and permeability control.

Molecular diffusion is the fifth transport mechanism shown in Figure 4 and only minimally controllable. Daniel[10] has noted that diffusion is probably a more important transport mechanism than hydraulic convection when permeabilities are below 10^{-6} cm/sec. Continuous pore channel pathways, however small and tortuous, exist throughout the system. There would initially be little diffusion through the fill itself, as concentration gradients from one point to another would be minor. However, a very large concentration difference will exist between the stabilized fill and the natural subgrade.

One method to reduce diffusion is to immobilize the contaminants as much as possible to reduce the source concentration, and another is to increase the liner diffusion path length. These serve to decrease the concentration gradient. Reducing the diffusion coefficients in both the fill and the liner are also worthy efforts, e.g., increasing channel tortuosity, but this is a relatively unstudied area. Both earthen and synthetic liners are under study, but the former has two desirable features: thickness and a potential for contaminant retardation within the liner material. Practical construction considerations also favor a compacted earth liner: a synthetic liner requires considerable bottom preparation and protection of the membrane against puncture, tearing, etc., whereas placement of clay liner can commence immediately upon grading the base of the excavation and impacts to the top few inches during filling tend to further remold, not damage, the material.

In any case, long-term molecular diffusion from the stabilized waste into the subgrade can be substantially mitigated, but not prevented entirely. Consequently, the goal is to reduce the diffusive flux to a rate within the dispersion and dilution capacity of the local groundwater, resulting in acceptably low concentrations at the site boundary.

SOURCE CONTROLS

The principle of stabilization beyond mechanical improvement is to reduce both the contaminant mobility and the internal transport ease within the waste fill itself. These features minimize either the volume or the concentration of fluid which must be prevented from degrading the local environment. The values and problems presented by using an industrially modified earth material, the spent clay, to provide both of these are illustrated in Figures 6 and 7.

In a field study of various stabilization alternatives, 50 yd³ cells of trial mixtures were placed on the site. One contained a sludge/clay/pozzolan mixture (Mix D), and another a sludge/portland cement/pozzolan blend (Mix E), each with a proportioning that had been optimized in the laboratory. Ongoing monitoring was conducted, as will be described later, and after a month of hardening, samples were recovered for mechanical testing.

Samples for one-dimensional consolidation study were placed in 1" high, 2.5" diameter stainless-steel rings and subjected to incremental loading as per standard compression procedures. Each mixture displayed about the same stiffness.[11] However, to saturate each sample for the test, the rings were placed in cylindrical plexiglass vessels flooded with distilled water. As a marked difference in clarity between the vessels was immediately seen, it was decided to replace the water at 3.5-day intervals and test the fluid, thus simulating (but not exactly) formal dynamic leaching tests,[12] which were not possible with the porous samples.

Figure 6, Total Organic Carbon (TOC) vs time, illustrates that the cement-bound Mix E was much less effective for immobilization than the clay-based Mix D. The observed mobilization and transport of carbonaceous material is a combination of compression, expulsion, and diffusion into the water bath. However, volumetric estimates indicate that the amount of contaminant migrating out of

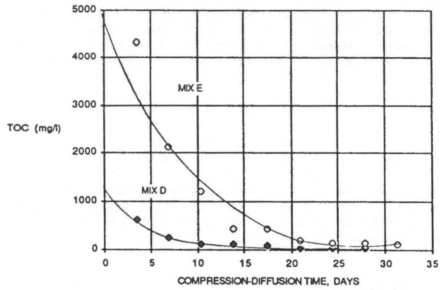

Mix D - Field clay-pozzolan mix, med. viscosity sludge
Mix E - Field cement-pozzolan mix, med. viscosity sludge

Figure 6. Comparison of static immobilization capability of clay- and cement-based stabilization mixtures.

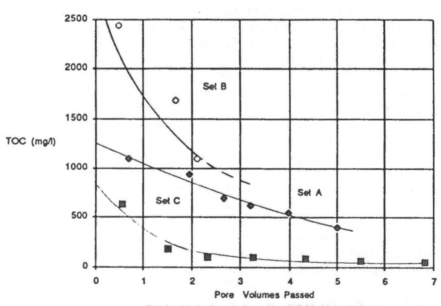

Set A: Med. visc. sludge, K = 2*E-08, high gradient
Set B: Med. visc. sludge, K = 2*E-08, low gradient
Set C: High visc. sludge, K = 2*E-06, low gradient

Figure 7. Composition and gradient influences on mobilization and convective of TOC.

the portland cement sample exceeded the volume change in the sample. There-fore, water entered the pore space to replace hydrocarbon loss out of the pores, in an attempt to achieve a high-concentration equilibrium. In contrast, much smaller amounts of contaminants exited the Mix D sample, due to either mobili-zation (dissolution) or transport (expulsion-diffusion) limitations. The average pH of the Mix E effluent was about 8, whereas the clay Mix D was approxi-mately neutral.

The equipment used was designed for mechanical-hydraulic experimentation, so that the implications of both the diffusion boundary conditions and the water-sludge ratio (about 10:1 by volume) are difficult to interpret. Consequently, the value of the tests is comparative, not an absolute indicator of expected field leachate concentrations. In terms of the sequential mobilization-transport model shown in Figure 3, it was concluded that the clay mix provided restrictions on both stages. The cement/pozzolan skeleton provided excellent mechanical stabilization, but did not significantly restrict either mobilization or diffusion.

To model both rainfall infiltration and groundwater seepage through the stabi-lized fill, many permeameter tests were run. These not only illustrated the in-fluence of various proportioning and in situ condition parameters, but also illustrated the control of either mobilization or convection, depending upon the flow rate.

Plexiglass permeameters were packed (usually compacted) with 200 g of freshly mixed material, and allowed to cure for two weeks before saturation, and then permeation commenced. Distilled water was used in most tests (pH=6.6), but solutions of 0.01 N sulfuric acid and 0.1 N acetic acid were also employed. These represented long-term conditions, the intermediate by-products of in situ biodegra-dation of the neutralized organic material.

Figure 7, TOC vs pore volumes passed, illustrates selected results of tests on the effluent from distilled-water permeation. Set C was done with a fairly vis-cous sludge, which made mixing without heating quite difficult.[13] A fairly dry clay, about 50% moisture content, was used. The minimum lime content required to assure uniform distribution (about 5% by weight) provided pH values of 7.5 to 8.0. One specimen was statically compacted, to represent the easiest construc-tion process, with a resulting permeability of 2×10^{-6} cm/sec. However, heavy kneading compaction to remold the clay structure decreased the permeability up to two orders of magnitude without significant changes in either porosity or ef-fluent quality.

The TOC vs pore volume results for Set C are quite good, and higher lime contents (and higher pH) improved them still further, but at the expense of in-creased permeability. The conclusion was that most of the hydrocarbons were in entrapped globule form as illustrated in Figure 3, but not very soluble. The effect of the clay was primarily physical, to provide the micro-encapsulating structure.

To provide data on less favorable materials, more liquid sludge and wetter clay (about 65% water content) were used, and a host of variables were encountered.

Sets A and B were identical samples employing the same lime proportion as Set C, but were tested at different hydraulic gradients. Both displayed 10^{-8} cm/sec permeability with minimal kneading, indicating easier dispersion of the medium viscosity sludge, but produced effluent samples with pH in the 6 to 7 range.

It can be seen from Figure 7 that the effluent quality is not nearly as good, especially for Set B. The lower gradient for the same permeability represents a slower velocity, but more importantly, a longer residence time. Consequently, the adsorbed or immobilized hydrocarbons could diffuse across the double layer into the main pore channel carrying the permeating water, as shown in Figure 3. For Set A, with the higher velocity, such mobilization was "washed out."

The lower hydraulic gradients used in Set B (about 15) still exceed those which would be expected in situ. Hence, it was concluded that the localized mobilization across the liquid/liquid interface shown in Figure 3 would dominate seepage leachate results, not internal convective transport. The focus was thus on improving adsorption, not on further reducing permeability. Major improvements in the effluent quality were found with either higher lime addition or drier clay. The results with the nonclay, cement stabilized mixtures (Figure 5) and the high viscosity sludge-clay mixtures (Set C, Figure 6) showed that pH was not in itself a major control; rather, high pH assisted the spent clay in immobilization of otherwise soluble fractions.

LOCAL SOIL INTERACTIONS

Despite the decades of petroleum refining waste storage on the site, the monitoring wells show very little groundwater impact except in areas immediately adjacent to lagoons or the spent clay piles. Whether this situation is the result of slow water movement or geochemical processes is unknown, but it raised the possibility of using the local soil as a liner, beneath the leachate collection system and the stabilized waste as indicated in Figures 2 and 4. It is a fairly simple matter to excavate a cell, bring some of the clayey silt to the proper moisture content, and recompact it on the base and side slopes. The results of two sets of tests are described herein, one involving extended incubation or hydration of several soil/liquid slurries, and permeameter tests.

It is hoped that an earth liner will adsorb or otherwise retard contaminants. The cation exchange capacity of the natural soil is low: 14.5 meq/g. To test the effects of aqueous solution contact, four liquids were used: distilled water (pH=6.6), 0.01 N sulfuric acid (pH=3.0), 0.1 N acetic acid (pH=3.0), and a synthetic leachate, actually a blend of the effluent collected from all of the permeability tests described earlier.

Slurries were made with 150 g of air-dried soil and an equal amount of leach liquid, and the specimens were sealed in jars for 70 days. Full contact between the soil particles and the liquid was assured, and it was reasoned that sufficient time for near-equilibrium interactions was allowed. At the end of the incubation, the supernatant from the settled slurry was decanted and analyzed. A separate liquid sample was taken by vacuum extraction through a filter at a suction pressure of 12 psi.

The low but finite buffering capacity of the soil was indicated by the pH of 6.23 for the vacuum extract from the acetic acid slurry, and a pH of 4.92 for the extract from the sulfuric acid hydration. More heartening was the evidence of organic material uptake (see Table 1).

Table 1. Organic Material Uptake.

Mixture	Original TOC	Supernatant TOC	Extract TOC
Soil-Acetic	3669 ppm	587 ppm	790 ppm
Soil-Leachate	1432 ppm	437 ppm	467 ppm

The reduction in free liquid organic content is substantial in each case, with the organic solute decreasing by two-thirds with the synthetic leachate, and a reduction of over 80% with the acetic acid. This is a favorable indication that the quality of leachate will be improved during seepage through both a compacted clay liner and the natural subgrade. The antecedent soluble organic content of the tested soil was quite low, with the extract after distilled water hydration showing a TOC of about 38 ppm, and the sulfuric acid solution incubation extract displayed a TOC of only 19 ppm.

With this data in hand, the next step was to determine hydraulic capabilities. Four permeameter samples were heavily kneaded near the optimum compaction moisture content and saturated to represent the compacted liner. To replicate the natural soil structure, the silty clay was aggregated to crumbs passing the #10 sieve, but retained on the #40 sieve. Six more permeameters were packed with light static effort on this material.

Consistent hydraulic results were obtained within each set. The heavily kneaded samples yielded hydraulic conductivities in the range of 2.0×10^{-8} to 3.5×10^{-8} cm/sec. This is quite acceptable for a liner, and further reduction would be expected in the field with the overburden pressure of 40' or so of stabilized waste. The aggregated, static-packed samples showed permeabilities from 4.2×10^{-5} to 8.3×10^{-5} cm/sec, similar to the results determined with the undisturbed samples. This contrast indicated that the two to three orders of magnitude difference between the liner and the subgrade required to minimize groundwater infiltration (Figure 5) can be obtained.

The effluent from the hydraulic conductivity tests was analyzed, indicating a similar pH as the influent (6.6+/−), and fairly low total dissolved solids. Following distilled water permeation, the other fluids were introduced. The aqueous acid permeation displayed an increase in permeability to the 10^{-7} cm/sec range, but the permeability of the sample passing the synthetic leachate did not vary from that found previously with water. The hydraulic stability of the liner to the range of fluid types to which it may be subjected in the field appears to be suitable, so that a synthetic liner, such as high density polyethylene, does not appear to be necessary from a chemical compatibility standpoint.[14]

Effluent quality results were less dramatic than the very high uptake observed with the extended hydration jar tests. There were substantial reductions in the organic content whether measured by TOC or COD, but not to the great extent found in the slurry equilibrium experiments. This was attributed to the effects of exposure and detention time, the inverse of the influences on organics mobilization shown in Figure 7. In the jar tests, there was full contact between soil particle surfaces and the liquid, and plenty of time for adsorption to occur. In the aggregated permeability test samples, the detention time was less by several orders of magnitude. The permeant passed through channels around aggregates and thus did not directly contact particles in the interior of the crumbs, which contained substantial bound water. Intermediate conditions (slower, more complete contact) existed in the low-permeability kneaded samples, which contained more tortuous and narrower pathways. Fortunately, there is no basis for expecting seepage through a liner at 10^{-3} cm/sec, which was the seepage velocity in the aggregated sample tests. Therefore, detention time is not expected to be a factor in limiting contaminant attenuation directly beneath the waste fills.

FIELD TEST RESULTS

Laboratory tests inherently represent some idealization, acceleration, or artificial boundary conditions. To more confidently predict worst-case field leachate quality, among other purposes, the field test site described earlier was installed in November 1986. The water quality sampler arrangement is shown conceptually in Figure 8.

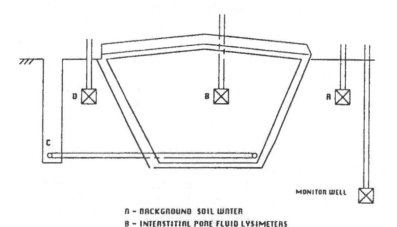

A – BACKGROUND SOIL WATER
B – INTERSTITIAL PORE FLUID LYSIMETERS
C – RUNOFF-COMPRESS LEACHATE
D – NATIVE SOIL LYSIMETER

Figure 8. Schematic of water quality sampler locations.

Background samples of both groundwater and soil water were obtained from a monitoring well and a lysimeter, respectively, 200′ upstream of the test cells, and labeled "A." To sample the porewater within the stabilized fill, lysimeters were placed in both the clay-based and nonclay, cement-based cells (Samples B-1 and B-2). To assess the consolidation/capillary drainage quality, leachate collectors were installed, leading to sample point "C" shown in Figure 8. Liquid draining from the cells was accumulated in the collection pit for 10 months, until deformation ceased. This is assumed to represent liquid expelled during consolidation and capillary drainage. Further accumulation is assumed to consist of leakage around the sides, in contact with the edge of the stabilized material, and thus represents, however approximately, runoff that has contacted the stabilized waste surface.

Finally, to assess interaction between leachate seeping through the unlined sides and the natural soil structure, lysimeters and monitoring wells were placed around the cells. However, no effect of the field test deposits has been seen to date in the groundwater monitoring wells, which were placed at some distance from the cells. Data shown in Table 2 is for lysimeters with the porous cups about 3′ below the surface. Lysimeter D-1 is in the soil 12′ away from the cement-sludge cell and D-2 is about 24″ from the clay-based sludge cell.

Table 2.   Results of Field Tests on Lysimeters and Monitoring Wells.

Sample	Date	Medium	Liquid	TOC (mg/L)	pH
A-1	6/88	Soil, bkgrnd	Soil H_2O	3.5	6.7
A-2	11/86	Soil, bkgrnd	Grndwtr	—	6.1
B-1	4/87	Clay-sludge	Pore liquid	2,027	6.8
B-1	6/88	Clay-sludge	Pore liquid	474	8.2
B-2	4/87	Cement-sludge	Pore liquid	13,462	6.9
B-2	4/87	Cement-sludge	Pore liquid	15,630	8.2
C	12/86	Collector pit	Drainage	263	6.2
C	4/87	Collector pit	Drainage	157	6.5
C	6/88	Collector pit	Runoff	38	6.7
D-1	6/87	Soil	Seepage	785	11
D-1	6/88	Soil	Seepage	235	7.4
D-2	6/87	Soil	Seepage	25	6.9
D-2	6/88	Soil	Seepage	4.0	8.0

Not only is the potentially transportable liquid in the clay-based mixture of much better quality than in the cement-bound sludge, but it appears that more immobilization occurred over a year's residence. The liquid in the earlier samplings of point C, the collection pit, is not as poor in quality as the pore liquid in the clay mixture. This may be due to dilution due to side permeation. However, the 6/88 sampling liquid is unquestionably seepage along the sidewalls and is of still better quality. In addition to runoff contact, this fluid could also represent the groundwater downstream of a buried but unlined monolith, i.e., marginal contact between the stabilized waste and water. The most encouraging results are those from sample point D-2, which captured capillary seepage from the clay-based cell. After 18 months, not only did little pollution appear in a region with high capillary suction, but the pattern of improved clay immobilization with increased pH appeared to work in situ also.

CONCLUSION

A technique has been developed to improve the onsite storage or disposal of hydrocarbon refining wastes that have little prospect of future reprocessing or feasibility for destructive (mineralization) treatment. The three basic controls on contaminant release from subsurface waste deposits or spill residuals are prevention of mobility, restriction of transport, and control of potential gradients. The focus in the stabilization process itself is on the first two.

The technique employed is to recycle one of the wastes, the spent clay, to mechanically, chemically, and hydraulically stabilize the other waste, acid hydrocarbon sludge. It is also necessary to control the pH and provide a pozzolanic cementing for the clay structure. The result is not unlike entrapment of hydrocarbons in an adsorptive, very fine-grained matrix, a micro-encapsulation. Movement from such a condition is restricted, but is certainly possible.

To better assess the likelihood, time frame, and consequences of contaminant release, a number of potential mechanisms of dissolved contaminant transport were identified. Four involved hydraulic gradients (capillary and consolidation drainage, and infiltration and groundwater seepage). The first two are expected to be important only during the first few years after deposition, while the risk of groundwater intrusion is a key long-term threat. To address these concerns, features such as a cap, a clay liner, a leachate collection system, and internal dewatering are under consideration, all employing gradient control in some way.

The last mechanism of release, molecular diffusion, can only be limited indirectly, by emphasizing contaminant fixation and liner thickening to reduce the concentration gradient. However, it was also demonstrated that the liner displayed significant capacity for contaminant attenuation as well as providing a reasonably stable hydraulic barrier. Because the liner is comprised of the same material as the local subgrade, the capacity for contamination retardation as well as groundwater dilution extends well beyond the man-made containment.

REFERENCES

1. Bagchi, A., "Design of Natural Attenuation Landfills," *J. Environ. Eng. Div. ASCE*, 109 (EE4), pp. 859–866 ASCE (1983).
2. Borden, R. C., and P. B. Bedient, "Transport of Dissolved Hydrocarbons Influenced by Oxygen-limited Biodegradation; 1. Theoretical Development," *Water Resour. Res.*, 22: pp. 1973–1982, (1986).
3. Dietz, D. N., "The Intrusion of Polluted Water Into a Groundwater Body and the Biodegradation of a Pollutant," in *Proceedings of the National Conference on Control of Hazardous Material Spills*, (Louisville, Ky, 1980) pp. 236–245.

4. Malone, P. G., and R. J. Larson, "Scientific Basis for Hazardous Waste Immobilization," in *Hazardous and Industrial Solid Waste Testing: Second Symposium*, ASTM STP 805, R. A. Conway and W. P. Gulledge, eds., 1983, pp. 168–177.
5. "Handbook for Stabilization/Solidification of Hazardous Wastes," EPA/540/2-86/001, U.S. Environmental Protection Agency, 1986.
6. Webster, W. C., "Role of Fixation Practices in the Disposal of Wastes," *ASTM Standardization News*, March 1984, pp. 23–25.
7. Sitar, N., J. R. Hunt, and K. S. Udell, "Movement of Nonaqueous Liquids in Groundwater," Geotechnical Practice for Waste Disposal '87, *Proceedings of the ASCE Specialty Conference for Waste Disposal*, (Ann Arbor, MI: June 1987), pp. 205–223.
8. Martin, J. P., A. J. Felser, and E. L. Van Keuren, "Hydrocarbon Waste Stabilization for Landfills," Geotechnical Practice for Waste Disposal '87, *Proceedings of the ASCE Specialty Conference for Waste Disposal*, (Ann Arbor, MI: June 1987) pp. 668–682.
9. Martin, J. P., W. T. Robinson, and E. L. Van Keuren, "Large-Scale Stabilization of Hydrocarbon Wastes with Spent Clays and Pozzolans," *Proceedings of the Oak Ridge Model Conference*, NTIS 871075—Vol.1-Pt 3., 1987, pp. 313–333.
10. Daniel, D. E., and C. D. Shackelford, "Disposal Barriers That Release Contaminants Only by Molecular Diffusion," *Proceedings of the Oak Ridge Model Conference*, NTIS 871075—Vol.1-Pt 3., 1987, pp. 313–333.
11. Van Keuren, E. L.,"Advanced Laboratory and Pilot Field Study of Hydrocarbon Refining Waste Stabilization with Pozzolans," thesis submitted to Drexel University, June 1987.
12. Bishop, P. L. "Prediction of Heavy Metal Leaching Rates from Stabilized/Solidified Hazardous Wastes," Toxic and Hazardous Wastes, *Proceedings of the 18th Mid-Atlantic Industrial Waste Conference*, Virginia Polytechnic Institute.2, 1986, pp. 236–252.
13. Martin, J. P., R. M. Koerner, A. J. Felser, and K. J. Davis, "Load Bearing Properties and Durability of Stabilized Waste," *Proceedings of the 38th Canadian Geotechnical Conference*, Edmonton, Alta., 1985, pp. 375–380.
14. Mitchell, J. K., and F. T. Madsen, "Chemical Effects on Clay Hydraulic Conductivity," Geotechnical Practice for Waste Disposal '87, *Proceedings of the ASCE Specialty Conference for Waste Disposal*, (Ann Arbor, MI: June 1987), pp. 87–117.

PART IV

Remedial Options

CHAPTER 17

Enhanced Bioremediation Techniques for In Situ and Onsite Treatment of Petroleum Contaminated Soils and Groundwater

Samuel Fogel, Margaret Findlay, and Alan Moore

INTRODUCTION

Bioremediation is a process by which organic contaminants are destroyed by the action of soil bacteria. Since the constituents of petroleum products are naturally occurring chemicals, soil bacteria capable of degrading them are relatively ubiquitous. These bacteria are capable of obtaining energy by breaking down petroleum hydrocarbons to carbon dioxide and water, as well as incorporating portions of the hydrocarbon to support their own growth. The types of petroleum products that we have treated by biological degradation include gasoline, jet fuel, #2, #4, and #6 heating fuels, diesel fuel, waste oil, and cutting oil.

Bioremediation involves increasing the numbers of these organisms in the contaminated soil by adding mineral nutrients and oxygen, which they require for growth. During petroleum degradation it is typical for several different kinds of bacteria to cooperate in the breakdown of the hydrocarbons.

Bioremediation can be carried out in situ for saturated soils located below the water table. Unsaturated soils can also be treated in situ or onsite. The same principles are involved, but the methods of delivering the oxygen and mineral nutrients differ. In this chapter, two case studies will be presented, in order to illustrate both types of soil treatment.

In Situ Aquifer Treatment

When saturated soils are contaminated with petroleum hydrocarbons, a portion of the contaminant becomes dissolved in the groundwater. However, since most of the constituents of petroleum products are relatively insoluble, a major portion of the contaminant will be trapped between soil particles or adsorbed to the soil. Depending on the spill and on hydrogeological conditions, there may be free product on the water that can be removed.

Traditional Approach

If the groundwater is pumped to the surface, the pumped groundwater can be treated above ground. This approach is not cost-effective, however, since it would take years of pumping to dissolve all of the petroleum trapped in the subsurface soils.

In Situ Approach

Compared to the "pump and treat" approach mentioned above, biodegradation in place is a faster and less costly method for the treatment of contaminated soil below the water table. In this approach, the oxygen and mineral nutrients are delivered to the contaminated area so that the bacteria can degrade the dissolved, adsorbed, and trapped hydrocarbons in place. The delivery of the nutrients is accomplished by dissolving them in groundwater that is recirculated through the contaminated area. Figure 1 illustrates this process: groundwater is withdrawn from wells located down-gradient of the contaminant, amended, and reinjected up-gradient using wells or trenches.

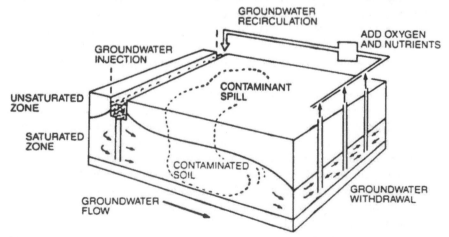

Figure 1.   In situ treatment of contaminated saturated soil by recirculating amended groundwater through the contaminated region.

Oxygen Amendment

The biodegradation of petroleum requires about 3 pounds of oxygen for every pound of petroleum hydrocarbons degraded. Thus an efficient means of oxygen transfer to the groundwater must be devised. Sparging with gaseous oxygen can deliver only 40 ppm at the injection point. Hydrogen peroxide, however, can be dissolved and injected at concentrations above 500 ppm and will gradually break down to oxygen during transport through the contaminated area. Therefore, although the use of hydrogen peroxide requires special design steps and handling procedures, it is employed by Cambridge Analytical Associates Bioremediation Systems (CAA) to achieve this more efficient aquifer treatment, since its use shortens treatment time and overall cost.

Case Study: In Situ Saturated Soil Treatment

CAA, working jointly with W. W. Irwin, a Southern California geotechnical firm, has completed remediation of contamination at a former retail gasoline site in Southern California. This project is one of several that W. W. Irwin and CAA are working on together.

In Situ Design

The design process is the critical element of any biological treatment approach. For in situ aquifer remediation, hydrogeological testing and geochemical analysis is necessary to determine the location and total amount of contaminant, the direction and rate of groundwater flow, and to characterize the aquifer sufficiently to control the groundwater flow. In addition, numerous laboratory tests are carried out on the contaminated soil and groundwater to determine bioremediation design parameters.

At this site, soil borings defined an area of unsaturated soil contamination of approximately $40' \times 50'$, which was to be remediated by soil vapor extraction. The water table existed at a depth of about 60 feet. Soil in the saturated zone below that depth consisted of sandy clay having a hydraulic conductivity of 10^{-5} to 10^{-4} cm/second, which was judged sufficient for groundwater recirculation. The groundwater in this area had concentrations of total petroleum hydrocarbons as high as 220 ppm, including 24 ppm of BTX (benzene, toluene, plus xylenes). Pumping tests carried out at monitoring wells in this area demonstrated the distance that could be influenced by pumping, and the flow that could be expected.

Laboratory Tests

Groundwater and soil from the contaminated saturated zone were subjected to a number of laboratory tests to obtain data for the remedial design. First, the total bacterial population as well as the numbers of bacteria capable of using gasoline as their only food source were documented by standard plate counting techniques. The gasoline-degraders ranged from 20,000 to 300,000 per mL of groundwater, amounting to between 2% and 10% of the total bacteria count.

A series of tests was performed on the soils and groundwater to provide information for the formulation of the chemical nutrient mixture. These included analysis of the groundwater for bacterial nutrients such as nitrogen and phosphate, and for cations likely to interfere with nutrient addition. These were followed by experiments to determine the solubility in site groundwater of CAA's proprietary ACT™ bacterial nutrient mixture. Other experiments measured the adsorption of ACT nutrients onto site soil. Following this, the stability of hydrogen peroxide was measured in various ACT nutrient/soil mixtures to determine the adjustments necessary to achieve the desired rate in the soil of peroxide hydrolysis to oxygen.

Before installation, the bioremediation design was tested by a laboratory treatment simulation in which site soil and groundwater were placed in sealed containers with gasoline, ACT™ nutrients, and oxygen, and the rate of biodegradation determined by analysis of samples using gas chromatography with flame ionization detection (GC/FID). These data are represented in Figure 2. It indicates that

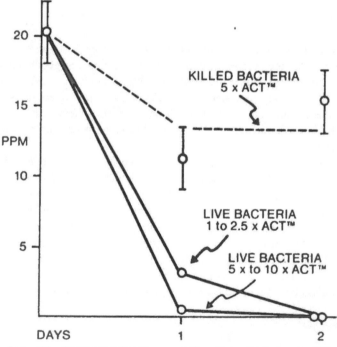

Figure 2. Laboratory Biodegradation Documentation. Groundwater and soil from the saturated zone of gasoline-contaminated site, in a sealed container with gasoline, ACT™ mineral nutrients, oxygen, and naturally present gasoline-degrading bacteria. Compounds analyzed were benzene, toluene, ethylbenzene, xylenes, and trimethylbenzene.

satisfactory biodegradation of the gasoline will occur with nutrient additions that
were determined to be required for optimal soil/groundwater interaction.

As a result of the hydrogeological and laboratory testing, the treatment design
was prepared. This consisted of (1) specifications for number and placement of
injection, recovery, and monitoring wells, (2) groundwater pumping rates, (3)
schedule and rate of nutrient and peroxide addition, and (4) sampling/monitoring
schedule. The design document was presented to the State of California Regional
Water Quality Board, which issued the permit. Figure 3 illustrates placement of
wells on the site.

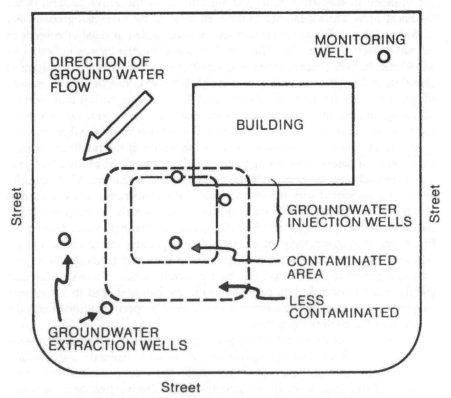

Figure 3. Site diagram for remediation of saturated soil and groundwater contaminated
with gasoline. Naturally occurring bacteria in the contaminated region are stimulated to degrade the contaminants in situ by recirculating groundwater amended
with mineral nutrients and an oxygen source through the contaminated region.

Field Data

Groundwater amended with mineral nutrients (ammonium and phosphate
salts), and hydrogen peroxide was circulated through the contaminated area
for 10 months, achieving approximately 3 cycles through the treatment zone.

Groundwater samples from the extraction wells were analyzed monthly to monitor the treatment process. Total Petroleum Hydrocarbon (TPH) analyses during the first six months showed an average of 15 ppm. During the seventh month the value fell to 0.1 ppm, after which the TPH content of the groundwater dropped to below the detection limit and remained undetectable for the next three months. The remediation, as measured by TPH and BTX concentrations in groundwater samples, was complete within six months of operation. Operation was continued for an additional four months to provide documentation for closure.

The mineral nutrients, nitrogen and phosphorus, added to the reinjected water, are adsorbed to subsurface soils, and are utilized by the active bacteria in the treatment zone. The appearance of these elements in the extracted groundwater is evidence that sufficient concentrations are being added to satisfy the needs of the active bacteria. The "breakthrough" of mineral nutrients was indicated by an increase in their concentrations in the extraction wells. The concentration of phosphate in the extraction wells averaged only 0.5 ppm during the first six months of operation, then increased to an average of 4 ppm during the last four months. Similarly, ammonium concentrations in the extraction wells were not detectable during the first seven months but increased to 0.7 ppm during the last three months.

Additional data which served to confirm the nature of the treatment process were counts of gasoline-degrading bacteria and measurements of dissolved oxygen in groundwater samples from the extraction wells. Although the majority of bacteria associated with the degradation of the contaminant are expected to be attached to soil particles in the treatment zone, bacteria in the groundwater samples can give an indication of conditions in the treatment zone. During the first month of treatment the number of gasoline-degrading bacteria in the extraction wells averaged 20,000 per mL. This value increased gradually during the treatment period to 300,000 per mL by the eighth month. These data indicate that the addition of hydrogen peroxide to the groundwater, and its subsequent conversion to oxygen in the treatment zone, were properly controlled and did not damage the bacterial population.

Measurements of the dissolved oxygen concentrations in the extraction wells confirm that sufficient hydrogen peroxide was added to maintain a satisfactory oxygen level in the subsurface treatment zone. The values obtained before the initiation of treatment were approximately 2 ppm. During treatment, although the rate of oxygen use was high, 3 to 4 ppm of dissolved oxygen was maintained.

Conclusion

Three months of nondetectable hydrocarbons in the groundwater indicate that the site has been cleaned up. Borings in the original contaminated area are being carried out to confirm that the soil no longer contains gasoline.

This case history is significant because it demonstrates that bioremediation can be brought about in slowly permeable soils and can result in reduction of contamination in groundwater to concentrations below the detection limit.

Excavated Soil Treatment

Soil which is contaminated by petroleum hydrocarbons may be excavated from a spill site, either to prevent further contamination of underlying strata or to permit excavation of a leaking tank. Such soil has traditionally been hauled to a "disposal site" and replaced with clean backfill. Recently the cost of soil disposal has risen sharply, making biological treatment a cost-effective alternative. Furthermore, biological treatment results in the destruction of the contamination and the elimination of future liability.

The general approach to excavated soil remediation is similar to saturated soil treatment described above, in that mineral nutrients and oxygen must be delivered to the soil bacteria. In addition, the moisture content of the soil must be maintained at an optimal level for bacterial activity. To accomplish this, a carefully engineered soil mound is constructed containing a network of perforated pipe for forcing air through the soil. A second network of pipes is provided for water delivery. The mound is usually constructed on a pad and underdrained to collect runoff. A generalized diagram of a soil treatment mound is given in Figure 4. The following case study will illustrate the CAA approach to this type of remedial challenge.

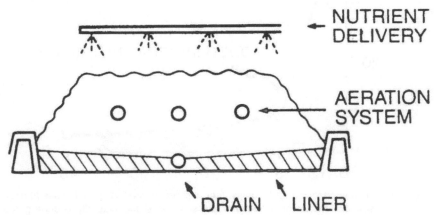

Figure 4. On site treatment of excavated contaminated soil by CAA Bioremediation System's Forced Aeration Contamination Treatment FACT™ process.

Case Study, Excavated Soil Treatment

CAA is currently conducting onsite treatment of 3,000 cubic yards of soil contaminated with diesel fuel. Our activities were initiated with composite sampling and analyses of the soil to determine the amount of contamination and the extent of "weathering" that had occurred. Following this, a number of laboratory tests were carried out to determine soil characteristics, moisture holding capacity, air permeability, and mineral nutrient content. The soil was found to be fine sand with silt having sufficient permeability to allow treatment.

A biodegradability test was performed to document the presence of an active microbial population and to investigate nutrient requirements. For this test, samples were placed in water in sealed bottles with excess mineral nutrients, such as nitrogen and phosphate, and with excess oxygen. The bottles were analyzed at different times to measure the rate of destruction of the diesel contaminant. Under these conditions, 93% of the diesel fuel disappeared in 10 days. Similar bottles, lacking mineral nutrients, showed no significant change in contamination in 10 days. These data are shown in Figure 5. This test demonstrated that a healthy population of fuel-degrading bacteria existed in the contaminated soil.

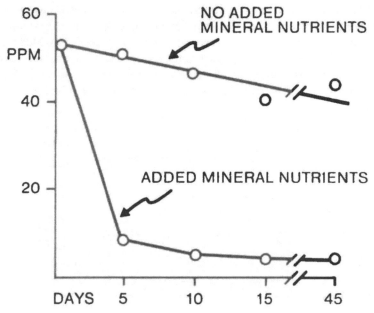

Figure 5. Biodegradability test for diesel fuel in contaminated soil from the site of onsite soil pile treatment. Contaminated soil containing naturally occurring fuel-degrading bacteria was placed in sealed containers with excess mineral nutrients, oxygen, and water. Samples were analyzed after different amounts of time to show the progress of biodegradation. The data is given in terms of the sum of the 18 major constituents of diesel fuel, analyzed by gas chromatography with flame ionization detection.

The Treatment System

A six-foot-high soil pile which covered an area of 200' ×60' was constructed on site. Forced aeration ductwork and water/nutrient delivery systems were incorporated in a manner similar to the design in Figure 4. The treatment system also included a leachate recovery and recirculation system and optional carbon treatment of off-gasses. The initial concentration of total petroleum hydrocarbons (TPH) was approximately 2,800 ppm. During pile construction and the first several weeks of treatment this value dropped to 1,100 ppm TPH. After the forced aeration system and the nutrient delivery system became fully operational, the TPH dropped to 800 ppm. Completion is expected to be achieved in a few months. During this time, project monitoring will be carried out in order to obtain data for process control. This will include enumeration of fuel-degrading bacteria and analyses for mineral nutrients in soil samples from the treatment pile.

CONCLUSION

The data from these two case studies demonstrate the efficacy of bioremediation of contaminated soil. To our knowledge, this aquifer remediation is the first published documentation of a commercial in situ hydrocarbon remediation using hydrogen peroxide as the oxygen source. Previously a pilot demonstration of this process was carefully documented and published by the American Petroleum Institute.[1] Both in situ aquifer treatment and onsite soil pile treatment processes are promising new cost-effective alternatives to costly traditional approaches.

REFERENCE

1. API Publication No. 4448. American Petroleum Institute, 1220 L Street, N.W., Washington, DC 20005.

CHAPTER 18

Biodegradation of Dissolved Aromatic Hydrocarbons in Gasoline-Contaminated Groundwaters Using Denitrification

Richard M. Gersberg, W. John Dawsey, and Harry F. Ridgway

INTRODUCTION

The four aromatic hydrocarbons commonly found in gasoline—benzene, toluene, xylene, and ethylbenzene—are among the top 20 of those hazardous substances most often found in sites on EPA's National Priority List. Many of the gasoline hydrocarbons are hydrophobic molecules, and hence possess low solubilities in water. They exhibit a marked tendency toward adsorption onto the soil phase of an aquifer, so the conventional removal of the "free product" from a contaminated site is largely ineffective at removing the majority of the spilled gasoline.

Serious risk to public health and the environment can occur when the aromatic hydrocarbons, benzene, toluene, and xylene (BTXs), being relatively soluble in water, form a plume of contaminated groundwaters spreading downgradient from the original site of contamination. Once contaminated, an aquifer is extremely expensive to clean up by conventional means.

In situ biodegradation of toxic organic compounds has received increasing attention in recent years as a site cleanup option. It appears to be a cost-effective alternative to conventional "pump and treat" methods. Moreover, biodegradation can result in the complete degradation of hazardous substances, while technologies such as carbon adsorption and air-stripping simply transfer the pollutants to a different medium. Indeed in the Superfund Amendments and Reauthorization Act (SARA) of 1986, a provision is included which imposes cleanup standards emphasizing permanent remedies, rather than simply moving the wastes and burying them somewhere else. In situ bioremediation of contaminated soils and groundwaters has the potential for providing just such a permanent remedy.

Presently, the great majority of bioremediation techniques involve the addition of oxygen (O_2) to the groundwaters in order to enhance the rate of biodegradation. However, given the relatively low solubility of oxygen in water (about 8.4 mg l^{-1} at $23°C$), and the reported high cost and technical difficulties of uniformly delivering large amounts of O_2 (either by air/O_2 sparging or by H_2O_2 addition) into an aquifer, using denitrification represents an innovative approach. Denitrification is an anaerobic microbial process whereby nitrate is used as an electron acceptor instead of oxygen. Since nitrate is very soluble in water, it can be easily and uniformly distributed throughout an aquifer.

Although an increasing body of evidence points to the relative ease at which BTX compounds can be degraded by denitrification in the laboratory (Major et al.[1]; Zeyer et al.[2]; Kuhn et al.[3,4]), there is little information available on the factors that may regulate or limit the process in the subsurface environment. This chapter presents data on BTX biodegradation in gasoline-contaminated groundwaters incubated in the laboratory both aerobically (using H_2O_2), and also under anoxic conditions with nitrate serving as the principal election acceptor. The specific effects of phosphorus and trace element additions on the rate of denitrification are evaluated.

METHODS AND MATERIALS

Groundwaters were obtained from a gasoline-contaminated site in San Diego, California at monitoring well #1 (MW1) by bailing. The water was distributed to 120-mL glass bottles fitted with gas-tight Teflon™ septa. Transfer of the groundwater into these bottles was carried out so as to minimize the introduction of oxygen, and indeed oxygen measurements (using the Winkler method[5]) showed levels below 1 mg^{1-} after transfer, this about the same value as the in situ level. At these low O_2 levels, denitrification will proceed if an alternate electron acceptor such as nitrate is available. According to stoichiometry, small amounts of oxygen (1 mg l^{-1}) present at the start of incubation will be used up rapidly after the consumption of less than 1 mg l^{-1} of BTX. The samples were incubated at room temperature. In the case of aerobic incubations, H_2O_2 was added by syringe to the samples, and then allowed to supply O_2 through its decomposition. Samples received nutrient amendments by syringe. Nutrients used were KNO_3 and/or K_2HPO_4. Unamended bottles served as the control. The trace element solution used for enrichments consisted of $(NH_4)_6(MO_7)_{24}$, 3×10^{-6}mM; H_3BO_3, 4×10^{-4}mM; $CoCl_2$, 3×10^{-5}mM; $CuSO_4$, 10^{-5}mM; $MnCl_2$, 8×10^{-5}mM; $ZnSO_4$, 10^{-5}mM and $FeCl_3$, 1×10^{-1}mM.

BTX disappearance was followed by peak attenuation of benzene, toluene, and total xylene isomers on a purge-and-trap GC with FID (modified method 602),[6] BTX levels were at natural in situ values, and no organic carbon enrichments were used.

RESULTS AND DISCUSSION

Values for the concentration of BTX compounds in groundwaters from the most contaminated well (MW1) at the research site are shown in Table 1. Levels were in the range of 13.3–13.7 mg l^{-1} for benzene, 33.7–39.5 mg l^{-1} for toluene, and 15.4–23.2 mg l^{-1} for total xylene isomers. The mass loss of BTX during an anoxic incubation (without added O_2) of these groundwaters is shown in Figure 1. During the 54-day incubation period, the nutrient enriched samples with 500 mg l^{-1} NO_3^-–N (acting as both a nutrient and an electron acceptor) and 50 mg l^{-1}

Table 1. Concentration of Benzene, Toluene, and Xylene Isomers in Groundwaters from Gasoline-Contaminated Test Site at Monitoring Well #1 (All values in μg 1^{-1}).

Compound	Nov. 4, 1987	Feb. 23, 1988
Benzene	13,307	13,725
Toluene	33,791	39,505
Total Xylene Isomers	15,368	23,243

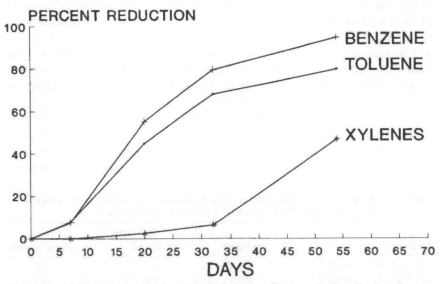

Figure 1. Percent removal of benzene (+), toluene (square sign), and xylene isomers (diamond sign) in nutrient enriched groundwaters incubated anoxically with 500 mg 1^{-1} NO_3^- – N and 50 mg 1^{-1} PO_4^{-3} – P. All removal values are calculated as compared to controls (no nutrients added).

$PO_4^{-3} - P$, showed significant decreases in BTX concentration. Levels decreased 80%, 95%, and 47%, respectively for benzene, toluene, and total xylene isomers, as compared to the unamended controls. This difference between the experimental treatments and the control to which neither nitrogen (N) nor phosphorus (P) had been added, reflects solely the stimulation of BTX biodegradation by nutrient enrichment, and would not encompass physical loss processes such as BTX volatilization or adsorption to the glass sample bottle.

Major et al.[1] in an investigation of the biodegradation of BTX in anaerobic batch microcosms containing shallow aquifer material, found that BTX loss occurred due to denitrification and followed zero-order kinetics. These investigators showed that mean removal rates were similar whether oxygen or nitrate was added as the election acceptor. They measured overall percent losses of 95%, 98%, and 12–15% for BTX compounds, respectively (during a 62 day incubation period), compared to our loss values of 80%, 95%, and 47% for BTX (in 54 days of incubation). However, it is important to note here that the initial concentration of benzene, toluene, and xylene were 3 mg 1^{-1} (for each) in the study by Major et al.,[1] while in our study the initial levels were much higher (in the range reported in Table 1). Of course, the amount of nitrate we added was also much higher (500 mg 1^{-1}) than that added by Major et al. (10 mg 1^{-1}), and we also added the important nutrient phosphorus. Due to such differences, an examination of the percent BTX loss solely does not always reveal the ultimate capacity of a given system for BTX removal.

In order to further examine the effect of phosphorus on denitrification and BTX biodegradation, an experiment was performed with groundwaters incubated for 51 days anoxically with 500 mg 1^{-1} $NO_3^- - N$, and varying levels of phosphorus, from 1–50 mg 1^{-1} (Table 2). The results show that rates increased only slightly with increasing P levels from 1 mg 1^{-1} to 10 mg 1^{-1}, and then decreased somewhat at the highest level of P(50 mg 1^{-1}). Although water chemistry analyses of groundwaters from the contaminated site show phosphorous levels to be very low, <0.1 mg 1^{-1}, apparently final concentrations of between 1–10 mg 1^{-1} are enough to satisfy assimilatory demand during denitrification. At the enrichment combination of 500 mg 1^{-1} of $NO_3^- - N$ and 10 mg 1^{-1} of $PO_4^{-3} - P$ where the rate of BTX loss was highest, the percent BTX removals of 93%, 97%, and 63%, respectively (as compared to unamended controls) were similar to those shown

Table 2.   The Effect of Phosphorus on the Removal of BTX (Benzene, Toluene, Xylene Isomers) During Anoxic Incubation with Nitrate.

Treatment	% Loss[a]		
	Benzene	Toluene	Xylene Isomers
500 mg/1 N; 1 mg/1 P	84	91	39
500 mg/1 N; 5 mg/1 P	91	96	59
500 mg/1 N; 10 mg/1 P	93	97	63
500 mg/1 N; 50 mg/1 P	88	93	40

[a]All values are calculated as percent loss as compared to controls (with no N or P addition).

in Figure 1 above. In both experiments (Figure 1 and Table 2) it is apparent that xylene is degraded more slowly than either benzene or toluene. This finding agrees with the results of Batterman[7] who found removals after 17 days of 92% and 95% for benzene and toluene, but only 76% for total xylene isomers when ground-waters were incubated anoxically with nitrate.

Bacteria that are capable of denitrification contain the dissimilatory nitrate reductase which contains iron (Fe) and molybdenum (Mo). The dissimilatory nitrite reductase contains molybdenum and copper (Knowles[8]). Availability of these and other trace elements might limit the rate of denitrification when nutrient enrichment (with N and P) is carried out. In order to investigate the role of trace elements in denitrification in these groundwaters, we enriched samples with a trace element solution containing iron, molybdenum, and copper as well as cobalt, manganese, boron, and zinc. Results show a significant ($P < 0.005$) stimulation of BTX loss in the trace-element-enriched groundwaters after 80 days of incubation, as compared to nonenriched samples (Table 3). Since iron in these groundwaters is at very low levels (approximately 10 μg l^{-1}), it is possible that this element may play an important role in such stimulation. Future experiments with individual trace element enrichments plus metal chelators will be performed to better delineate the specific element(s) involved.

Table 3. Percent Mass Loss of Benzene, Toluene, and Xylene Isomers in Trace Element-Enriched and Unenriched Groundwaters.[a]

Compound	% Loss[b] Trace Element Enriched[c]	% Loss[b] No Trace Element Enrichment
Benzene	85	66
Toluene	96	86
Total Xylene Isomers	76	39

[a]All samples were also enriched with 500 mg l^{-1} NO$_3$-N and 10 mg l^{-1} PO$_4$-P.
[b]Percent loss as compared to control (no N or P).
[c]Trace element enrichment included the following final concentrations: (NH$_4$)$_6$ (MO$_7$)$_{24}$, 3×10^{-6}mM; H$_3$BO$_3$, 4×10^{-4}mM; CoCl$_2$, 3×10^{-5}mM; CuSO$_4$, 10^{-5}mM; MnCl$_2$, 8×10^{-5}mM; ZnSO$_4$, 10^{-5}mM and FeCl$_3$ 1×10^{-1}mM.

In addition to the BTX loss experiments with no added oxygen, we performed incubations of groundwaters to which H$_2$O$_2$ was added as an oxygen source, with and without the nutrients nitrate and phosphate (Figure 2). Over a 51-day incubation period, total BTX losses in the samples enriched with both H$_2$O$_2$ (50 mg l^{-1}) plus nitrate and phosphate, were 87%, 95%, and 35%, respectively (as compared to unamended control values). The peroxide treatment with no N and P addition showed removal values as low as the control, indicating that the nutrients N and P may be limiting aerobic microbial activity in situ.

An experiment conducted thereafter showed that at a peroxide level of 250 mg l^{-1} (with N and P), overall percent loss of benzene and toluene in just 21 days could be greatly increased to 91% and 94%, respectively. Apparently, aerobic

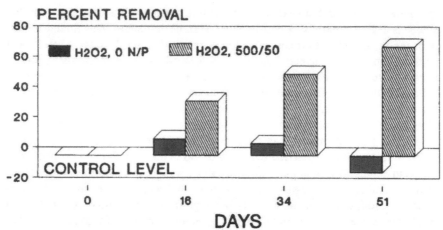

Figure 2. Percent removal of total BTX (benzene, toluene, and xylene isomers) in ground-waters incubated aerobically with 50 mg 1^{-1} H$_2$O$_2$ (as an O$_2$ source) plus either nutrients (500 mg 1^{-1} NO$_3$ – N and 50 mg 1^{-1}PO$_4$$^{-3}$ – P) or no nutrients. All removal values are calculated as compared to controls (no H$_2$O$_2$ and no nutrients).

loss rates for these monoaromatics can be very rapid when peroxide distribution (and subsequent breakdown to O$_2$) is uniform, as exists in these laboratory incubations. The difficulty of distributing H$_2$O$_2$ uniformly throughout an aquifer and also controlling its decomposition in the subsurface environment makes an extrapolation of these laboratory results to an actual field-scale situation problematic.

The data presented above show that the microorganisms in groundwaters have the ability to degrade BTX under anoxic (denitrifying) as well as aerobic conditions. The ability of phosphorus to enhance such activity was demonstrated. Rates of BTX loss under denitrifying conditions were high enough to suggest that the addition of nitrate to BTX-contaminated groundwaters may prove to be a cost-effective alternative to conventional aerobic processes for bioremediation of aquifers contaminated with monoaromatic hydrocarbons.

ACKNOWLEDGMENTS

Funding was provided by the Electric Power Research Institute (EPRI). Services-in-kind were provided by San Diego Gas and Electric Company and the Orange County Water District. We thank M. Bradley, S. Lyon, and W. Hayhow for laboratory assistance.

REFERENCES

1. Major, D. W., C. I. Mayfield, and J. F. Barker. "Biotransformation of Benzene by Denitrification in Aquifer Sand," *Ground Water* 26:8–14 (1988).
2. Zeyer, J., E. P. Kuhn, and R. P. Schwarzenbach. "Rapid Microbial Mineralization of Toluene and 1, 3-Dimethylbenzene in the Absence of Molecular Oxygen," *Appl. Environ. Microbiol.* 52: 944–947 (1986).
3. Kuhn, E. P., P. J. Colberg, J. L. Schnov, O. Warner, A. J. B. Zehnder, and R. P. Schwarzenbach. "Microbial Transformations of Substituted Benzenes During Infiltration of River Water to Groundwater: Laboratory Column Studies," *Environ. Sci. Tech.* 19:961–968 (1985).
4. Kuhn, E. P., J. Zeyer, P. Eicher, and R. P. Schwarzenbach. "Anaerobic Degradation of Alkylated Benzenes in Denitrifying Laboratory Aquifer Columns," *Appl. Environ. Microbiol.* 54:490–496 (1988).
5. *Standard Methods for the Examination of Water and Wastewater*, 16th ed. (Washington, DC: American Public Health Association, 1984).
6. *Methods for Organic Chemical Analysis of Municipal and Industrial Wastewater*, U.S. Environmental Protection Agency, EPA-600/4-82-057, 1982.
7. Batterman, G. "Decontamination of Polluted Aquifers by Biodegradation," in *Contaminated Soil* (Dordrecht: Martinus Nijhoff, Publishers, 1986).
8. Knowles, R. "Denitrification," *Microbiol. Rev.* 46:43–70 (1982).

CHAPTER 19

Bioremediation of Petroleum Contaminated Soils Using a Microbial Consortia as Inoculum

Barry A. Molnaa and Robert B. Grubbs

INTRODUCTION

In the past few years, as landfills have become more and more scarce and concomitantly more and more cost prohibitive, interest in biological methods to treat organic wastes has increased. One area, in particular, that has received increased attention is the biological treatment of petroleum contaminated soils.

The term bioremediation has been given to describe the process by which the use of living organisms (in conjunction with or independent from other technologies) is employed to effectively decontaminate a polluted system. In most cases the organisms employed are bacteria; however, work is being conducted using fungi and plants. Water hyacinths have been utilized in water systems to effectively remove trace organics and trace metals.

There are two techniques for utilizing bacteria to degrade petroleum in the soil. One method uses the bacteria that can already be found in the soil. These bacteria are stimulated to grow by introducing nutrients into the soil and thereby enhancing the biodegradation process. This process is known as biostimulation. The other method involves culturing the bacteria independently and adding them to the site. This process is known as bioaugmentation.[1]

One advantage of bioremediation is that the process can be done onsite with a minimum amount of space and equipment. By treating onsite, costs and liability are greatly reduced while extending the life of the current landfills by reducing the amount of waste they would normally receive.

Onsite treatment may involve excavation of the contaminated soil and construction of a lined treatment cell. If excavation is impractical the treatment may be conducted without disturbing the contaminated site by using a recirculating injection well system. This process is considered in situ treatment.[1,2]

Both onsite and in situ treatment have their advantages and disadvantages and the decision to use one method of treatment or the other is often dictated by various factors at the site.

ONSITE VS IN SITU TREATMENT

Onsite treatment, whereby the contaminated soil is excavated and placed into a lined treatment cell, has some distinct advantages. It allows for better control of the system by enabling the engineering firm to dictate the depth of soil as well as the exposed surface area. By controlling the depth and exposed surface area of the soil one is able to better control the temperature, nutrient concentration, moisture content, and oxygen availability.[1] The presence of the liner is an added benefit since the liner prevents the migration of the contaminants, there is no possibility of contaminating the groundwater. After treatment the liner is picked up and properly disposed of, generally by incineration.

Onsite treatment has an added benefit in that it is much easier to demonstrate the site is clean than in an in situ cleanup. By isolating the contaminated soil in the treatment cell it is possible to sample the site in a more thorough and therefore representative manner. This may prove a necessity if the regulating agency or the customer desire to optimize the reliability of sampling and analysis.

The excavation of the contaminated soil adds to the cost of a bioremediation project, as does the liner and the landfarming equipment. In addition to these costs it is necessary to find enough space to treat the excavated soil onsite. In some states areas are now being set aside to provide the needed space to treat these soils.

In situ treatment is advantageous in instances where the excavation of the contaminated soil is cost-prohibitive or impossible. The method of in situ treatment generally involves establishing a hydrostatic gradient through the area of contamination. Water is placed on the site so that it will flow through the area of contamination, carrying nutrients and possible organisms to the contaminants. Once the water has passed through the site it is pumped up through wells and returned to the beginning of the system. This continuous recirculation is carried on until the site has been determined to be clean (Figure 1).

Recovery of the percolating water is the most difficult aspect of this treatment method. Sites may contain a natural clay or rock barrier which collects the percolating water, in which case extraction wells can be placed in this collection zone. Other sites may require the construction of collection trenches or numerous recovery wells at the bottom of the contaminated soil horizon. Given the various geologic/hydraulic conditions that exist at a site, the application of this

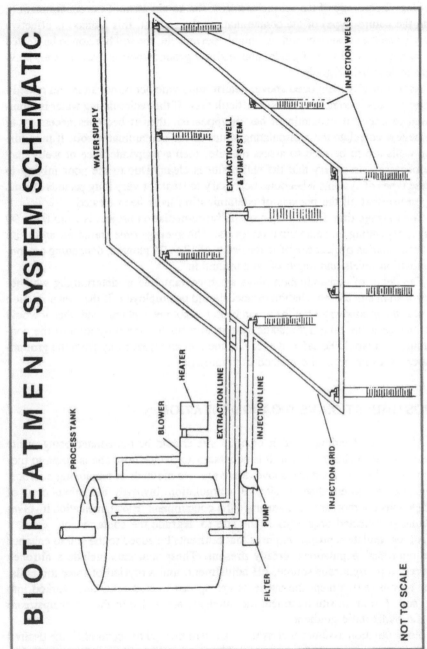

BIOTREATMENT SYSTEM SCHEMATIC

PROCESS TANK

BLOWER

HEATER

EXTRACTION LINE

INJECTION LINE

WATER SUPPLY

EXTRACTION WELL
& PUMP SYSTEM

INJECTION WELLS

FILTER

PUMP

INJECTION GRID

NOT TO SCALE

Figure 1. Continuous recirculation.

technology may be limited and would depend on whether regulatory agencies would consider this to be an appropriate and feasible alternative.

The most effective means of implementing these principles depends on the geology/hydrology of the subsurface area, the areal extent of the contamination, and the nature (type) of the contamination. In general, this method is effective only when the subsurface soils are highly permeable, the soil horizon to be treated is within 20 to 30 feet of grade, and shallow groundwater is present, i.e., at 30 feet or less below grade.

As was briefly mentioned above, determining whether or not an in situ remediation process is complete can be a difficult task. If the recirculating water is monitored to check if contaminant has disappeared, then it becomes necessary to somehow correlate the recirculating water to the contaminated soil. If monitoring wells are to be used to assess the site, then a preponderance of wells may be necessary to satisfy that the entire site is clean. Due to the poor mixing in these types of systems it becomes necessary to treat for very long periods of time to ensure that all the pockets of contamination have been treated.

The average time frame for an onsite bioremediation project is from 60 to 90 days, depending on contamination levels. The average time frame for an in situ bioremediation project can be in the order of 12 to 24 months, depending on contamination levels and depth of contamination.

The depth of contamination plays an important role in determining whether or not an in situ bioremediation project should be employed. If the contamination is near the groundwater but the groundwater is not yet contaminated, then it would be unwise to set up a hydrostatic system and further the migration of the contaminant. It would be safer to excavate the soil and treat away from the groundwater by using an onsite method of treatment.

BIOSTIMULATION VS BIOAUGMENTATION

Along with deciding whether or not a site should be remediated using onsite treatment or in situ treatment, it is necessary to decide how one is going to bioremediate the site. As stated above, there are two methods of employing microorganisms to bioremediate a site. Biostimulation involves the stimulation of indigenous microorganisms to degrade the contaminant. Bioaugmentation involves adding preselected organisms to the site to degrade the contaminant.

A biostimulation project requires that nutrients be added to the soil to enhance the microbial populations already present. These nutrients include a nitrogen source, a phosphorous source, pH adjustment, and a myriad of trace minerals. For an onsite treatment the nutrients are spread over the site and worked into the soil. For an in situ treatment the nutrients are added to the water upstream in the hydrostatic gradient.

Biostimulation assumes that every organism needed to accomplish the desired treatment results is, in fact, present. Therefore, all that is required to achieve effective biodegradation is to provide (or enhance) an ideal environment for these ubiquitous microorganisms to live and work.[1]

There are numerous shortcomings with this hypothesis. For example, how can we be certain that those organisms present are the most suitable to degrade all materials present? Secondly, what if the only organisms stimulated are those that eliminate the primary substrate, but do not cometabolize the *specifically targeted* substrates? At any given site, many of the problem substrates may not be able to be biodegraded directly. If they are the only food source available, the microbes may not be able to degrade these targeted organics, since they do not serve as primary food sources on which the microbes feed.

To ensure that the necessary organisms are present it is generally necessary to conduct a feasibility study on the soil from the site before any biostimulation project is undertaken. The cost of such a study can range from $5,000 to $40,000, depending on the extent of contamination and the characteristics of the contaminants.

Bioaugmentation is the controlled addition of specially formulated biocultures to assist those found naturally in the soil. It is done in conjunction with the development and monitoring of an ideal growth environment in which these selected bacteria can live and work.

In most cases, the targeted organic contaminants either serve as the food source or are cometabolized. Essential elements are added to the "food source" to provide the required nutrient levels, and water provides the media in which the bacteria function.

The mere addition of bacteria will not, in itself, solve the problem. Studies conducted in 1979 by Dibble and Bartha clearly demonstrated that *sewage sludge* actually inhibited hydrocarbon biodegradation in soil, and the use of yeast extract had no effects whatsoever.[3] The selected microorganisms must be carefully matched to the waste contamination present in the soil, as well as the metabolites formed. They must favorably compete with the ubiquitous organisms found in the expected environmental conditions.

Bioaugmentation allows one to control the nature of the biomass. It provides an element, heretofore not available; that of *predictability*. Bioaugmentation ensures that the proper team of microorganisms is present in the soil in sufficient type, number, and compatibility to effectively and efficiently attack the waste constituents and break them down into their most basic compounds.

One objection to bioremediation has been that it takes an inordinate amount of time for the process to work. In the case of biostimulation this is true, however the addition of specially selected microbial consortia allows one to control the biomass of the contaminated site. The additional control of the biomass enables one to increase the kinetic rates of removal from the contaminated site by selecting a more efficient consortia of microorganisms than might be present at the site.

By increasing the kinetic rates it has been possible to remediate sites in 60 to 90 days using addition of selected consortia of microorganisms.

By selecting the microbial consortia beforehand it is possible to select for organisms that will not produce nuisance odors such as hydrogen sulfide. Petroleum degradation can create anaerobic conditions within the soil. Once anaerobic conditions are present it becomes possible to generate phytotoxic compounds such as hydrogen sulfide.[4] If one augments the soil with organisms that do not possess the ability to generate these phytotoxic compounds, a potential hazard to onsite petroleum degradation can be averted.

The cost of the selected microorganisms has been mentioned as a disadvantage in treating contaminated soils, but if one considers the cost of a feasibility study to ensure that a biostimulation project will work, the cost is considerably less for the bioaugmentation products.

THE PROCESS

There is far more involved with bioremediation projects than simply adding microorganisms. Various factors need to be considered to ensure the success of these programs. The proper engineering to facilitate biological growth is a crucial step in the process of bioremediating a site.

An electron acceptor is required for breakdown of hydrocarbons. Oxygen, nitrate, and sulfate are the most common. In a bioremediation project the presence of oxygen is one of the most crucial factors to the rate of reaction. This is especially true early in a project, before any oxygenated intermediates are formed. Sporadic reports of anaerobic degradation in vitro remain controversial, and convincing proof of significant anaerobic hydrocarbon biodegradation is still outstanding.[4] Sulfates are a potential electron acceptor, but are not abundant in soils. Nitrate is not energetically favorable for this purpose in soils.[5]

In soils, aeration depends on the total amount of air-filled pore space. Elimination of air-filled pore space by waterlogging or compaction reduces oxygen transfer. Large amounts of biodegradable organics in the top layers will deplete oxygen reserves in the soil and slow down oxygen diffusion rates to the deeper layers.

Oxygen can become a limiting factor in all types of petroleum degradation, so aeration is required in most applications. In aqueous systems aeration and agitation also provide more surface area of hydrocarbons to the bacteria which live only in the aqueous phase of the system and work at the oil to water interface.

Another essential parameter in a bioremediation process is moisture. Bacteria rely on water to exchange everything through the cell. At 100% moisture in soils, however, all pore spaces are filled with water. At only 10% moisture, osmotic and matrix forces reduce metabolic activity to marginal levels. Moisture levels in the range of 20% to 80% of saturation generally allow suitable biodegradation in soils.[4]

The addition of large quantities of hydrocarbons in a system usually creates a nutritional imbalance which needs to be corrected by the application of inorganic fertilizers containing nitrogen and phosphorous. Biosludges from refinery and petrochemical treatment facilities normally contain enough nitrogen and phosphorous.

For landfarming operations the American Petroleum Institute recommends a C:N ratio of 160:1. Laboratory experiments by Dibble and Bartha showed a C:N ratio 60:1 and a C:P ratio of 800:1 to be optimum.[4] The expense of fertilizer and the potential for groundwater contamination encourage more conservative application rates. Most agricultural fertilizers contain excessive P and K for microbial use. Urea and ammonium compounds can be added to such fertilizer to bring up the nitrogen levels. Nitrates can pose leaching problems and encourage denitrification under anaerobic conditions. The ammonium ion being positively charged binds to the negatively charged soil particles. However, in well aerated soils with neutral pH values above 50°F the ammonium ion is nitrified to nitrates in one to two weeks after application.[6]

In cleanup situations one frequently cannot do a mass balance of pollutants. Sufficient nitrogen and phosphorous must be present to start off microbial activity and must be monitored continually to assure that they don't become too low due to assimilation into cell mass, leaching, nitrification, or volatilization. We recommend maintaining nitrogen levels in excess of 5 ppm at all times and phosphorous levels of 1 ppm or more. These levels will ensure that microbial activity is not lost.

Temperature affects the rates of microbial metabolism as well as the physical state of hydrocarbons. It also affects the solubility of the substrates. Some small alkanes are more soluble at 0°C than at 25°C.[7] Elevated temperatures can influence nonbiological losses, mainly evaporation. In some cases the decreased evaporation of toxic components at lower temperatures has been reported to have inhibited degradation.[8] In general most mesophilic bacteria perform best at about 35°C but their performance can be affected by these other factors. Consequently researchers have reported different optimums and considerable variance in activity at different temperatures, little change in activity over given temperature ranges, and other superficial contradictions. Huddleston and Cresswell reported petroleum degradation in soils as low as −1.1°C as long as the soil solution remained liquid.[9] Degradation rates were quite slow. In natural habitats shifts in microbial populations due to temperature changes have been reported.[10] As one might suspect from such shifts, as well as changes in solubilities, there are reports showing the types of hydrocarbons being degraded may vary with temperature.

While the pH of the marine environment is uniform, steady, and alkaline, the pH of various soils covers a wide range. The marine environment is well buffered. In soils and poorly buffered treatment situations, organic acids and mineral acids from the various metabolic processes can significantly lower the pH. The overall biodegradation rate of hydrocarbons generally is higher under slightly alkaline conditions. Therefore, appropriate monitoring and adjustments should be made to keep such systems in the 7.0 to 7.5 pH range. Variations or swings in pH in treatment systems can have a very deleterious effect on the performance of the biomass.

Since oils and most petroleum hydrocarbons are only sparingly soluble in water, the relatively small interfacial area of oil in contact with water can limit the microbial degradation of oil. Microbes colonize the surfaces of oil droplets and the undersides of slicks. Many hydrocarbon-using microorganisms produce emulsifying agents which greatly enhance their effectiveness in handling the oil. It is widely held that emulsifiers can be involved in the entry of hydrocarbons into the cells, but degradation can occur without emulsification. Emulsifiers have proven useful in some cleanup operations but various sources indicate that not all dispersants enhance biodegradation.[6,11]

Most of the parameters that need to be monitored in a bioremediation project are a function of good environmental application. Once the environment has been made conducive to bacterial growth and a satisfactory monitoring system has been established, the programs are not very labor or capital intensive.

SUCCESSFUL BIOREMEDIATION PROGRAMS

Several innovative and successful bioremediation programs have been conducted by Solmar Corporation in conjunction with various environmental engineering firms and remediation contractors.

Case #1

Bioremediation was selected as the method of choice to clean up an abandoned refinery site in southern California. The 32-acre site was located in a prime industrial area and the goal was to clean the site to a level sufficient that commercial buildings could be built.

The initial contamination levels for the site ranged from a low of 1,500 ppm to a high of 30,000 ppm. The site was sectioned off into several treatment zones and a bioremediation program was begun using a consortia of microorganisms supplied by Solmar Corporation of Orange, California. Since the site had been contaminated on and off for a period of 40 years with little or no sign of decontamination by indigenous organisms, it was concluded that a bioaugmentation program could accelerate the remediation process.

The treatment was conducted over a period of six months. While areas were being treated, other areas were being taken out of service until the entire tank farm was dismantled. As areas were taken out of service, treatment was begun to remediate those sections of the property.

The entire area was certified as clean within a period of one year and construction of office buildings has begun.

Case #2

The city of Carson decided to exercise its redevelopment powers and condemned a site that had been used as a petrochemical tank storage site and salvage operation. The site had been an eyesore. Rather than seal the contaminants at the site under buildings and parking lot, the city decided to get rid of the contaminants. The site had been earmarked as a park and the city officials were concerned that if the contaminants were left in place they might endanger the health of children using the park.[12]

The price for hauling away the contaminated soil for proper disposal was estimated to be $2 million. The estimated amount of contaminated soil was approximately 10,000 cubic yards. A bioaugmentation program was proposed and adopted at the site.

The cost of the cleanup was less than $132,000 and the city began seeking bids for its most elaborate recreation facility.

Case #3

When the Sacramento Utilities District purchased a small parcel of land to expand their existing parking lot they were unaware that the land had been previously contaminated with diesel fuel. Once the contamination had been detected the Utilities District decided to take it upon themselves to clean up the site rather than attempt to pursue any legal action.

The District realized that merely excavating and hauling the contaminated soil to a dump site was just transferring the problem to another site. In keeping with the District's policy of concern with the environment, other alternatives to land disposal were sought.

Upon examination of treatment options the District decided to implement a bioremediation program using bioaugmentation as the source of organisms. The bioremediation of the 2000 cubic yards of contaminated soil reduced the Total Petroleum Hydrocarbon levels from 2800 ppm to less than 38 ppm (Figure 2) in approximately 74 treatment days.[13] The cost of treatment was $360,000 less than the total price of disposal without the inherent liability.

Case #4

Bioremediation was the method of treatment opted for, to treat 1500 cubic yards of diesel-contaminated soil at the former King's Truck Stop in Sacramento, California. The project reduced the diesel contaminant levels from 3000 ppm to less than 30 ppm in approximately 62 treatment days.

Case #5

In situ bioremediation was necessary to clean up contamination from a ruptured transfer line that passed under a railroad track. A jumbo tank car had been moving on the track as solvents were being pumped through the line. The resulting rupture led to a loss of 300 to 400 gallons of solvent at a depth of 38 inches beneath the surface along 120 feet of the track.

A continuously recirculating ground injection system was designed and installed to treat the contaminated soil (see Figure 1). Following a cleanup program of

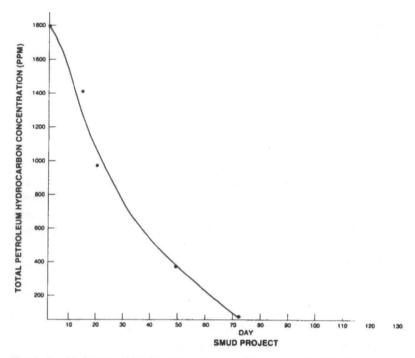

Figure 2.   Reduction of TPH levels in 74 days. (Courtesy of ProTek Environmental, Inc., Huntington Beach, CA).

Table 1.   Degradation Schedule of the Contaminants.

Component	09/24/84 (ppb)	10/31/84 (ppb)	04/04/84 (ppb)	% RED.
Benzene	N/A	96	31	67.7
Carbon Tet.	N/A	65	Nil	99.9
Chlorobenzene	9,050	227	37	99.6
1,1 DCE	N/A	508	341	32.9
Ethyl Benzene	154,000	1,119	382	99.8
Toluene	31,000	1,276	526	98.3
111 TCA	N/A	82	Nil	99.9
Xylene	1,249,000	16,825	1,979	99.8

N/A = not analyzed for

nine months with the bioaugmented system, a 99.5% degradation of the contaminants was achieved (Table 1).

Case #6

A bioremediation project involving 25,000 cubic yards of soil contaminated with various lubrication and form oils is currently ongoing. Preliminary results indicate that the contamination levels have been reduced from a high of 4800 ppm down to 125 ppm in the most contaminated cell (Figure 3). In a lesser contaminated cell the levels have been taken from a high of 1400 ppm down to below the action level of 100 ppm (Figure 4).

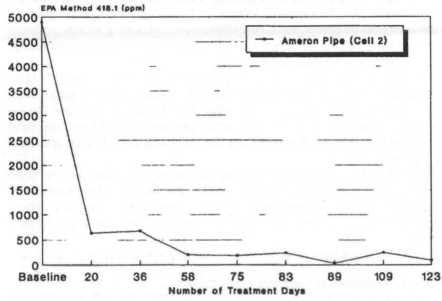

Figure 3. Contamination level reduction in the most contaminated cell. (Courtesy of ProTek Environmental, Inc., Huntington Beach, CA).

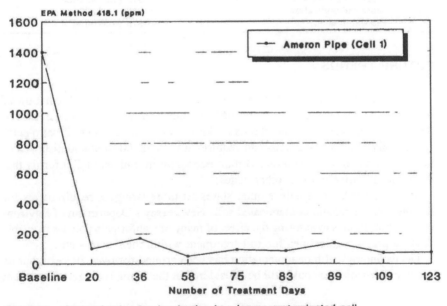

Figure 4. Contamination level reduction in a lesser contaminated cell.

COST OF TREATMENT

Cost effectiveness, it seems, plays only a small role in the agencies' pursuit of the elusive Best Demonstrated Alternative Technology (BDAT). The facts are that economics do govern, and if cost-effective ways of dealing with the problems can be found, then more sites will be cleaned up and fewer generators will resort to legal delays in effecting cleanups.

Feasibility studies conducted on the projects discussed above found that bioremediation is a most cost-effective means of dealing with contaminated soils. As with most technologies, cost is directly related to the size of the site and extent of contamination. However, bioremedial approaches tend to have lower fixed costs and therefore are able to compete favorably with other technologies from a cost standpoint.

When looking at a bioaugmentation project one must consider the cost of the cultures. Generally, the cost of the cultures is less than 1% of the total cost of the project. When one weighs the cost of the organisms versus the assurance of mind in knowing the correct organisms have been provided, this is a small price to pay.

Table 2 gives a breakdown of various technologies and their costs per ton.

Table 2. Breakdown of Technologies and Costs.

Treatment Process	Cost Per Ton
Landfill disposal fees:	$140 to $120/ton + Taxes + Transportation
Mobile Incineration:	$150 to $400/ton
Stabilization/fixation:	$100 to $200/ton
Bioremediation:	$15 to $70/ton

FUTURE TRENDS

At the present time, California seems to be pushing for bioremediation of petroleum-contaminated soils more than any other state. This is due in part to the stringent regulations within the state. Since California classifies all petroleum-contaminated soil as hazardous and requires it to be manifested and disposed of in a class one landfill, there are certain economic incentives in California that do not at this time exist in other states.

It will be not be long before other states set more stringent requirements for dealing with petroleum-contaminated soil. New Jersey's Department of Environmental Protection is monitoring the effect of many of California's policies to enable them to prepare guidelines for soil treatment within their own state.

The technology of bioaugmentation has been around for over 20 years but its use in bioremediation could still be considered in the formative stage. Treatment of "simple" wastes such as waste oil are fairly straightforward and other more complex wastes are being treated every day. Solmar Corporation is currently looking at being able to handle several troublesome substrates that had not been considered as candidates for bioremediation in the past. Concurrently, we are also assessing the possibility of selective removal of contaminants. There is some evidence that it may be possible to selectively remove PNAs from a site before removing the other contaminants, the advantage being that it would then be possible to delist certain contaminated sites by removing the characteristic that made them hazardous in the first place.

Regulatory obstacles to bioremediation are becoming fewer and fewer. In certain regions of California it is now required that a site must explore the possibility of bioremediation before any other technology can be adopted. We see this as a trend that will continue to grow as landfill alternatives and economic constraints limit the number of viable alternatives for hazardous waste disposal.

BIBLIOGRAPHY

1. Mathewson, J. R., and R. B. Grubbs. "Innovative Techniques for the Bioremediation of Contaminated Soils," presented at the second annual CWPCA Industrial and Hazardous Waste Information Exchange, Oakland, CA, 1988.

2. Grubbs, R. B., and B. A. Molnaa. "In Situ Biological Treatment of Troublesome Organics," presented at the first annual CWPCA Industrial and Hazardous Waste Information Exchange, Fresno, CA, 1987.

3. Dibble, J. T., and R. Bartha. "Effects of Environmental Parameters on the Biodegradation of Oil Sludge," *Appl. Environ. Microbiol.* 37:729–739 (1979).

4. Bossert, I., and R. Bartha. "The Fate of Petroleum in Soil Ecosystems," in *Petroleum Microbiology*, R. M. Atlas, Ed. (New York: Macmillan Publishing Company, 1984) Chapter 10.

5. Hansen, R.W., and R.E. Kallio. "Inability of Nitrate to Serve as a Terminal Oxidant for Hydrocarbons," *Science*, 125:1198 (1957).

6. Robichaux, T. J., and N. H. Myrick. "Chemical Enhancement of the Biodegradation of Crude Oil Pollutants," *J. Petrol. Tech.* 24:16–20 (1972).

7. Polak, J., and B. C. Lu. "Mutual Solubilities of Hydrocarbons and Waters at 0° and 25°C," *Can. J. Chem.* 51:4018–4023 (1973).

8. Floodgate, G. D. "The Fate of Petroleum in Marine Ecosystems," in *Petroleum Microbiology*, R. M. Atlas, Ed., (New York: Macmillan Publishing Company, 1984).

9. Huddleston, R. L., and L. W. Cresswell. "Environmental and Nutritional Constraints of Microbial Hydrocarbon Utilization in the Soil," in Proceedings of 1975 Engineering Foundation Conference: The Role of Microorganisms in the Recovery of Oil, NSF/RANN, Washington, 1976, pp. 71–72.

10. Westlake, D. W. S., A. Jobson, R. Phillippe, and F. D. Cook, "Biodegradability and Crude Oil Composition," *Can. J. Microb.* 20:915–928 (1974).

11. Mulkin-Phillips, G.J., and J.E. Stewart. "Effect of Four Dispersant on Biodegradation of Growth of Bacteria on Crude Oil," *Appl. Microbiol.* 28:547–552 (1974).

12. Stein, G., "Oil-Gobbling Bacteria Clean Soil at Site of Park in Carson," *Los Angeles Times*, October 11, 1987.
13. Rittenhouse, R.C., "Quality–the Critical Element in Liquid Fuel Handling," *Power Eng.* July, 1988.
14. Grubbs, R. B. "Enhanced Biodegradation of Aliphatic and Aromatic Hydrocarbons through Bioaugmentation," presented at the fourth annual Hazardous Materials Management Conference/Exhibit, Atlantic City, NJ, 1986.

CHAPTER 20

Cutoff Walls to Contain Petroleum Contaminated Soils

Marco D. Boscardin and David W. Ostendorf

INTRODUCTION

Cutoff walls and other barrier systems have a long history of use in traditional geotechnical construction for the control of water seepage, flows, and pressures. With the increased awareness of the hazard posed by contaminated soils and groundwater, it was a natural step to employ barrier wall systems to contain and prevent the spread of contaminants. However, the criteria for successful performance of a barrier to contain a contaminant plume are, in general, much stricter than the criteria for a barrier to control water in traditional geotechnical construction. For example, in traditional geotechnical construction, cutoff walls are typically required to reduce water flows or control water pressures so that they are not problems during construction or operation of a facility. As a consequence, small seeps and flows may be acceptable and measures to accommodate small flows and bleed off slight pressures are incorporated in the design. In contrast, a barrier called upon to contain a contaminant plume should ideally permit no passage of contaminants or at least permit passage of contaminants at such a slow rate that the material escaping does not constitute a hazard. The impact of the chemistry of the permeant on the properties, particularly hydraulic properties, of the barrier must also be assessed when the barrier is required to contain contaminants. Most cutoff walls in traditional geotechnical construction must only deal with water with a relatively narrow range of chemistries. On the other hand,

233

barriers used to contain pollutants can encounter a much wider range of pore fluid chemistries which are often much more chemically aggressive than unpolluted water. This chapter examines various cutoff wall systems available from the perspective of construction-related defects and interaction with contaminants.

The impacts of composition and method of construction on performance for barriers based on portland cement, bentonite, asphalt, and chemical binders or grouts are discussed, as are synthetic membrane barriers. Slurry wall, permeation grouting, jet grouting, and vibration beam methods of construction are considered. The effects of organic-based contaminants on material properties and barrier integrity are examined for several cases and a breakthrough model that quantifies the time available for true remediation as a function of chemical compatibility of the contaminant and the barrier is presented.

CUTOFF WALL MATERIALS

Hydraulic barriers have been constructed using a variety of materials. These materials include bentonite, asphalt, portland cement, synthetic membranes and liners, chemical grouts, and composite materials composed of two or more of the above. In general, these materials exhibit relatively low hydraulic conductivities ($< 10^{-9}$ m/s) for laboratory specimens with passive or nonaggressive permeants (see Table 1). In contrast, field test data, again for chemically passive permeants, indicate conductivities 10 to 10,000 times greater than the laboratory values, depending on the material and the method of construction.[3] When an incompatibility exists between the cutoff wall material and the permeant, further

Table 1.   Conductivity of Barrier Materials with Passive Permeant.

Material	K_p (Lab) m/sec $\times 10^{-10}$	K_p (Field) m/sec $\times 10^{-10}$
Bentonite slurry	5	50
Compacted soil bentonite	0.1	10
Cement bentonite	3 (filter cake)	100
Asphalt slurry	0.1	1
Soil cement	0.5	50
Cement grout	0.1 (single well)	1000 (curtain)
Concrete	0.1	—
Butyl rubber	0.001	0.01
Polyvinyl chloride	0.00007	0.01
Chlorinated polyethylene	0.0002	0.01
Ethylene propylene diene monomer	0.002	0.01
Chlorosulfonated polyethylene	0.0004	0.01
High density polyethylene	0.0001	0.01
Silicate grout	47	10000
Acrylamide grout	0.5	10000
Dynagrout T	1	10000
Resorcinol	1	10000

Based on data from References 1–6.

increases in conductivity can also occur, as discussed in a later section. General characteristics of each material are briefly described below.

Bentonite

Bentonite slurries, soil-bentonite mixes, and portland cement-bentonite mixes are commonly used in what is termed slurry wall construction to form vertical cutoff walls. An important aspect of the slurry wall construction is the formation of the bentonite filter cake along the sides of the trench. The filter cake forms along the sides of the trench during excavation as long as the level of the slurry is such that flow is directed out of the trench and into the surrounding soil. The filter cake is typically thin, 3×10^{-3} m, but has a low hydraulic conductivity, on the order of 10^{-10} m/s.[7] When sufficient excavation has been performed, the slurry trench is backfilled with either a soil-bentonite mix, a cement-bentonite mix, or a plastic concrete mix. D'Appolonia[8] indicates that the conductivity of soil-bentonite backfill depends on the quantity of bentonite added to the soil and on the grainsize and plasticity characteristics of the soil. For reasonable mixes, hydraulic conductivities of 5×10^{-10} to 5×10^{-9} m/s can be achieved for passive permeants.[8] In contrast, cement-bentonite mixes typically have conductivities on the order of 10^{-7} to 10^{-8} m/s.[1,5] Hydraulic conductivity of plastic concrete is typically less than 1×10^{-9} m/s.[1]

Asphalt

The use of asphalt slurries for cutoff walls is relatively new. The slurries are typically proprietary mixtures that include emulsified asphalt, sand, water, and portland cement.[9,10] Laboratory conductivities for asphalt slurries begin at 1×10^{-10} m/s, but rapidly decrease with loss of fluid and curing to less than 1×10^{-12} m/s within a few hours.[5,11] The low conductivity, high cost, and greater resistance to degradation by aggressive permeants makes this an attractive material for use in thin barriers.

Portland Cement

Portland cement-based systems include grouts and conventional concretes, as well as the cement-bentonite mixes described previously. Kosmatka and Panarese[4] indicate passive conductivities in the range of 10^{-12} to 10^{-11} m/s for conventional concretes. However, additives such as polymers and fly ash can reduce the conductivities and increase chemical stability.[12,13] Passive hydraulic conductivities for cement grouts are typically 1×10^{-12} m/s or less.[14]

Synthetic Membranes and Liners

Synthetic barrier materials are typically polymer-based and take the form of sheets. There are three basic families of synthetics used in practice; elastomers, thermoplastics, and crystalline thermoplastics (see Table 2).[2] These materials are characterized by very low passive conductivities, $\ll 10^{-12}$ m/s, making their use in thin sheets viable. To form an effective barrier, the sheets must be bonded together to form a continuous unit and installed in such a manner as to prevent tears and punctures from forming and compromising its integrity. In general, the elastomer family is the weakest and most difficult to seam of synthetic barrier materials.[3,15] The thermoplastic family is easier to seam in the field and tends to have somewhat better tear and tensile strength properties than the elastomers.[3] The crystalline thermoplastics, particularly high density polyethylene, tend to resist mechanical tearing and puncturing better and tend to resist aging and exposure related degradation better than the other synthetic barrier materials though seaming may be more complicated.[3,15]

Synthetic barriers are subject to aging and physical and hydraulic property degradation, due to exposure to ultraviolet radiation, ozone, thermal variations, microbial attack, and volatilization.[3] In addition, these materials may be attacked by aggressive permeants and, therefore, compatibility of the barrier with the exposure conditions and the contaminant needs examination. However, the variety of synthetics available means that materials compatible with the various conditions and the various aggressive permeants can be found.

Table 2. Synthetic Membranes.

Class	Typical Materials
Elastomers	Butyl rubber (IIR)
	Chlorinated rubber (CR)
	Ethylene propylene diene monomer (EPDM)
Thermoplastics	Polyvinyl chloride (PVC)
	Chlorinated polyethylene (CPE)
	Chlorosulfonated polyethylene (CSPE)
	Elasticized polyolefin (ELPO)
Crystalline thermoplastics	Low density polyethylene (LDPE)
	High density polyethylene (HDPE)

After Folkes [2].

Chemical Grouts

Chemical grouts are liquid solutions designed to harden after injection into a porous medium. The grout, which in theory permeates the voids in the medium, increases the strength and reduces the permeability of the medium as it hardens. Injection of particulate grouts such as cement and bentonite grouts are typically limited in use to rather coarse-grained soils such as gravels and coarse sands with large interparticle voids. Chemical grouts typically have low viscosities and no particulate solids and are therefore able to permeate much finer materials, fine sands to coarse silts, than the particulate grouts.[16] Chemical grouts include silicates, lignosulfites, phenoplasts, aminoplasts, acrylamides, polyacrylamides, acrylates, and polyurethanes.[17] Laboratory conductivities of these materials are typically in the 10^{-10} to 10^{-9} m/s range (see Table 1). Areas of concern with chemical grouts are degree of permeation of the medium, reaction of the grout with the present and future pore fluids, and the potential release of toxic or corrosive materials due to improper mixing or reaction with the soils and its pore fluid.[3,16]

CONSTRUCTION OF A CUTOFF WALL

The goal during construction of a cutoff wall is to erect a continuous, impermeable barrier (usually but not always vertical) to isolate a contaminated mass of soil from the surrounding, uncontaminated soil and water. This is typically accomplished by extending the barrier to intersect continuous, impermeable (very low permeability) soil or rock layer to effect closure. Equally important, but not considered here, is the construction of an impermeable cap to prevent recharge of the contaminated groundwater and the attendant flows.

Materials with the desired hydraulic and chemical resistance properties can be identified in the laboratory. Unfortunately, many materials do not perform as well in the field as the laboratory studies would indicate (see Table 1), due to the presence of defects in the field situation. In this section, the propensity for defects (windows, cracks, or high permeability zones) to occur during and after construction of the various types of barriers is examined. The barrier construction techniques considered include slurry wall methods, vibrating beam methods, injection grouting, and jet grouting.

Slurry Wall Methods

Slurry wall-based methods of creating a vertical barrier are among the most common methods in use. In general, the method consists of excavating a narrow trench, typically 0.6 to 1.5 m thick, down into an aquiclude while using a bentonite-water slurry to keep the trench open and the side walls stable. Slurry trenches in excess of 120 m deep have been constructed,[14] though for contaminant control, slurry trench cutoffs are typically less than 50 m deep.[7] Up to 15 m, backhoes are generally most efficient. Draglines are typically restricted to depths less than 25 m, while clamshells are used for deeper excavation.[8] Once a sufficient length of trench is excavated, the bentonite-water slurry can be replaced by either a soil and bentonite backfill,[8] a plastic, portland-cement concrete,[18] an asphalt slurry,[19] or a synthetic membrane and backfill composite.[20] A variation is the cement-bentonite slurry method,[21] that incorporates portland cement in the slurry which eventually hardens in the trench and does not require backfilling. D'Appolonia,[8] Millet and Perez,[22] Ryan,[23] and Xanthakos[19] describe the construction and specification of slurry walls in more detail.

Evans[1] describes common construction and postconstruction defects encountered in slurry walls. He defines defects as zones of the wall that do not provide the same resistance to groundwater flow and contaminant migration as the good, intact portions of the cutoff. Construction defects include poorly mixed backfill, trapped pockets of slurry, trapped pockets of material that has spalled from the sides of the trench, incomplete keying into the low permeability layer, and loss of the filter cake in portions of the trench wall. Postconstruction defects include cracking due to changes in moisture, temperature, consolidation and stress, as well as increases in conductivity due to chemically aggressive permeants.

A case where a composite, synthetic membrane (high density polyethylene, HDPE) and soil barrier was installed using a modification of the slurry trench technique is described by Druback and Arlotta.[20] The unique feature of this system is the granular material placed within the membrane that lined the top, bottom, and sides of the trench which would permit checking the effectiveness of the interior layer of the liner and allow collection and extraction of any contaminant that passes through the inner layer. Defects in such a system can be associated with separations at seams, and tears and punctures created during installation, backfilling, and postconstruction deformation.[24]

Vibrating Beam Methods

The vibrating beam method of cutoff wall construction, as described by Leonards et al.[5] and Jepsen and Place,[11] consists of driving an injection beam into the ground to a preselected depth, up to 25 m, with a vibratory driver so that the soil is consolidated and displaced during the driving. During extraction of the beam, the void space is pressure grouted with either a cement-bentonite slurry or a slurry based on an asphalt emulsion which hardens in place. The injection beam is a wide flange section with a web on the order of 0.85 m and of sufficient stiffness to minimize bowing under the applied loads. The tip of the beam is fitted with wear plates to protect the beam and injection system, and to create a void of the desired width. Full length grout pipes are attached to the beam to inject the slurry. The beam is also fitted with a fin that acts as a guide to ensure continuity of the wall by running in the grouted void created by the previous penetration. In addition, each beam penetration overlaps the previous penetration by about 10%. In this manner, a thin cutoff wall (on the order of 0.1 m) is constructed. This method is generally used to create vertical walls that intersect an aquiclude for closure. However, Leonards et al.[5] describe a test installation in sandy soil utilizing cement-bentonite walls inclined 45° from the vertical. The use of inclined walls permitted closure without intersecting an aquiclude. The test installation was checked via pumping tests and the effective conductivity for the test cell was 2×10^{-8} m/s, which is approximately the permeability of the cement-bentonite slurry. After the pumping tests, the test section was excavated and inspected. The vertical walls were consistently 0.089 to 0.1 m thick. The inclined walls were thinner, 0.051 to 0.064 m thick, and the intersections of the walls were complete.

Jepsen and Place[11] describe potential defects for cutoffs constructed by the vibrating beam method. They include narrowing, spalling, or collapse of the sides of the wall due to vibrations of the beam as adjacent portions of the wall are constructed, uncertainty associated with intersection of the aquiclude, and control of alignment and verticality of the beam to ensure continuity and to prevent unfilled or ungrouted windows. Jepsen and Place[11] imply that the effective depth for vibrating beam walls is in the range of 14 m. Below 14 m, the potential for windows is greater without special controls on alignment and verticality. However, Leonards et al.[5] and discussions of Jepsen and Place[11] indicate that this method can be used successfully. The presence of numerous boulders or very hard or dense zones is likely to compromise the ability of this method to construct a continuous wall, particularly at depth.

Injection Grouting Methods

Permeation grouting has been used since the early 1800s to control water flow and to increase the strength of the soil.[25] The fluid grout material is typically injected under relatively low pressures to avoid creating fractures, with the aim of filling the void spaces in the soil. After it is in place, the grout will set to bind the soil particles together into a coherent mass and to block water flow through the soil. Grouting to control contaminant transport is especially attractive in several situations including: cases where the barrier must extend to depths greater than those conventionally possible with other techniques, cases where it is desirable to form an impermeable horizon when the natural one is relatively deep, and cases where ground conditions may reduce the practicality of other systems (e.g., many boulders present).[26]

Grouts are usually classified as particulate grouts (e.g., cement and bentonite grouts) and solution grouts (chemical grouts). The particulate grouts are usually restricted to clean, coarse sands and gravels soils,[27] whereas the chemical grouts are capable of penetrating soils with up to 20% coarse silt.[17,27] The grouts are usually injected using parallel, offset rows of holes. The outer rows of holes are grouted first and the inner row(s) is grouted last to ensure complete grouting. The spacing of the holes depends on type of grout, grout viscosity, injection pressure, soil permeability, and rate of grout take.[25]

Defects in grouted barriers usually take the form of zones that have not been permeated with grout and zones where conditions are such that the grout will not set up. In addition, postconstruction defects can develop due to aging of the grout and chemical interaction with the surrounding soil and fluids.[28] The continuity of the grouted barrier is of particular concern. Grout tends to flow along the easiest path available and it is very difficult to achieve the symmetrical, intersecting grout bulbs postulated as illustrated by the cases presented by May et al.[29]

Jet Grouting Methods

Jet grouting is a form of grouting where, rather than permeate the soil voids with minimal disruption of the soil structure, the jetting process is used to mechanically mix the soil and the grout. In general, cement grout is used, but other types of grouts may be employed.[30,31] The method consists of drilling a guide hole to the desired depth, inserting the jetting tool to the bottom of the hole, and then the fluid grout is injected at high pressure as the injector rod is lifted and rotated. The high speed fluid cuts and mixes the cementing agent with the soil to form cylindrical columns 0.7 to 3 m in diameter, depending on the soil type and the specific injection system used, and up to 50 m deep. To form a barrier, a row of intersecting columns must be constructed. Novatecna[30] recommends multiple rows of intersecting columns for depths greater than 15 m, or monitoring and correcting the verticality of the guide holes to ensure continuity. No information was available regarding conductivity of the cement jet grouted soil, but it is anticipated that it would be in the range of portland cement-soil mixes and cement concretes.

Defects in a jet grouted barrier fall into three basic categories: incomplete mixing of the soil and grout, inability of the grout to set up, and gaps or windows in the continuity of the wall. Incomplete mixing can be controlled by jetting pressure, injector rod rotation rate, size and number of injectors, lifting speed, and grout mix and injection rate.[31] Compatibility of the grout mix with the host soil and pore fluid can be checked and altered if need be. The continuity of the wall is the greatest concern. The presence of a guidehole at the location of each cylinder does permit checking the intersection with an aquiclude, and checking and correcting verticality.

The most attractive uses of this technique appear to be to create a floor under a contaminated area in cases where an aquiclude is rather deep and to construct a barrier in areas where space to excavate and mix backfill is very restricted. However, due to the apparent lack of data, caution must be exercised if using jet grouting to create a barrier to contain contaminants.

IMPACT OF AGGRESSIVE CHEMICALS

The above discussion has only considered the hydraulic conductivity of the barriers for passive permeants. However, it is well known that contaminated permeants can react with barrier materials and potentially cause significant increases in conductivity. In this section, the effects of petroleum-based and other organic contaminants on the barrier conductivity are examined briefly.

In general, bentonite-based barriers appear to fare poorly when faced with petroleum products and other organic contaminants.[32,33] However, proprietary bentonites reported to be more resistant to contaminants are available,[11] and low concentration solutions and certain types of immiscible organic contaminants may not cause as large changes in the bentonite mixes.[1,34,35] Portland cement-based mixes appear to be able to resist oily wastes and sludges.[32,36,37] Of the synthetic membrane materials, high density polyethylene and polyvinyl chloride appear least affected by oily materials.[32,37,38] Asphalts tend to fare poorly in the presence of oily and aromatic hydrocarbon contaminants,[5,37] but Anderson et al.[9] indicate that an asphalt slurry composed of asphalt emulsion, sand, water, and portland cement performed well in tests where xylene, methanol, and creosote oil were used as permeants. Lord et al.[39] indicate resistance of a silicate grout to several organic contaminants and Bodocsi et al.[21] found an acrylamide grout resistant to acetone. However, Malone et al.[28] found that gasoline and oil, as well as other organic permeants, increase the gel times for an acrylate and a silicate grout, thus indicating a potential for compatibility problems.

BREAKTHROUGH MODEL

In a strict sense, any barrier system is temporary. All the barrier systems described above are considered to transmit fluids, though some at very slow rates. The role of a breakthrough analysis is to estimate the time required for a contaminant to breach the barrier. This information can then be used in the planning and timing of more permanent remediation.

Following Ostendorf et al.,[3] the breakthrough model proposed rests on an empirical relationship between cutoff wall permeability, K, and number of contaminated pore volumes, N, discharged through the barrier.

$$K = K_b[1 + (\frac{K_b}{K_p} - 1)(1 - N)^\alpha]^{-1} \qquad (N < 1) \qquad (1a)$$

$$K = K_c[1 + (\frac{K_c}{K_b} - 1)(N)^{-\beta}]^{-1} \qquad (N > 1) \qquad (1b)$$

As suggested by Figure 1, the empirical conductivity relation simulates the initial decrease of permeability from its passive K_p value to a minimum breakthrough level, K_b, upon the discharge of a single pore volume. The prebreakthrough behavior may be attributed to initial dissolution of the solid matrix by the chemically aggressive permeant with subsequent clogging of the pores. This behavior is modeled using the exponent α. The post breakthrough behavior is marked by a rise to a higher contaminated conductivity, K_c, due to piping and further dissolution of the solid matrix. The post breakthrough behavior is modeled using a second exponent β.

The conductivity determines the specific discharge, v, of permeant under a hydraulic head gradient, S, across the cutoff wall.

$$v = KS \qquad (2a)$$

$$v = \frac{discharge}{gross\ area} \qquad (2b)$$

The specific discharge may also be cast in terms of the rate of change of pore volume, leading to an integral relation between time, t, and N

$$v = nL \frac{dn}{dt} \tag{3a}$$

$$\frac{S}{nL} \int_0^t dt' = \int_0^N K^{-1} \, dN' \tag{3b}$$

with effective porosity, n, and cutoff wall thickness, L. A straightforward substitution of the empirical conductivity relation (Eq. 1) into Eq. 3b yields the following algebraic relationships between time and pore volume number

$$t = \frac{nL}{K_b S} \left\{ N + (\frac{K_b}{K_p} - 1) \frac{[1 - (1 - N)^{\alpha + 1}]}{\alpha + 1} \right\} \qquad (N < 1) \tag{4a}$$

$$t_b = \frac{nL}{K_b S} \left(\frac{\alpha + K_b/K_p}{\alpha + 1} \right) \qquad (N = 1) \tag{4b}$$

$$t = t_b + \frac{nL}{K_c S} \left[N - 1 + (\frac{K_c}{K_b} - 1)(\frac{1 - N^{1 - \beta}}{\beta - 1}) \right] \qquad (N > 1) \tag{4c}$$

The actual time of breakthrough, t^b, coincides with the discharge of one pore volume through the wall, and is given by Eq. 4b. The expression simplifies considerably for those permeants that do not alter the conductivity of the cutoff wall.

$$t_b = \frac{nL}{K_p S} \qquad \text{(passive only)} \tag{5}$$

Eq. 5 is used to assess the time of breakthrough for essentially pure water or dilute solutions.

Eqs. 1, 2, and 4 may be combined to yield an implicit account of the temporal variations of the specific discharge of the cutoff wall under the action of a chemically aggressive permeant. Figure 2 shows the expected behavior for the following parameter values:

Hydraulic gradient (S) = 10
Barrier thickness (L) = 1 m
Passive conductivity $(K_p) = 1.6 \times 10^{-11}$ m/s
Breakthrough conductivity $(K_b) = 2.6 \times 10^{-12}$ m/s
Contaminated conductivity $(K_c) = 1 \times 10^{-10}$ m/s
Effective porosity (n) = 0.152

In this example, the actual breakthrough time of 2×10^9 s (63 years) is followed by another 60-year period of slow discharge until a rapid rise of discharge occurs. Thus, the discharge of relatively significant amounts of contamination occurs abruptly at an effective breakthrough time, t_e, about 120 years after the initial exposure of the barrier to the waste. A working definition of this time

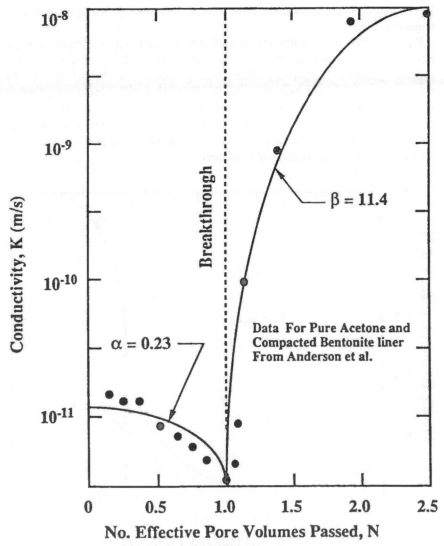

Figure 1. Conductivity vs pore volume, aggressive permeant.

follows by setting t equal to t_e when the barrier conductivity (and specific discharge) achieves half of its contaminated value.

$$K = \frac{K_c}{2} \quad \text{(at } t=t_e) \tag{6a}$$

$$t = t_b + \frac{nL}{K_cS} \left[\frac{(\frac{K_c}{K_b} - 1)^{1/\beta}(\beta - 2) + \frac{K_c}{K_b} - \beta}{\beta - 1} \right] \tag{6b}$$

Eq. 6b is used in the assessment of the time of breakthrough for aggressive contaminants.

Eqs. 5 and 6b suggest that the time of breakthrough varies directly with the effective porosity and wall thickness, and inversely with the conductivity and hydraulic gradient. The porosity effect is most pronounced in channeling, whereby the majority of flow proceeds through a small fraction of the void space, leading to short t_b values. The thickness dependency implies that the synthetic barriers provide no lead time if they fail due to the extreme thinness typically associated with this class of barriers. Larger hydraulic gradients and higher conductivities lead to shorter breakthrough times as well.

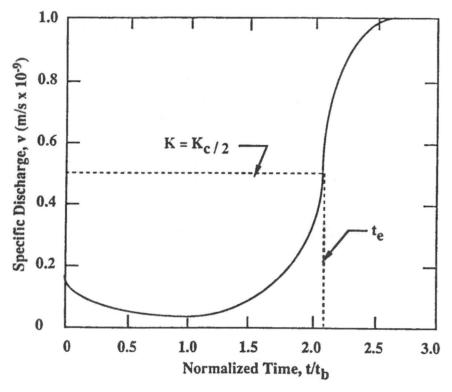

Notes: For Liner Thickness of 1 m and Hydraulic Gradient of 10.
Breakthrough Time $t_b = 2.0 \times 10^9$ s. Kp $= 1.6 \times 10^{-11}$ m/s,
$K_b = 2.6 \times 10^{-12}$ m/s, $K_c = 10^{-10}$ m/s, n = 0.15.

Figure 2. Specific discharge vs time.

TIME OF BREAKTHROUGH ASSESSMENT

Table 3 presents breakthrough times computed for passive permeation through a 1-m-thick cutoff wall under a constant hydraulic gradient of 10, corresponding to a high buildup of mildly contaminated groundwater behind the containment barrier. The alternative materials may be compared on constitutive grounds by contrasting laboratory values for the time of breakthrough. Soil cement, cement grout, lean portland cement concrete, and asphalt slurry (Cases 7, 9, 11, and 6) all offer relatively effective retention of chemically inert wastes in the absence of field imperfections, providing from 15 to 22 years of lead time before the breakthrough of pollution. The bentonite slurry wall (fine backfill), conventional portland cement concrete, acrylamide grout, Dynagrout T, and resorcinol grout alternatives (Cases 1, 12, 14–17) are less effective in passive retention, with breakthrough times of about 1.6 years in magnitude. The bentonite slurry (course backfill), cement bentonite, and silicate grout materials (Cases 2–4 and 13) are relatively ineffective, and yield passive breakthrough times ranging from 2 to 5 months.

Table 3.   Breakthrough Times for Passive Permeants.

Case	Barrier (Condition)	K_p m/s $\times 10^{-10}$	n %	t_b s $\times 10^8$	Ref.
1	Bentonite slurry wall (fine backfill)	5	70	1.40	[8]
2	Bentonite slurry wall (coarse backfill)	50	70	0.14	[8]
3	Cement bentonite slurry wall (slag)	10	5	0.05	[18]
4	Cement bentonite slurry wall (no slag)	100	5	0.005	[18]
5	Asphalt slurry (laboratory)	0.1	5	5.00	[9]
6	Asphalt slurry (field)	1	5	0.50	[3]
7	Soil cement (laboratory)	0.5	35	7.00	[2]
8	Soil cement (field)	50	35	0.07	[2]
9	Cement grout (single well, laboratory)	0.1	5	5.00	[40]
10	Cement grout (curtain, field)	1000	5	0.0005	[41]
11	Portland cement concrete (lean mix)	0.1	5	5.00	[42]
12	Portland cement concrete (conventional)	1	5	0.50	[42]
13	Silicate grout (laboratory)	50	5	0.01	[43]
14	Acrylamide grout (laboratory)	0.5	5	1.00	[44]
15	Dynagrout T grout (laboratory)	1	5	0.50	[45]
16	Resorcinol grout (laboratory)	1	5	0.5	[43]
17	Grout (field)	10000	5	0.00005	[46]

Based on passive permeation through a 1-m-thick wall under a hydraulic gradient of 10.

It is important to note the dramatic fall in cutoff wall effectiveness when field improprieties are considered. The chemical grout alternative breakthrough time falls to a scant 1.4 hours (Case 17) due to leaks in the curtain, while the cement grout and soil cement alternatives (Cases 10 and 8) also suffer with t_b values of 14 hours and 2.7 months, respectively. Only the asphalt slurry alternative demonstrates a moderately effective breakthrough time of 1.6 years under field conditions. The decline of field effectiveness underscores the need for detailed specifications of materials and methods, thorough quality assurance procedures, and careful monitoring of cutoff wall performance for leaks and imperfections. Additional field testing of all alternative cutoff wall systems is certainly warranted as well.

Tables 4 and 5 summarize the effects of chemically aggressive permeants on the performance of some of the materials. The varying porosity values cited in

Table 4 indicate channeling as a mode of barrier failure, particularly for the bentonite and asphalt slurry walls (Cases 18, 19, and 22–24). Relatively effective retention is exhibited by the latter barrier in contact with the three organic wastes (effective breakthrough times of 8.6 to 20 years). The bentonite slurry wall resists attack by inorganic acids and bases (Cases 20 and 21), while the soil cement and acrylamide grout alternatives are compatible with separator waste (Case 25) and acetone (Case 33), respectively. Dynagrout T and plastic concrete solid matrices are clogged without piping when subjected to sulfuric acid (Case 31) and scrubber waste (Case 27) for the duration of their testing periods; the longest t_e values (300 and 50 years) are simulated as a consequence. The bentonite slurry wall is relatively ineffective in the retention of pure xylene (Case 18), while the soil cement and silicate grout are degraded by municipal leachate and organics, respectively (Cases 26 and 28). These latter three effective breakthrough times range from 12 days to 5 months. Moderate degrees of effectiveness are exhibited by the remaining cases.

The widely varying response of the cutoff walls, as modeled, to different aggressive contaminants has several implications for the design of effective retention facilities. It is essential that the composition of the waste be known so that a chemically compatible material may be elected. In this regard, particular attention should be paid to immiscible contaminants which will exist at full strength in discrete, density determined locations in the aquifer. The barrier material selected should be tested in a permeameter with its waste to establish data comparable to Figures 1 and 2 with a duration sufficient to observe a rise of conductivity after the breakthrough of contamination. Finally, the aggressive permeant simulations are all based on laboratory data and do not reflect the potentially serious loss of effectiveness inherent in field construction procedures.

Table 4. Chemically Aggressive Permeant Cases.

Case	Barrier	Permeant (Concentration)	n %	Ref.
18	Bentonite slurry wall	Xylene (pure)	6.8	[33]
19	Bentonite slurry wall	Methanol (pure)	14	[33]
20	Bentonite slurry wall	Hydrochloric acid (pH = 0)	69	[8]
21	Bentonite slurry wall	Sodium hydroxide (pH = 14)	69	[8]
22	Asphalt slurry wall	Methanol (pure)	0.4	[9]
23	Asphalt slurry wall	Xylene (pure)	3.2	[9]
24	Asphalt slurry wall	Creosote oil (pure)	4.5	[9]
25	Soil cement	Separator waste	5.0	[46]
26	Soil cement	Municipal leachate	5.0	[15]
27	Plastic concrete	Scrubber waste	5.0	[18]
28	Silicate grout	Organics	5.0	[39]
29	Dynagrout T grout	Phenol (8%)	5.0	[45]
30	Dynagrout T grout	Potassium ferrocyanide (15%)	5.0	[45]
31	Dynagrout T grout	Sulfuric acid (5%)	5.0	[45]
32	Dynagrout T grout	Perchloroethylene (0.01%)	5.0	[45]
33	Acrylamide grout	Acetone	5.0	[21]

Table 5. Breakthrough Times for Aggressive Permeants.

Case	K_a m/s x 10^{-10}	K_b m/s x 10^{-10}	K_c m/s x 10^{-10}	α	β	t_b s x 10^8	t_c s x 10^8
18	5.0	5.0	100	—	10.6	0.14	0.15
19	5.0	5.0	280	—	3.73	0.28	0.38
20	5.0	5.0	35	—	5.00	1.38	1.69
21	5.0	5.0	35	—	5.00	1.38	1.69
22	0.051	0.004	0.016	0.28	7.74	2.80	3.87
23	0.13	0.019	0.034	0.36	4.00	6.27	6.27
24	0.29	0.078	0.29	0.18	8.64	2.20	2.71
25	0.5	0.1	0.1	0.23	—	1.75	1.75
26	150	8.0	40	0.23	5.00	0.01	0.02
27	1.8	0.006	0.006	0.23	—	15.84	15.8
28	50	50	8000	—	5.00	0.01	0.01
29	1.0	1.0	100	—	5.00	0.50	0.63
30	1.0	1.0	30	—	5.00	0.50	0.63
31	8.0	0.001	0.001	0.23	—	93.5	93.5
32	1.0	1.0	3000000	—	5.00	0.50	0.50
33	0.5	0.045	0.045	0.23	—	2.89	2.89

For purposes of breakthrough time estimation, representing exponential values ($\alpha = 0.23$ and $\beta = 5.00$) were used in cases where reported data were absent. Based on aggressive permeation through 1-m-thick wall with hydraulic gradient of 10.

SUMMARY AND CONCLUSIONS

A number of barrier materials and methods of construction were examined and their strengths and weaknesses were briefly discussed. Passive hydraulic conductivities were examined and compatibility with organic permeants were noted. In addition, the ability of the various construction methods to create a continuous barrier with few defects was examined. A breakthrough model was also presented to assist in estimates of the time available before contaminants breach the barrier.

Based on the above discussions, several barrier types appear more suited to retain petroleum related contaminants than the others. In general, permeation grouting appears to be a poor choice due to the difficulties in ensuring continuity of the barrier. Slurry wall construction techniques with the appropriate backfill or liner composition should perform satisfactorily. Plastic concrete backfill or a composite of backfill and membrane liner (HDPE or PVC) may be attractive alternatives. With the latter, difficulties with seaming and installation without tears or punctures will require special construction techniques and may hinder its use. Asphalt slurries, particularly those that include portland cement, may also provide satisfactory performance as a slurry trench backfill. However, the high cost of the asphalt slurries has tended to counteract their use in the relatively wide slurry walls. In general, the bentonite based backfills tend to fare poorly when faced with organic contaminants; however, proprietary mixes of treated bentonite may provide satisfactory performance under the appropriate conditions. The vibrating beam method of barrier construction creates a much thinner wall than the traditional slurry trench method. Still, use of a low conductivity backfill such as an asphalt slurry may permit satisfactory performance if compatibility problems do not exist. Jet grouting techniques appear to have potential for use in contaminant control, particularly for formation of a horizontal barrier beneath contaminated zones, but more information is needed regarding the continuity of a barrier constructed by this technique.

Crucial factors for success of barriers installed in situ to surround a zone of contamination are compatibility of the materials and the permeant (at the appropriate concentration), compatibility of the construction technique with the barrier materials and the ground conditions encountered, and good specifications and quality assurance controls for construction.

REFERENCES

1. Evans, J. C., "Slurry Trench Cutoff Walls for Waste Containment," *International Symposium on Environmental Geotechnology*, Vol. I, Envo Publishing Company (1986), pp. 303–311.
2. Folkes, D. J., "Fifth Canadian Geotechnical Colloquium: Control of Contaminant Migration by the Use of Liners," *Can. Geotech. J.*, Vol. 19, No 3 (1982), pp. 320–344.
3. Ostendorf, D. W., R. R. Noss, A. B. Miller, and H. S. Phillips, "Hydraulic Containment of Low-Level Radioactive Waste Disposal Sites," Completion Report, UADOE, Environmental Engineering Program, University of Massachusetts (1987).
4. Kosmatka, S. H., and W. C. Panarese, *Design and Control of Concrete Mixtures*, 13th Ed. (Skokie, IL: Portland Cement Association, IL, 1988).
5. Leonards, G. A., J. L. Schemednecht, J. L. Chameau, and S. Diamond. "Thin Slurry Cutoff Walls Installed by the Vibrating Beam Method," *Hydraulic Barriers in Soil and Rock*, ASTM STP 874 (1985) pp. 34–43.
6. Davidson, R. R., and J. Y. Perez. "Properties of Chemically Grouted Sand at Lock and Dam No. 26," *Grouting in Geotechnical Engineering*, ASCE (1982) pp. 433–449.
7. Spooner, P., R. Wetzel, C. Spooner, C. Furman, E. Tokarshi, G. Hunt, V. Hodge, and T. Robinson. *Slurry Trench Construction for Pollution Migration Control* (Park Ridge, NJ: Noyes Publications, 1985).
8. D'Appolonia, D. J. "Soil Bentonite Slurry Trench Cutoffs," *J. Geotech. Eng. Div.*, ASCE, Vol. 106, GT4 (1980), pp. 399–417.
9. Anderson, D. C., K. W. Brown, and J. Green. "Effects of Organic Fluids on the Permeability of Clay Soil Liners," *Proc. Eighth Annual Research Symposium*, USEPA (1982), pp. 179–190.
10. Jogis, H., and R. Bell. "Vibrating Beam Asphaltic Slurry Wall – A Case of Sealing Pond Dikes," National Conference on Hazardous Wastes and Environmental Emergencies, Houston, TX (1984).
11. Jepsen, C. P., and M. Place. "Evaluation of Two Methods for Constructing Vertical Cutoff Walls at Waste Containment Sites," *Hydraulic Barriers in Soil and Rock*, ASTM STP 874 (1985), pp. 45–63.
12. Office of Nuclear Waste Isolation, "Evaluation of Polymer Concrete for Application to Repository Sealing," Report No. ONWI-410, Columbus, OH (1982).

13. Vesperman, K. D., T. B. Edil, and P. M. Berthouex. "Permeability of Flyash and Fly Ash-Sand Mixtures," *Hydraulic Barriers in Soil and Rock*, ASTM STP 874 (1985), pp. 289–298.
14. Littlejohn, G. S. "Design of Cement Based Grouts," *Proc. of Conf. on Grouting in Geotechnical Engineering*, ASCE, (1982), pp. 35–48.
15. Haxo, H. E., R. S. Haxo, N. A. Nelson, P. D. Haxo, R. M. White, S. Dakessian, and M. A. Fong. *Liner Materials for Hazardous and Toxic Wastes and Municipal Solid Waste Leachate* (Park Ridge, NJ: Noyes Publications, 1985).
16. Karol, R. H. *Chemical Grouting* (New York: Dekker, 1983).
17. Karol, R. H., "Chemical Grouts and Their Properties," *Proc. of Conf. on Grouting in Geotechnical Engineering*, ASCE (1982), pp. 359–377.
18. Adaska, W. S., and N. J. Cavalli. "Cement Barriers," *Management of Uncontrolled Hazardous Waste Sites*, HMCRI (1984), pp. 126–130.
19. Xanthakos, P. P. *Slurry Walls* (New York: McGraw-Hill Book Co., 1979), 622 pp.
20. Druback, G. W., and S. V. Arlotta. "Subsurface Pollution Containment Using a Composite System Vertical Cutoff Barrier," *Hydraulic Barriers in Soil and Rock*, ASTM STP 874 (1985), pp. 24–33.
21. Bodocsi, A., I. Minkarah, and B. W. Randolph. "Reactivity of Various Grouts to Hazardous Wastes and Leachates," *Proc. of the Tenth Annual Research Symposium*, USEPA (1984), pp. 43–51.
22. Millet, R. A., and J. Y. Perez. "Current USA Practice: Slurry Wall Specifications," *J. Geotech. Eng. Div.*, ASCE, Vol. 107, GT8 (1981), pp. 1041–1056.
23. Ryan, C. R. "Slurry Cutoff Walls; Application in the Control of Hazardous Wastes," *Hydraulic Barriers in Soil and Rock*, ASTM STP 874 (1985), pp. 9–23.
24. Nuclear Regulatory Commission. "Trench Design and Construction Techniques for Low-Level Radioactive Waste Disposal," Report No. CR-3144, Washington, D.C. (1983).
25. Herndon, J., and Lenahan, T. *Grouting in Soils, Vol. 1: A State-of-the-Art Report*, FHWA Report No. FHWA-RD-76-26, Washington, D.C. (1976).
26. Malone, P. G., R. J. Larson, J. H. May, and J. A. Boa. "Test Methods for Injectable Barriers," *Hazardous and Industrial Solid Waste Testing: Fourth Symposium*, ASTM SPT 886 (1986), pp. 273–284.
27. Baker, W. H., "Planning and Performing Structural Chemical Grouting," *Proc. of Conf. on Grouting in Geotechnical Engineering*, ASCE (1982), pp. 515–539.
28. Malone, P. G., J. H. May, and R. J. Larson. "Development of Methods for In Situ Hazardous Waste Stabilization by Injection Grouting," *Proc. of the Tenth Annual Research Symposium*, USEPA (1984), pp. 33–42.
29. May, J. H., R. J. Larson, P. G. Malone, and J. A. Boa. "Evaluation of Chemical Grout Injection Techniques for Hazardous Waste Containment," *Proc. of the Eleventh Annual Research Symposium*, USEPA (1985), pp. 8–18.
30. "High Technology in Jet Grouting," Novatecna Company Brochure, 1988.
31. Guatteri, G., J. L. Kauschinger, A. C. Doria, and E. B. Perry. "Advances in the Construction and Design of Jet Grouting Methods in South America," *Proc. Second International Conf. on Case Histories in Geotechnical Engineering*, Vol. II (1988), pp. 1037–1046.
32. Sharma, H. D., and P. Kozicki. "The Use of Synthetic Liner and/or Soil Bentonite for Groundwater Protection," *Proc. Second International Conf. on Case Histories*

in Geotechnical Engineering, Vol. II (1988), pp. 1149–1157.

33. Anderson, D. C., W. Crawley, and J. D. Zabcik. "Effects of Various Liquids on Clay Soil-Bentonite Slurry Mixtures," *Hydraulic Barriers in Soil and Rock*, ASTM STP 874 (1985), pp. 93–103.

34. Fernandez, F., and R. M. Quigley. "Hydraulic Conductivity of Natural Clays Permeated with Simple Liquid Hydrocarbons," *Can. Geotech. J.*, Vol. 22, No. 2 (1985), pp. 205–214.

35. Foreman, D. E., and D. E. Daniels. "Permeation of Compacted Clay with Organic Chemicals," *J. Geotech. Eng.*, ASCE, Vol. 112, No. 7 (1986) pp. 669–681.

36. Haxo, H. E., Jr. "Durability of Liner Materials for Hazardous Waste Disposal Facilities," *Proc. 7th Annual Research Symposium on Landfill Disposal: Hazardous Waste*, USEPA (1981) pp. 140–156, as cited by Folkes (1982).

37. Stewart, W. S. *State of the Art Study of Impoundment Techniques*, "USEPA Report No. EPA-600/2-78-196, Cincinnati, OH (1978).

38. Haxo, H. E., N. A. Nelson, and J. A. Miedema. "Solubility Parameters for Predicting Membrane-Waste Liquid Compatibility," *Proc. of the Eleventh Annual Research Symposium*, USEPA (1985b), pp. 198–212.

39. Lord, A. E., R. M. Koerner, and E. C. Lindhult. "The Hydraulic Conductivity of Silicate Grouted Sands with Various Chemicals," *Management of Uncontrolled Hazardous Waste Sites*, HMCRI (1983), pp. 175–178.

40. Gulick, C. W., J. A. Boa, and A. D. Buck. "Bell Canyon Test Cement Grout Development Report," Report No. 80-1928, Sandia National Laboratories, Albuquerque, NM (1980).

41. Powell, R. D., and N. R. Morganstern. "The Use and Performance of Seepage Reduction Measures," *Seepage and Leakage from Dams and Impoundments*, ASCE (1985), pp. 158–182.

42. Troxell, G. E., H. E. Davis, and J. W. Kelly. *Composition and Properties of Concrete* (New York: McGraw-Hill, 1968).

43. Spalding, B. P., L. K. Hyder, and I. L. Munro. "Grouting as a Remedial Technique for Problem Shallow Land Burial Trenches of Low-Level Radioactive Solid Wastes," *J. Environ. Qual.*, Vol. 14 (1985), pp. 100–130.

44. Clarke, W. J. "Performance Characteristics of Acrylate Polymer Grout," *Grouting in Geotechnical Engineering*, ASCE (1982), pp. 418–432.

45. Muller-Kerchenbauer, H., W. Friedrich, and H. Hass, "Development of Containment Techniques and Materials Resistant to Groundwater Contaminating Chemicals," *Management of Uncontrolled Hazardous Waste Sites*, HMCRI (1983), pp. 167–174.

46. Davis, K. E., and M. C. Herring. "Laboratory Evaluation of Slurry Wall Materials of Construction to Prevent Contamination of Groundwater from Organic Constituents," *Proc. Seventh National Groundwater Quality Symposium*, NWWA (1984), pp. 491–512.

47. Anderson, D. C., A. Gill, and W. Crawley. "Barrier-Leachate Compatibility: Permeability of Cement/Asphalt Emulsions and Contaminant Resistant Bentonite/Soil Mixtures to Organic Solvents," *Management of Uncontrolled Hazardous Waste Sites*, HMCRI (1984), pp. 126–130.

CHAPTER 21

Thermal Desorption of Hazardous and Toxic Organic Compounds from Soil Matrices

C. Peter Varuntanya, Michael Hornsby, Arun Chemburkar, and Joseph W. Bozzelli

INTRODUCTION

Thermal desorption is a physical separation process, although chemical transformations may occur, depending on thermal stability of the organic compounds and operating temperature.[1] Nearly complete desorption of many organic pollutants from soil is possible for relatively small desorption oven residence times at temperatures near the boiling point of the pollutant. The principal advantage of the technique is that moderate temperatures, and therefore moderate energy inputs to the system with an inexpensive carrier gas such as CO_2 exhaust or N_2, are sufficient to place the organic compounds into the vapor phase. Subsequent incineration of the organic compounds in the vapor phase could then be accomplished in an afterburner.[2] This two-stage technique requires significantly less energy input and therefore less cost than direct incineration of the entire soil mass. It is, therefore, intended to be considered as an alternative to incineration for soil cleanup, and is in fact "the" method of choice. It is important that knowledge of the thermal decomposition behavior of hazardous organic materials is first obtained under controlled, environmentally safe laboratory conditions before expensive scale-up and permanent construction.[3-5] Then, many problems associated with full-scale thermal destruction may be avoided. Conventional equipment such as an asphalt kiln or a rotary kiln may be utilized during full-scale operations.

The laboratory scale research described here was designed to study and demonstrate the feasibility of the removal of selected toxic organic compounds from soil by thermal desorption and gas purge; to develop and utilize a data base on time and temperature requirements for complete thermal desorption; to formulate an engineering model which will eventually predict a flow period and temperature required to remove toxic species from soil matrices for proper re-use of soil, and so more expensive removal procedures. Two series of experiments were performed: Plug Deposition Experiments and Desorption Experiments with uniformally contaminated soils. The selected compounds and their physical constants studied in the Plug Deposition Experiments and the desorption experiments are reported in Table 1. The compounds were selected on the basis of their

Table 1.   Hazardous and Toxic Organic Compounds.

Chemical Species	Formula	FM	BP	D	HVAP	DM
1,2,4-Trichlorobenzene	$C_6H_3Cl_3$	188.45	214.0	1.446 (25)	10227 (TR)	NIA
Hexachlorobenzene	C_6Cl_6	284.78	323–6	2.044 (24)	12558 (TR)	NIA
1-Chloronaphthalene	$C_{10}H_7Cl$	162.62	259.3	1.1938 (24)	11178.3 (TR)	1.59
1,1,1-Trichloroethane	$C_2H_3Cl_3$	133.41	74.0	1.3376 (20/4)	7960	1.79
Chlorobenzene	C_6H_5Cl	112.56	131.7	1.1063 (20)	8730	1.69 1.60*
Acetone	CH_3COCH_3	58.08	56.24	0.7908 (20/4)	6240	2.88 2.90*
Benzene	C_6H_6	78.114	80.3		7352	0.0
Toluene	$C_6H_5CH_3$	92.14	110.6	0.8660 (20/4)	7930	2.20 0.40*
Naphthalene	$C_{10}H_8$	128.17	217.7	1.162 (20/4)	10304.7	0.0
Anthracene	$C_{14}H_{10}$	178.23	340	1.25 (27/4)	13500	0.0 —
Dichloromethane	CH_2Cl_2	84.933	40.0		6690	1.8*
Chloroform	$CHCl_3$	119.39	61.7	1.4985 (15)	7028.7	1.1*
Tetrachloromethane	CCl_4	153.8	77.0	1.59	7160	0.0

Abbreviations:
    FM   = Formula Weight, gm/mole
    BP   = Boiling Point, °C
    D   = Density, gm/cc
    HVAP = Heat of Vaporization, cal/mole
    DM   = Dipole Moment, Debye unit
Note:
    TR   = calculated from Trouton's Rule
    NIA = No Information Available
    20/4 = Measured at 20°C
    *   = Reference No. 8

frequency of occurrence in actual waste streams and predicted range of thermal stability.[6,7] The thermal desorption behavior of the compounds studied is statistically analyzed and related to an elementary first-order differential equation involving compound mass and boiling point.

The mathematical model developed predicts temperature, residence time, and flow rate for known fraction removal of organic pollutants from soil in this type of desorbing apparatus.

EXPERIMENTAL

Two basic types of experiments were performed: Plug Desposition Experiments, which consisted of syringe injections of selected organic compounds into a heated column containing one of the four soil matrices we used; and secondly, quantitative desorption experiments utilizing uniformly precontaminated soil.

Plug Deposition Experiments

The objective of the plug deposition experiments was to establish a relationship between the temperature and the required time for desorption plus transport of organic compounds from different types of soil. Organic contaminants in natural soils often migrate as vapors. They sorb onto soil particles, then desorb and migrate to other soil particles. This behavior is very much like the behavior of organic compounds in a gas chromatograph (GC) column.[9] Plug deposition refers to the entrance, passage, and exit of a singular mass of organic vapors through a soil column. The column packings and organic compounds were chosen to model that behavior. The four packings were chosen to represent natural soils with a range of retentive tendencies as shown in Table 2.

Table 2.   Column Packings for Plug Deposition Experiments.

Packings	Mass (grams)	Mesh	Material
Soil	2.95	35–45	dried top soil
Sand	3.84	45–80	silicone oxide
Poropak T	2.38	100–120	ethylene glycol dimethacrylate
(A very retentive polymer for organics.)			
Gaschrom R	2.14	60–70	aluminum and silicone oxides

One microliter (μL) samples of 1,1,1-trichloroethane, carbon tetrachloride, chlorobenzene, 1,2,4-trichlorobenzene, dichloromethane, benzene, and toluene were injected into the injector port. This was an injection port from a GC that had a coaxial-parallel flow design, and carrier flow is 30 cm³/min. The compounds were selected because they presented a range of boiling points and chemical properties.[10]

The times required for contaminant removal levels, i.e., 90% removal and 50% removal, were established at various operating temperatures at a known carrier gas flow. For 50% removal of the compound from the column the time required was taken as the peak time indicated on the integrator report. The time for 90% removal of the compound from the soil was taken as the point where the flame ionization detector (FID) output signal dropped to 10% of its maximum value.

The desorption curves are presented in Figures 1–6. Two types of graphs are shown: Temperature vs 50% Retention Time; and Temperature vs 90% Retention Time. These data are specifically relevant to in situ desorption at hazardous waste or landfill sites where steam can be injected in the ground to flow up through soil and decontaminate and purge.

Desorption Experiments With Uniformly Contaminated Soil

The quantitative desorption experiments utilized soil that had been precontaminated with known quantities of 1,2,4 trichlorobenzene (TCB), hexachlorobenzene (HXCB), and 1-chloronaphthalene (CNAP). The soil was packed into columns, installed in a nitrogen gas line, and placed in a preheated oven. The desorbed gases were analyzed by online GC. Gases were either directed to the FID, or to activated carbon absorber adsorption tubes. Solvent extractions and GC analysis were further performed on the contaminated soil, the desorbed soil, and the activated carbon to provide a mass balance through the system.

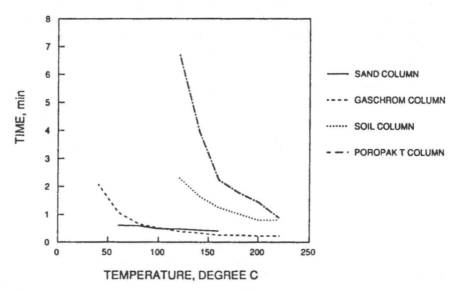

Figure 1.   Chlorobenzene desorption times—RT50.

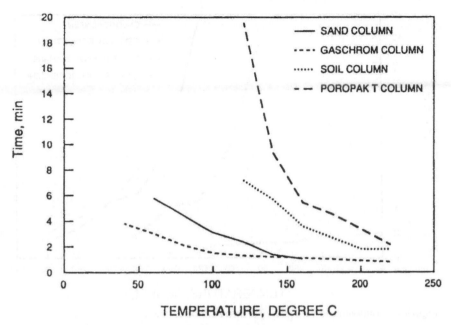

Figure 2. Chlorobenzene desorption times – RT90.

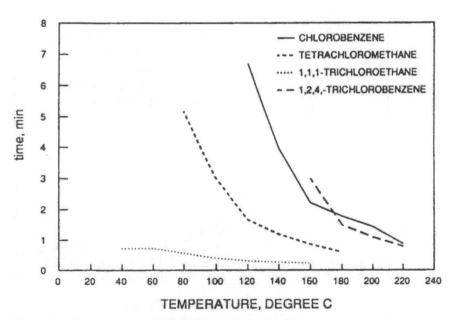

Figure 3. Temperature vs 50% RT (Poropak T column).

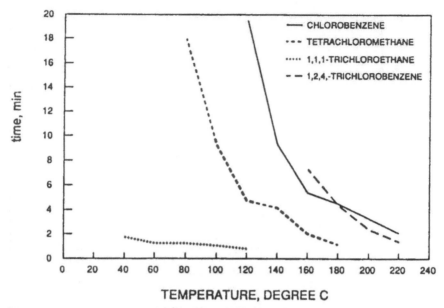

Figure 4. Temperature vs 90% RT (Poropak T column).

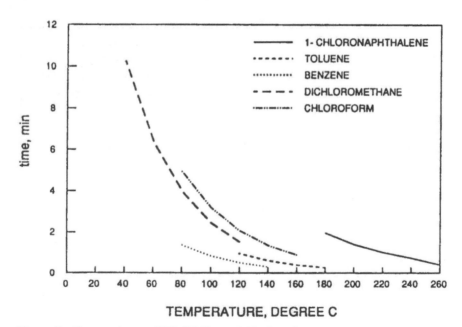

Figure 5. Temperature vs 50% RT (Poropak T column).

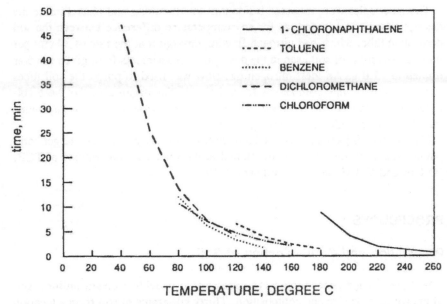

Figure 6.   Temperature vs 90% RT (Poropak T column).

A diagram of the desorption system is shown in Figure 7. The tubing and fittings were stainless steel #316. Glass materials were used for inertness and to ensure that they do not catalyze the decomposition of the organic compound at the temperatures being utilized in the desorption furnace. The quartz tube was filled with 15.0 g of loosely packed soil confined by quartz wood plugs at either end.

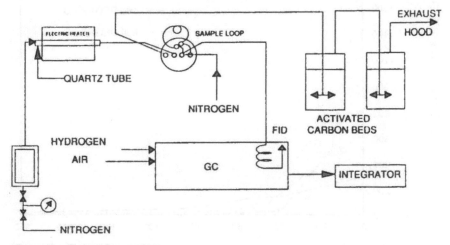

Figure 7.   Desorption system.

Thermocouples were mounted inside thin quartz tubing and placed inside the desorption tube to compensate for any temperature difference between the soil desorption tube, which had nitrogen flowing through it at the rate of 30 cm³ per minute. The mass of soil used in the desorption experiments for a given run was limited to 15.0 g. An inert nitrogen gas flow was used to purge the soil tubes and carry the organic compounds from the soil. The organic vapors were collected on two 26-g activated carbon tubes and placed in series or in a known volume sample loop on a GC gas-sampling valve.

The periodic injections produced a series of chromatograms. The target compound area units and the time are plotted at the selected temperatures of TCB, HXCB, and CNAP shown in Figures 8–10.

PROCEDURES

Preparation and Contamination of Soil

Soil preparation procedures were necessary to provide a clean, uniform soil of known particle size and composition. Thirty kilograms of soil from a location within the New Jersey Institute of Technology campus at Newark, NJ was put in large aluminum trays, placed in the laboratory sink, and tap water was allowed to run onto the trays. This caused the finest particles, salts, and possibly polar

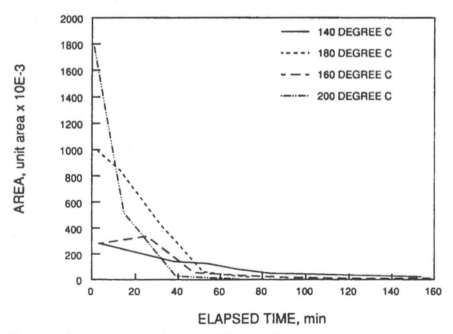

Figure 8. Elapsed time vs unit area for TCB desorption.

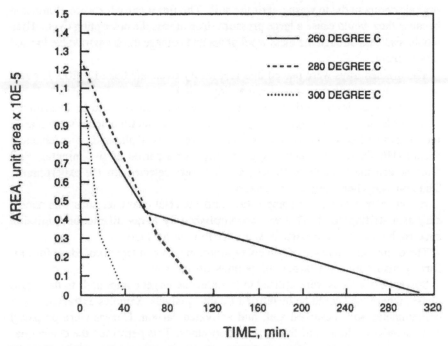

Figure 9. Elapsed time and unit areas for HXCB desorption.

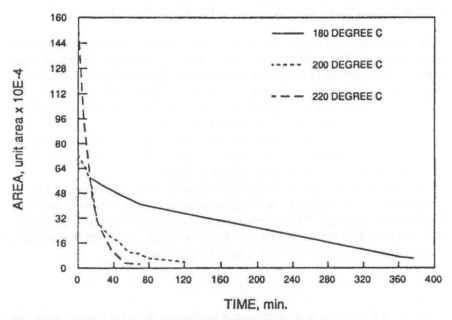

Figure 10. Elapsed time vs unit areas for CNAP desorption.

organic compounds, to wash out of the soil. The finest particles were eliminated because they would cause a large pressure drop across the desorption oven. They would also pass through the glass wool plugs and damage the six-port valve located downstream.

The drained trays were placed in a laboratory oven and baked at 200°C overnight. This procedure removed remaining water.

The soil was sieved so that the particle size distribution could be determined and in order to separate aggregated particles. Two kilogram batches were placed into a standard sieve set (Soil Test). The sieve set was placed in a mechanical shaker (Humboldt Manufacturing Company) for a period of five minutes. Soil from screen mesh numbers 40, 60, and 70 were selected for the experiments. The soil was chemically characterized.

A soil sample of known composition and material characteristics was necessary as a starting point. This was accomplished by X-ray diffraction, emission spectrophotometry, and particle size distribution analysis.

The evaporative method of soil contamination was a suitable procedure for uniformly distributing the target compounds on the soil.[11]

Solvent extractions were utilized to transfer the target compounds to the liquid phase, where they could be quantified by GC analysis. Samples extracted were: contaminated soil, desorbed soil, and activated carbon. Curves were prepared from standard solutions of the target compounds. This permitted the determination of the concentration of the target compound in the extract, and therefore its mass, in the sample.

Mathematical Modeling for Desorption Experiments

A mathematical model was developed to describe the desorption processes of the experiments. The models were based on the chemical properties heat of vaporation or boiling point (a rough empirical relationship between the normal boiling point and the heat of vaporization is by Trouton's Rule: $\Delta H_v/T_b=21$ cal.mol^{-1}.K^{-1}) and dipole moment, or total molecular mass. Residence time, oven temperature, initial and final soil concentrations also were parameters of the model. The models are designed to aid in any future efforts to scale up the system to a pilot plant study. Also, they can be very useful in the prediction of required conditions for nonstudied chemical species.

The selected model was a direct function of the heat of vaporization in the exponential part of the equation, and of dipole moment or polarizability in the linear portion of the equation. The overall form of the equation expresses concentration removed from the soil matrix as a function of time at a given temperature. The change in concentration with time was expressed as:

$$dC_i/dt=kC_i \tag{1}$$

where C_i is concentration of species, i, adsorbed on the soil. This equation can be integrated to yield:

$$C(t)/C(o) = \exp(kt) \tag{2}$$

where $C(t)$ is concentration remaining on the soil at time, t.

Clearly a model based upon these fundamental chemical properties will be of value in predicting removal rates and conditions, for compounds not included in the lab experimental data base. The empirical desorption coefficient k_i is a function of temperature, dipole moment, heat of vaporization, purge flow rate and soil particle shape and size distribution, and soil organic content. However, at a constant purge gas flow rate, it may be expressed energetically in a Arrhenius-type correlation:

$$k_i = a_i \,(\text{mass})\, \exp\,[(-b_i\, \Delta H_{vap})/RT] \tag{3}$$

or

$$k_i = a_i \,(\text{mass})\, \exp\,[(-b_i'\,T_b)/RT] \tag{4}$$

where: a_i and b_i coefficients are fitted empirical constants
      R is the ideal gas constant $= 1.98$ Cal/K.mole
      T is temperature in degrees Kelvin
      T_b is the boiling point temperature in degrees Kelvin
      H_{VAP} is heat of Vaporization in calories/mole

Since the model was based upon fundamental chemical properties as boiling point and mass, it will be of value in predicting removal rates and conditions for compounds not included in the lab experimental data base.

The curves shown in Figure 11 illustrate the curve fitting and optimization capabilities of the procedure and model. This in turn will lead to suggestions for the

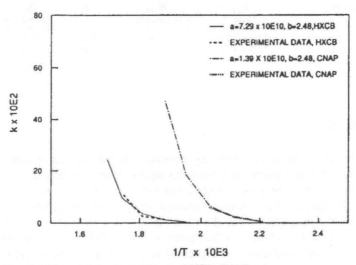

Figure 11. Rate constant vs temperature (HXCB, CNAP).

design of a mobil unit, such as a rotary kiln for the onsite decontamination of soil matrices.

Analysis of the experimental results (Tables 3–5) provide numerical values for the physical parameters of the mathematical model. The parameters are presented in Table 6 for TCB, HXCB, and CNAP.

Table 3. Summary of Parameters for Mathematical Modeling 1,2,4-Trichlorobenzene Desorption.

T, °C	C_0	k_i	R
140	0.384	1.73×10^{-2}	0.981
160	0.500	2.98×10^{-2}	0.958
180	1.52	6.03×10^{-2}	0.984
200	2.06	8.23×10^{-2}	0.979

Table 4. Summary of Parameters for Mathematical Modeling Hexachlorobenzene Desorption.

T, °C	C_0	k_i	R
260	0.03141	1.47×10^{-2}	0.999
280	0.0591	2.56×10^{-2}	0.964
300	0.2643	11.56×10^{-2}	0.994

Table 5. Summary of Parameters for Mathematical Modeling Chloronaphthalene Desorption.

T, °C	C_0	k_i	R
180	0.0365	0.62×10^{-2}	0.999
200	0.1513	2.65×10^{-2}	0.975
220	0.3343	6.13×10^{-2}	0.985

Table 6. Parameters for the Second Modeling.

Compound	a_{i-1} min	a_{i-1} mol.gm.min^{-1}	b_i	b_i' cal/mol·K	R
Trichlorobenzene	4.68×10^{10}	2.58×10^{10}	2.30	48.30	0.99
Hexachlorobenzene	7.29×10^{10}	2.55×10^{10}	2.48	52.07	0.96
Chloronaphthalene	1.39×10^{10}	0.85×10^{10}	2.28	47.95	0.99
Average	4.51×10^{10}	2.00×10^{10}	2.35	49.44	

RESULTS

The results of the preliminary experiments verified predictions regarding the parameters that will affect the thermal desorption of organic pollutants from soil. The sand column shows the weakest, and the Poropak T column the strongest, affinity for the pollutants. Also, an increase in operating temperature drastically decreases the retention time of the compound. The results also were consistent with the calculations based on the mathematical model. All the parameters for the model were established and presented in Figures 1–6. The average correlation coefficient was 0.975.

Desorption curves for TCB, HXCB, and CNAP were plotted in terms of the remaining mass of the target compound versus time. These curves were prepared for TCB at temperatures of 140, 160, 180, and 200°C, and for CNAP at 180, 200, and 220°C. The curves are presented in Figures 8–10, respectively.

A mass balance was also performed. It shows that 96% of the TCB and 102% of the HXCB was accounted for. The percent removals were 95% for TCB, and 62% for HXCB. Final TCB soil concentration was 84 ppm after desorption at 200° for 69 min. Final HXCB soil concentration was 143 ppm after desorption at 300° for 39 min. Increasing the desorption oven temperature to 320° should result in a significant drop in the HXCB concentration, as its boiling point is 322°C.

The best fit exponential decay mathematical model was determined, as well as its accompanying physical parameters. The data fit the equation with an average correlation coefficient of 0.98 for TCB, HXCB, and CNAP. The desorption curves for these compounds at each studied temperature were plotted in Figures 11.

The model which is based upon the fundamental chemical properties for the empirical coefficient, k_i, is expressed in a correlation with two average constants, a_i and b_i as:

$$k_i = 4.51 \times 10^{10} \exp [(-2.35 \, \Delta H_{vap})/RT] \tag{5}$$

or

$$k_i = 4.51 \times 10^{10} \exp [(-49.44 \, T_b)/RT] \tag{6}$$

The constant b_i is a direct function of heat of vaporization or boiling point temperature and a_i is a function of dipole moment (a linear portion of the equation).

The equation above can now be used to predict desorption conditions and parameters for all chlorinated aromatic species, e.g., PCBs (polychlorinated biphenyl), chlorinated PAHs (polycyclic aromatic hydrocarbon) and chlorinated benzene species which are known to be difficult to remove from soil matrices, due to low volatility. Some of the fitted curves are shown in Figure 12.

CONCLUSION

We conclude that thermal desorption of soil matrices contaminated with organic compounds (compounds such as chlorinated dioxans not tested yet) is the method of choice when considering for remediating hazardous waste sites. Moderate desorption oven temperatures, in the vicinity of the boiling point of the target compound, can be suitable for reducing concentrations to acceptable levels. The moderate temperature requirements can be equated to moderate energy inputs and costs. Cost is a major criteria in the selection of a remedial technology. Modified conventional equipment such as rotary kilns, asphalt kilns, or incinerators operated at lower than normal temperatures may be used in scaled-up operations. This will serve to further lower costs.

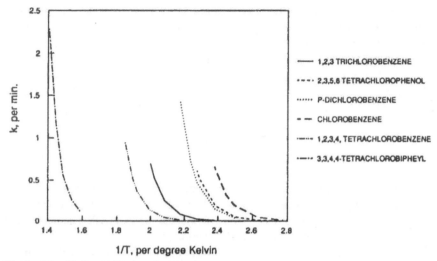

Figure 12. Rate constant vs temperature.

Thermal desorption also satisfies two Environmental Protection Agency (EPA) policies established in the 1986 Superfund Amendments and Reauthorization Act (SARA): onsite treatment and permanent solutions. The thermal desorption system could be portable (flat bed truck convoy) and transported to a site, then removed when all the soil has been processed. Desorption can be a part of a permanent solution. Once the vapors have been desorbed, they could easily be incinerated in an afterburner, thus reducing long-term risks and liabilities.

EPA's Record of Decision,[4] stating that the technology has been selected for cleaning of a Superfund site in the USA (Lipari Landfill, NJ), demonstrates that thermal desorption is a feasible method for cleaning up hazardous waste sites.

ACKNOWLEDGMENT

The authors gratefully acknowledge funding from the NJIT-NSF University/Industry IHTW Research Center.

REFERENCES

1. Miller, R. A., R. W. Helsel, E. S. Alperin, A. Groen, D. M. Catalano, J. L. Flemming, and D. M. Pitts, "Thermal Desorption and Heat Characteristics of Soil Contaminated with 2,3,7,8-Tetrachlorodibenzo-p-dioxin," presented before the Division of Environmental Chemistry, American Chemical Society, New York, New York, April 1986.
2. Office of Technology Assessment, Technologies and Management Strategies for Hazardous Waste Control, U.S. Government Office, Washington, D.C., pp 7–23, 1986.

3. Jury, W. A., W. F. Spencer, and W. J. Farmer, *J. Environmental Quality*, "Behavior Assessment Model for Trace Organics in Soil: I. Model Description," 12 (4), 558 (1983).
4. Jury, W. A., W. F. Spencer, and W. J. Farmer, *J. Environmental Quality*, "Behavior Assessment Model for Trace Organics in Soil: II. Application of Screening Model," 13 (4), 573 (1984).
5. Garbarini, D. R., and L. W. Lion, "Influence of the Nature of Soil Organics on the Sorption of Toluene and Trichloroethylene," *Environ. Sci. Technol.*, 20 (12), 1263 (1986).
6. Chemical Substances Index, American Chemical Society, 1986. (Chemical Abstract)
7. United States Environmental Protection Agency, New York, New York, News Release, 86 (57), June 1986.
8. Reid, R. C., J. M. Prausnitz, and T. K. Sherwood, *The Properties of Gases and Liquids*, McGraw-Hill, New York, 1977.
9. Levenspiel, O., *Reaction Kinetics*, John Wiley & Sons, New York, 1986, pp. 256–278.
10. Morrison, R. T., and R. Boyd, *Organic Chemistry*, Allyn and Bacon, Boston, 1983, pp. 573–621.
11. University of Minnesota, Department of Chemical Engineering, Chemistry 5139 Laboratory Manual, 1982, pp. 261–280.

CHAPTER 22

Hot Mix Asphalt Technology and the Cleaning of Contaminated Soil

Raymond C. Czarnecki

Three distinct but related phenomena have created the need for cost-effective and environmentally sound processes to decontaminate soil polluted with oil: (1) the legal mandate to remediate sites; (2) the progressive leakage of underground tanks; and (3) the continuing occurrence of oil spills. Taken together, these three conditions constitute a worthwhile and lucrative market for industrialists, and much progress has been made over the past several years. At Brox Industries, we've devoted considerable resources to understanding and improving performance in this emerging industry.

INCREASING DEMAND FOR CLEANUP

In the area of onsite remediation or environmental assessment, a recent *Wall Street Journal* article stated that "environmental exams for real estate are becoming common. In fact, the environmental inspection is becoming to industrial and commercial property what the termite inspection is to home sales." Some states are imposing regulations requiring sellers to notify buyers that waste has been stored or disposed of on the property. In other states, laws mandate that the property be inspected and given a "clean bill of health" before title can be transferred. At the federal level, a 1986 amendment to the Superfund toxic waste cleanup program now effectively obligates a buyer to make a diligent inquiry "into whether a property is contaminated" or be held responsible if contamination is subsequently found.

The law is entering the picture in other ways, too. A recent court decision has held that the mortgagee of a foreclosed property can be held responsible for cleanup of waste on the property. This issue is pervasive and continuing; in the words of a real estate analyst, only an "idiot would overlook the environmental risks before buying a piece of property." Many states have "buyer protection plans," and other states insist that an inspection be made of virtually all commercial and industrial real estate before it is sold. Property assessors Brox interviewed indicate that at least 20% of the inspections made in Brox's territory over the past year discovered oil or gasoline contamination in the soil. Recently, several real estate transactions were held up because of environmental problems. As a result, many owners are taking the initiative to have voluntary inspections made to ensure that their property is safe, or to take further action if it is not. Such inspections surely will create more opportunities for cleanup specialists.

Another source of oil-contaminated soil is the underground storage tank. Although state and federal regulations concerning the leaking of underground storage tanks have been on the books for some time, it was not until 1984 that a formal regulatory program for this problem was established (see amendments to the Resource Conservation and Recovery Act). The federal program enables states with their own programs to enforce environmental standards, provided each state's guidelines meet federal minimums. As a result, every state in Brox Industries' region has established a protocol for monitoring, testing, and replacing underground storage tanks. It has been found that fully 25% of all underground storage tanks at least 20 years old show signs of leakage; most states vigorously encourage the removal of all tanks after 25 years of service. It is estimated that each removal of a leaking tank will generate 30 to 50 cubic yards of soil.

Another significant source of contaminated soil is the "spill." Most states have well delineated cleanup procedures for spills of gasoline, oil, or kerosene: the spill is absorbed by sand or other solid absorbents, followed by removal and isolation of the total mass. Massachusetts estimates that the state suffers approximately 4,000 spills every year.

Having reviewed the data, Brox concluded that oil contamination is both a local and a national problem, and will continue as such for many years to come.

THE ADVANTAGES OF HMA TECHNOLOGY

Processing contaminated soil in a Hot Mix Asphalt (HMA) facility has many advantages:

1. The incineration technology required is related to HMA technology.
2. Capital investment is "in place" with only modifications required.
3. Decontaminated soil can be incorporated into HMA.

4. Small quantities of soil can be processed economically. That is, large quantities or continuous runs are not necessary.
5. With over 5000 HMA plants, a plant should be near the product, and thereby lower freight costs.
6. No continuing liability for the generator.

Early attempts to process soils in an HMA facility involved incorporating the soil with regular aggregate. Figure 1 is a flow diagram for a typical HMA batch plant. Conventionally, the aggregate moves from the cold feed bins via belt feeders to a collecting conveyor and then enters the dryer at the end opposite the burner (counter flow). After the water evaporates and the aggregate is heated to approximately 350°F it leaves the dryer and is carried via a bucket elevator to a set of screens and then is separated into hot bins. It is subsequently made into HMA. As mentioned earlier, first attempts to process contaminated soils involved incorporating the soil with the virgin aggregate, followed by normal processing. The end result was that hydrocarbon levels in the stack were elevated. Tests run in California on such systems showed that destruction of hydrocarbon was about 65%. Figure 2 compares dryer length with temperature, and helps to explain the mechanism involved and the reason for elevation of hydrocarbon in the stack. Hot gases at 2000°F enter at one end and cold, wet aggregate enters at the other. The wet aggregate immediately reaches 212°F, at which time evaporation of hydrocarbon and water occurs. The temperature stays at 212°F until all the liquids are evaporated into the exiting gas stream, then the aggregate is heated until it reaches 350°F, when it leaves the dryer and enters the bucket elevator. The hot gases enter at about 2000°F and leave the dryer at 300–350°F. Since the evaporated hydrocarbon is entering a low temperature gas stream (300–400°F), very little incineration takes place. The California tests confirm the proposed mechanism.

With this knowledge and background, we at Brox Industries chose to work with one of our batch plants—a 3-ton Madsen equipped with a 50,000 CFM baghouse for air cleaning and an 8' diameter×30' long dryer. By this time, we had seen enough contaminated soil to know that a ratio of 95% virgin to 5% contaminated soil would be a ratio we could use and still be able to produce a quality HMA product. We then called in Astec Industries (Don Brock and Gail Mize), giving them the following constraints: (1) that the hydrocarbon be incinerated, and (2) that we be able to process at a rate of 8-9 tons/hour (5%×180T/hr=9T/hr). The solution is shown in Figure 3. A concrete storage area with 6-ft. sides is used for holding the contaminated soil. The soil is loaded into a bin via a front-end loader and the material is fed in proportion to virgin aggregate to the process by a belt feeder. The soil is passed over a scalping screen to remove large particles, and then transferred via belt conveyor to a chute into a ceramic cylinder inserted between the burner and dryer.

The ceramic cylinder, the key to the process, is 5 ft in diameter and 5 ft long, and is equipped with flights. It rotates counter to the rotation of the dryer and

FLOW DIAGRAM

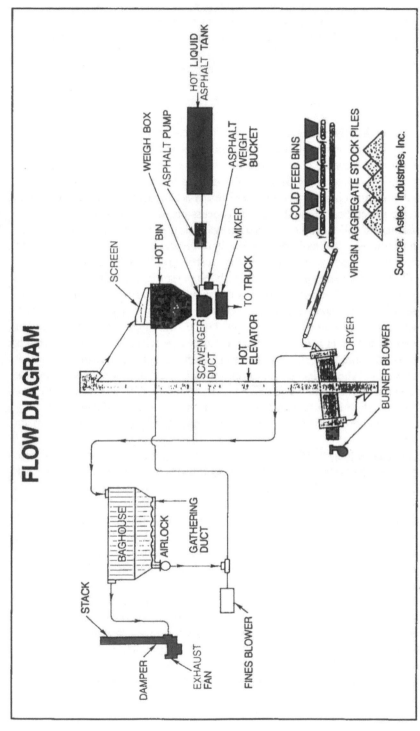

Figure 1. Flow diagram of Hot Mix Asphalt (HMA) process. Source: Astec Industries. Used with permission.

OIL CONTAMINATED SOIL
IN EVAPORATIVE TYPE UNIT
(Batch Plant)

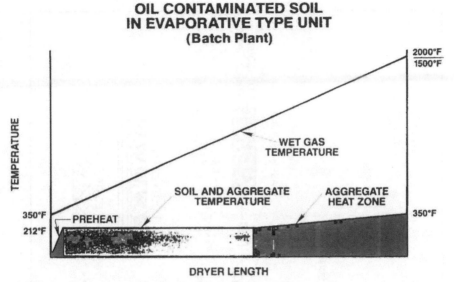

Figure 2. Comparison of batch plant dryer length and temperature.

can be adjusted in RPM from 0–6. Holdup time of the soil in the ceramic cylinder is at least one minute to assure that the soil is freed of hydrocarbon. The entire system from storage area to ceramic cylinder is covered by a roof and rests on concrete walkways with sumps.

The mechanism involved in this process is illustrated in Figure 4, which compares the length of the ceramic cylinder with temperature. The ceramic cylinder is filled with flame and acts as a combustion chamber. The flame extends beyond the cylinder for approximately 11 ft into the dryer. The soil enters the rotating cylinder and immediately flashes off the water and hydrocarbon into the 2000°F flame. The soil leaves the cylinder at approximately 800°F.

The vaporized hydrocarbon is incinerated and destroyed as it is carried along the 16 ft of flame into the dryer. Exposure to this 2000°F flame is calculated to be 1.2 seconds, based on a 16-ft flame length and a gas velocity of 800 ft/minute (16 ft/800 ft minute×60 sec/min=1.2 sec). Experts agree that this exposure time is sufficient to destroy the vaporized hydrocarbon.

Not all soils are suitable for recycling into Hot Mix Asphalt, owing to clay content and particle size. High clay content is objectional because it reduces the overall strength of HMA. Soils with a high content (20%) of any minus 200 mesh material may also deteriorate quality when incorporated into HMA by contributing to potential stripping problems. An average HMA composition contains 60% coarse aggregate, 40% fine aggregate (sand), and 5.5% liquid asphalt. The total acceptable minus 200 mesh material in the end product is approximately 6%.

From a practical point of view, Brox has generally tried to limit the minus 200 mesh material in soil to about 20% of the sample. Table 1 shows the screen analysis of a number of soil samples we have received; very few exceeded our

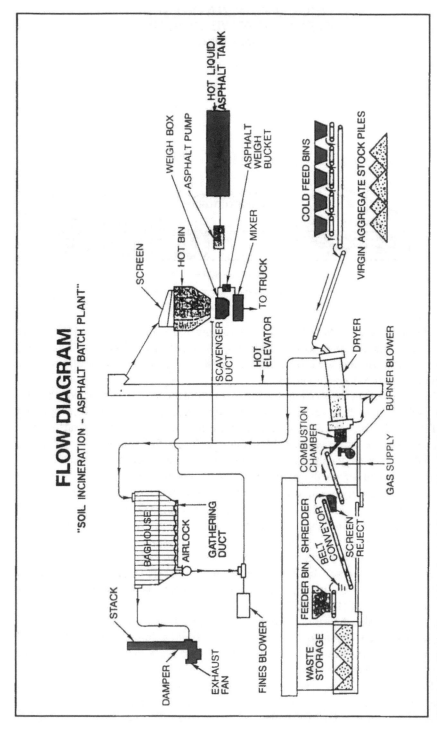

FLOW DIAGRAM
"SOIL INCINERATION - ASPHALT BATCH PLANT"

Figure 3.   Flow diagram showing solution provided by Astec Industries.

OIL CONTAMINATED SOIL
IN INCINERATION TYPE UNIT

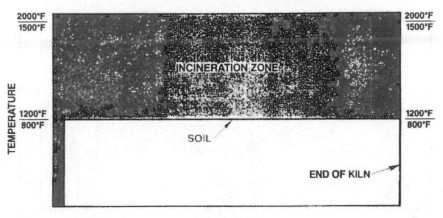

DRYER LENGTH

Figure 4. Comparison of ceramic cylinder length and temperature.

Table 1. Screen Analysis Oil Contaminated Soil Samples.

	NAT. SAND	C.S.-121	% Passing C.S.-42	C.S.-58	C.S.-90	C.S.-75
3/4	100	100	84	92	97	100
3/8	99	100	80	89	96	89
#4	95	97	73	85	94	82
#50	43	46	28	58	70	45
#100	13	12	13	42	45	22
#200	3	3	7	15	18	12

established limit. Frankly, Brox is having more of a problem with oversize material. There appears to be a substantial quantity of contaminated soil which contains large diameter particles.

Our current policy is to reject this material for direct processing. However, considerable effort has been expended to develop new techniques for handling these materials. An obvious solution is to screen all materials on site and ship the fine material to an asphalt plant and the coarse material to a landfill. The approach Brox chose was to develop a new process which would have a wider tolerance for the quality and particle size of the contaminated soil and yet use essentially the same equipment. Figure 5 illustrates the process flow of our new procedure. Virgin decontaminated aggregate is charged to a cold feed bin and metered into the dryer via a belt feeder-conveyor. The contaminated soil is fed, as in the earlier process, by means of a bin, belt feeder, and conveyor to the 5-ft diameter ceramic cylinder. The decontaminated soil then combines with the

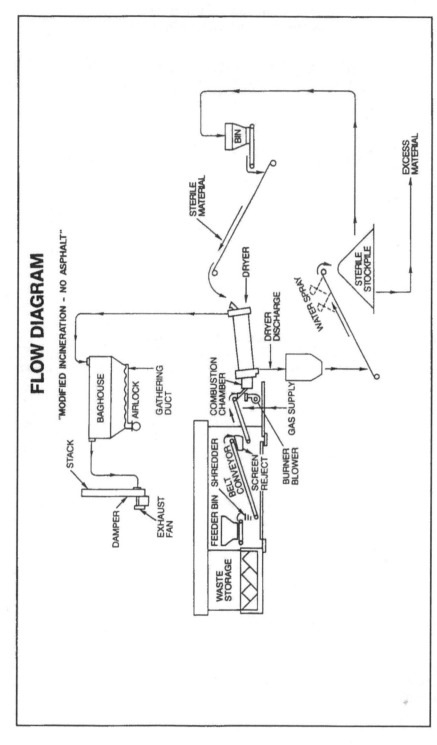

Figure 5. Flow diagram of the new Brox process.

virgin aggregate and the combined streams leave the system by means of a metal conveyor belt. The hot dry soil/aggregate is cooled with a water spray and deposited on a storage pile for recycling to the process. At regular intervals, decontaminated soil is removed from the process and sold as base material. The process offers considerable advantages over recycling in Hot Mix Asphalt:

1. A wide tolerance for gradation from a maximum of 1½ " in diameter to virtually unlimited amounts below 200 mesh.
2. Soil containing silt or clay are acceptable.
3. The ratio of soil to virgin aggregate can be increased and is limited only by the incineration capacity of the equipment.
4. The process can be run year-round in the cold and rain, in contrast to Hot Mix Asphalt, which is run only when the weather is favorable.

DEQE APPLICATION PROCESS

Solving the technological challenge is only part of the task confronting anyone seeking to enter the field of environmental engineering. Even though contaminated soil is not hazardous waste per se, it still falls under the legal mandate of the Massachusetts Department of Environmental Quality Engineering (DEQE), and thus demands a comprehensive application. Brox's application to the Department included the physical and chemical parameters of our proposed process: (1) Brox would not accept material with free liquids; (2) particle size must be less than ¾ " diameter; (3) flashpoint must exceed 140°F; (4) Brox's limits on heavy metals and chlorides were such that petroleum in the soil could be categorized as virgin materials.

The listing that follows will give an accurate impression of the information covered.

A. Project Details

- waste description including chemical and physical characterization
- quality control
- process description
- facility description
- environmental controls
- process or operation controls
- site description—elevation, 100+500 year flood lines, borings to demonstrate stability of base under facility.

B. Safety Procedures

- site security
- training

- containment structure
- emergency response
- closure procedures

C. Record Keeping and Reporting

- manifest
- receiving record
- bill of lading
- quality control
- production
- inventory
- safety and environmental inspection

D. Inspection and Maintenance

After months of negotiations and discussions, Brox finally received the permit on December 1, 1987, allowing the company to process the following:

a. soil that is contaminated with virgin oil, kerosene, or gasoline
b. soil that meets a specific chemical analysis and physical properties characterized as virgin hydrocarbon
c. soil that contains no more than 3% oil
d. soil processed in a required ratio of 95 virgin/5 contaminated soil
e. 95% destruction of hydrocarbon must be achieved.

In addition, we must meet all recordkeeping and reporting restrictions and demonstrate that we have not increased emissions from our plant. Demonstrating that we have achieved 95% destruction with no increase in emissions is both a procedural and an analytical challenge. From December 1, 1987, until recently, we have been working with DEQE to establish a procedure which is acceptable. It was agreed that the procedure involved making a material balance around the incineration system using pure chemicals such as benzene, toluene, xylene, and ethylbenzene in a sand base as a controlled substitute for oil contaminated soil. Three points in the system were picked for analysis: incoming soil (sand), soil stripped of contaminant, and stack. It was proposed that our analytical problem would be more reproducible and controlled if we used pure chemicals instead of oil.

Our analytical work is continuing and Table 2 shows the results of three runs we have made. In all of these runs, the stack opacity was below 10 and the solid samples of stripped soil contained less than 1.0 ppm of oil. The Hot Mix Asphalt made during these runs was of excellent quality, meeting all state specifications.

We recently ran a test burn using toluene and xylene in sand as simulated contaminated soil. The test was directed to making a material balance around the

Table 2.   Brox Paving Company Incineration Runs With Oil Saturated Sand.

Run#	Oil	H20%	Feed Rate	Inc. RPM	Virgin Agg. Rate	Opacity Stack	Ratio Vir./ Recycle
1-Orig.	1.1	4.3	8T/HR	3.5	160T/HR	<10	95/5
1-Ceramic	N.D.						
1-Hot bin	N.D.						
2-Orig.	1.5	4.1	9T/HR	5	170T/HR	<10	94/6
2-Ceramic	N.D.						
2-Hot bin	N.D.						
3-Orig.	1.3	5.0	4T/HR	5	150T/HR	<10	97.3/2.7
3-Ceramic	N.D.						
3-Hot bin	N.D.						

Table 3.   Analytical Results Xylene/Toluene Blend.

CONTAMINATED AGGREGATE—3% ORGANIC—50/50 BLEND OF
XYLENE AND TOLUENE IN NATURAL SAND
TOTAL SAMPLE—9 TONS
MOISTURE CONTENT—3.3%

DURATION OF RUN—1 HR.

TEMPERATURE—50°F

Sample Location	Xylene/Toluene
START	30,000 PPM
HOPPER	4,000 PPM
CONVEYOR	1,500 PPM
CYLINDER EXIT	0.9 PPM
ELEVATOR	.017 PPM
LOAD OUT (SAND BIN)	<.010 PPM
STACK (GAS CHROMATOGRAPH)	<0.200 PPM

% DESTRUCTION—99.25%

system, and resulted in 99.25% destruction of hydrocarbon fed to the incinerator (see Table 3).

In summary, we at Brox believe that the processing of soil contaminated with hydrocarbon in a modified Hot Mix Asphalt facility is a practical and economic technique. We believe the current process offers advantages to both the generator and the Hot Mix Asphalt producer, and provides the state DEQE with another disposal option. A new modified process offers the potential of lower cost, 12 months of operation, and a wider tolerance of soil quality.

CHAPTER 23

Removing Petroleum Products from Soils with Ozone, Ultraviolet, Ultrasonics, and Ultrapure Water

Lucas Boeve

For many years, chemical plants, refineries, oil drillers, fuel depots, service stations, and other petroleum handlers have been careless in their methods of waste disposal. Petroleum products have been dumped into the ground and water supplies, without any regard for environmental pollution. Most polluters thought the soil would absorb the petroleum products or the groundwater would wash them away. However, we all now know that most petroleum products do not break down in soil or water, and we are now faced with a huge pollution problem. This country, as well as all other industrialized nations, must find effective means to remove petroleum products from soil and water supplies, and keep the contaminants from reaching the food chain.

Traditional remedial programs to remove petroleum products from soil include incineration, landfilling, biological activity, and chemical treatment. Incineration is very expensive, can produce dioxins, and still results in a fly ash to dispose of. Landfilling only moves the problem from one location to another, and does not remove the contaminant. Biological activity is very sensitive to numerous environmental factors, slow to react, and produces unpredictable results. The use of chemicals often creates unwanted by-products, leaves residues, and sometimes is more toxic or hazardous than the target contaminant.

During the past several years, Excalibur Enterprises, Inc., has developed an effective and economical system to purify air, water, and inert materials, including soils and sediments. The OUUU process utilizes ozone, ultraviolet, ultrasonics, and ultrapure water to remove and destroy contaminants. The OUUU process is a two-stage system, where the contaminants are first removed from the soils with ultrapure water as the solvent, and then the aqueous solution is treated to convert the contaminants into carbon dioxide, water, and harmless salts. The process requires the use of ozone, ultraviolet, ultrasonics, and ultrapure water. The OUUU process requires no incineration, landfilling, or chemicals, is portable, can be operated onsite, and converts the contaminants into carbon dioxide, water, and harmless salts.

Contaminated soils are first put in a slurry (30–40% solids), and rinsed with ultrapure water. Ultrapure water is known as the "universal solvent," and for many years has been utilized by the pharmaceutical and electronics industries for rinsing purposes, to remove organics and other contaminants. Compared to chemical solvents, ultrapure water has much less molecular weight, can more readily penetrate inert materials, leaves no residues, is not a chemical compound, forms no by-products, and is not carcinogenic.

The problem with solvents is that they "leach out" petroleum products from inert materials and transfer the contaminants into the liquid solution, which then has to be properly treated. Chemical solvents have to be distilled, and the contaminants are transferred to the distillate, which can then be considered hazardous. The OUUU process, on the other hand, requires no distillation and removes the contaminants from the ultrapure water with ozone, ultraviolet, and ultrasonics.

After the water is separated from the solids through a centrifuge or cyclone, the aqueous solution is first treated with ozone, to break down hydrocarbon double bonds. Ozone (O_3), as an active form of oxygen (O_2), is a very strong oxidizer, and is produced by passing oxygen through a corona arc or a source of ultraviolet energy. Ozone, however, is a very unstable gas, must be generated onsite, and quickly reverts to oxygen, which is why ozone cannot be bottled or stored like other chemicals. The sweet smell in the air after a thunderstorm is actually ozone, which was generated by the lightning passing through the air and converting some of the oxygen into ozone.

The ozone is applied to the aqueous solution through a special reactor, which also contains an ultraviolet source and an ultrasonics transducer, since the interaction of ultraviolet and ultrasonics, used as catalysts, greatly enhances the oxidative capacity of ozone. Ozone in water decomposes to oxygen, free radicals, and hydrogen peroxide. Ozone and free radicals are particularly reactive with aliphatic carbon/carbon double bonds, and are capable of aromatic ring cleavage. Ozone alone seldom leads to complete oxidation or mineralization to carbon dioxide and water, but leads to partial oxidation products.

Ultraviolet light provides photons which enhance radical production. Ultrasonic waves produce cavitation, which also generates free radicals, and is capable of breaking chemical bonds. The use of ultraviolet and ultrasonics as catalysts with ozone can reduce petroleum products three times more rapidly than the use of ozone alone. It should also be noted that microscopic turbulence of the cavitational energy produced through ultrasonics is the reason for increased mass transfer of ozone into solution. Ultrasonics reduces the thickness of the liquid film, enhances gas transfer, and reduces bubble coalescence, which increases the interfacial area for gas transfer.

It should also be noted that the ultrasonics play an additional role by maintaining clean surfaces on the ultraviolet crystals. When dealing with the average wastestreams that contain petroleum products, the opacity of the fluid blocks the passage of the ultraviolet light, and the crystals become coated. However, with ultrasonics, the opacity is reduced and the ultraviolet crystals are kept clean, which allows the ultraviolet light to effectively penetrate the liquid media.

After the contaminated soils have been rinsed, and the contaminants destroyed with the OUUU process, the soil can then be returned to the original location, or recycled.

The OUUU process has been used to remove a number of organic contaminants including polychlorinated biphenols (PCBs), pentachlorophenols (PCPs), 2,4,5-trichlorophenols (TCPs), benzene, and cyanide from water and inert materials, such as soil, wood, paper, metals, and resins. The size of the systems utilized ranged from one cubic foot per hour of inert material with a water flow rate of one gallon per minute, to 27 cubic yards per hour (Figure 1) with a water flow rate of fifty gallons per minute. Figure 2 shows a typical system for treating 10 cubic feet per hour.

Figure 1.   System to remove contaminants from soil at the rate of 27 cubic yards per hour.

Figure 2.   System to remove contaminants from soil at the rate of 10 cubic feet per hour.

The most important feature of the OUUU process is that all decontamination is accomplished on the site and no landfilling, incineration, or chemicals are required and soils are suitable for reuse. No unwanted by-products are ever released into the environment, and operating costs are very low, compared to traditional methods. The U.S. Environmental Protection Agency (EPA) has already demonstrated that the OUUU process is the best practical method to remove and destroy PCBs found in soils and sediments (Contract #68-02-3992, dated September, 1987, and released in February, 1988).

CHAPTER 24

Cleanup of a Gasoline Contaminated Site Using Vacuum Extraction Technology

James J. Malot

INTRODUCTION

Cleanup of contaminated soils is often the most important aspect of rapid, cost-effective remediation at sites contaminated with petroleum hydrocarbons. A leaky tank or pipeline, surface spill, or other release can quickly contaminate a large volume of subsurface soil and rock. Frequently, underground pollution problems go undetected for years. Once contaminants seep underground, they spread continually; resulting groundwater and soil contamination is responsible for numerous abandoned water supply wells, inoperable utility systems, and deadly explosions.

An increasing awareness of the environmental impact of soils contaminated with hydrocarbons is apparent in various state and local regulations and in the Environmental Protection Agency (EPA). For example, the EPA is implementing an Underground Storage Tank (UST) program to control the estimated 100,000 to 400,000 leaking underground storage systems (EPA 625/6-87/015, 987). Several state agencies have implemented leak detection programs. The Florida Department of Environmental Regulation (FDER) has further taken responsibility to manage and finance assessment and cleanup efforts of more than 2000 petroleum contamination sites under their Early Detection Incentive program. Other states are following Florida's lead with regulations requiring that contaminated soils and groundwater be cleaned up in order to protect groundwater resources and the environment.

Most spills and leaks leave behind more residual hydrocarbons in the soils than ever enter the aquifer or that may be recovered as free product. Free product floating on the aquifer spreads out and often moves slowly in the direction of groundwater flow. However, as the water table fluctuates, free product rises and falls with it, and a great volume of soil becomes filled with immobile, residual hydrocarbons. Typically, a vertical "smear zone" of residual hydrocarbon is apparent within a few feet of the water table. This smear zone expands vertically with seasonal changes in the water table and contains the highest contaminant levels beyond the source area itself. As free product migrates, the trail of residual hydrocarbons that are left behind are a greater and more lasting threat to groundwater quality than the floating free product itself.

Even if the hydrocarbon leak is small and free product never is detected on the water table, residual hydrocarbons within the vadose zone will leach out, continuing to contaminate the aquifer for years. As a result, cleanup of residual hydrocarbons is often the most important aspect of subsurface remediation. Stringent cleanup goals for groundwater quality cannot be achieved without direct remediation of all of the residual hydrocarbons within the soils.

The high incidence of petroleum contamination from leaking underground storage systems and spills requires that efficient and cost-effective solutions be found to achieve cleanup of these sites. Experience with the remediation of petroleum hydrocarbons contamination has shown that where soils are contaminated and allowed to remain in place untreated, the cleanup process is costly, lengthy, and may never reduce contaminant levels to where the water resource can again be used for a potable supply.

Each contamination site must be evaluated with respect to potential exposure pathways, time and cost constraints, and cleanup objectives imposed by local regulators in order to optimize treatment strategies for soil contamination. Advantages and disadvantages of some of the more common soil remediation techniques are summarized here.

Vacuum extraction is an in situ cleanup process that removes petroleum hydrocarbons directly from the soils. Successful cleanup of residual hydrocarbons and soils using vacuum extraction has been demonstrated at numerous sites. This chapter details conditions and results at a demonstration site in Florida. Results of the in situ soil treatment pilot study are presented with respect to stringent soil and groundwater cleanup objectives that are enforced in Florida.

DELINEATION OF SOIL CONTAMINATION

Prior to any remediation, except emergency response to eliminate an active source, an assessment of the subsurface conditions is required. Subsurface hydrocarbons disperse into four distinct plume types or phases:

1. free product floating on the water table
2. dissolved hydrocarbons within the aquifer
3. residual hydrocarbons absorbed in the soil matrix
4. vapor plume emanating from each of the others

The initial focus of the subsurface investigation is often the delineation of the free product plume. However, if groundwater quality is of potential concern, delineation of the dissolved hydrocarbons and the residual hydrocarbon is necessary. Groundwater monitoring wells are used to determine free product or groundwater quality. Delineation of the residual hydrocarbon plume is often neglected, although quantification of the extent and magnitude of contamination in the soils is critical to effective remediation of the groundwater at the site.

The fourth type of plume, vapors, emanates from the other three sources of hydrocarbons. Due to the volatile nature of most hydrocarbons, vapors are generated from dissolved components within the aquifer, free product at the surface, and residual hydrocarbons above the groundwater. Driven by diffusion in response to concentration gradients, gravity, and advection caused by natural changes in subsurface pore pressures, vapors move away from source areas.

A soil vapor survey can be employed at the beginning of the investigation to provide useful data at low cost. Establishing the lateral concentration profile of hydrocarbon vapors extracted from a shallow probe can be useful in identifying source areas and the extent of free product plume. In some cases where stratigraphic interferences are minimal, soil vapor surveys can provide an indication of groundwater quality. Utilization of these data are most useful prior to installation of groundwater/product monitoring wells.

Quantification of the vertical extent and magnitude of soil contamination is essential to properly evaluate remedial activities. Vertical and areal profiles of soil contamination must be known before remedial options are evaluated. Onsite analysis of soils by headspace techniques is a rapid and effective method to quantify the three dimensional distribution of VOCs in subsoils.

SOIL REMEDIATION METHODS

Numerous methods for treating soils contaminated with hydrocarbons are available. Each method has limitations with respect to the contaminant, soil types, depth, time frame, and the residual concentration that can be achieved. Unfortunately, there is no single method that is applicable for all chemicals at all sites. For sites with a wide variety of chemicals present, two or more technologies may be required to adequately treat the soil. The following are some of the more common approaches to remediate soils:

• Excavation and Treatment
• Pump and Treat

- Soil Flushing
- Bioremediation
- Soil Venting
- Vacuum Extraction

Perhaps the most commonly used method to remediate soil contamination is excavation and disposal, although transporting the removed soils to a suitable disposal location is expensive and merely transfers the contamination, rather than eliminating it. Removal is also disruptive and may be impractical due to physical conditions at the site, such as the locations of buildings, roadways or utilities and the proximity to residents. Studies have shown that the process of soil excavation can release 60% to 90% of the volatile contaminants into the atmosphere.[1]

Options for excavation and treatment include offsite disposal, aeration, incineration, chemical treatment, and heap vacuum extraction. Aeration, spreading of soils, and allowing volatiles to disperse naturally can be a low-cost alternative if the transfer of contaminants to the atmosphere is of little concern. Incineration may be applicable to heavy hydrocarbons, pesticides, and PCBs with proper controls on emissions and disposal of residuals. Numerous methods for chemical treatment or stabilization of soils are available, depending on the type of contaminants. "Ex-situ" or "heap" vacuum extraction is another alternative for sites where excavation is necessary whereby volatiles are removed by a horizontal vacuum system underlying a pile of soil.

Pumping and treating groundwater is usually conducted at all sites where groundwater quality is significantly impacted. However, pump and treat systems do not treat the soils directly. Initial capital costs may be low to moderate, but high operations and maintenance costs are required as contaminants continue to leach from soils. Pumping and treating groundwater alone is very slow, taking decades to restore aquifers to common cleanup goals. Where free product is present, pump and treat systems can cause groundwater to become worse as residual hydrocarbons are spread throughout the cone of depression of a pumping well, causing groundwater concentrations to rise significantly whenever pumps are turned off.

Soil flushing may enhance contaminant recovery from pump and treat systems if contaminants are water soluble and contained in highly permeable, homogenous soils. Difficulties result with most types of hydrocarbon contamination since they are generally immiscible with water, and the surfactants or other additives may be difficult to remove from extracted groundwater. Uncertainties of flow conditions within the unsaturated zone due to stratified soils, heterogeneities, and the presence of silts and clays can severely limit the effectiveness of flushing systems.

Bioreclamation can provide in situ destruction of contaminants. However, difficulties with flow control in unsaturated zone, similar to flushing, are often encountered. Intermediate by-products may be more mobile and hazardous than the original components. Clogging of nutrient injection systems with biogrowth can limit the effectiveness of bioremediation systems. Successful applications are limited to high permeability soils, and high operations and maintenance costs are often incurred.

"Soil venting" is a term derived from passive venting of landfill gases, although there are references in the literature to "soil venting" systems that actually use vacuum extraction, causing substantial confusion. As it is referred to here, "soil venting" applies to low vacuum, fan-type systems that are commonly employed in wells and trenches. Distinguishing features with vacuum extraction are that soil venting systems typically induce a small radius of influence and produce low recovery rates. As a result, soil venting systems are slow and are not effective in low permeability soils, in contrast to vacuum extraction.

Advantages of vacuum extraction systems are that cleanup of contaminated soil and groundwater is in situ, rapid, and low cost. Vacuum extraction systems are not limited by depth to groundwater, with successful application demonstrated at sites with groundwater as deep as 300 feet and as shallow as 3 inches. In shallow groundwater applications "dual extraction," simultaneous extraction of groundwater and vapors from the same well, is particularly effective.

Vacuum extraction has been demonstrated effective in virtually all hydrogeologic settings: clays, silts, sands, gravel, alluvium, glacial till, and fractured rock. Vacuum extraction eliminates residual hydrocarbon within the unsaturated zone. Treatment of vapors is required at most sites, although risk assessment often indicates that dispersion is sufficient to render groundwater contaminants harmless. However, carbon adsorption and catalytic oxidation are effective if air quality restrictions apply.

For multimedia contamination sites, those sites with numerous types of compounds (i.e., VOCs, PCBs, pesticides, and metals) a phased approach is often required. In these cases, it is prudent to remove VOCs first, using vacuum extraction so that other technologies can then be applied more cost-effectively and safely. For example, with chemical treatment or incineration of soil, which require excavation, the health risk of excavation is minimized if the majority of VOCs are removed first, in situ, by vacuum extraction.

CASE HISTORY

A gasoline contamination site in Belleview, Florida came to the attention of the Florida Department of Environmental Regulation (FDER) in late 1982 when residents complained that water from the municipal well field had an objectionable taste and odor. Analyses of water samples from the municipal wells detected the total concentrations of benzene, ethyl benzene, toluene, and xylenes (BTX) ranging from 15 to 470 ppb.

Based on the compounds detected and the pattern of the chromatograms, FDER concluded that the contamination was a partially weathered gasoline. Since the wells all contained benzene in excess of Florida's Maximum Contaminant Level

(MCL) of 1 μg/L for the compound, the three public supply wells had to be abandoned. A subsequent inspection of the inventory records of gasoline stations in the area revealed that between October 1979 and March 1980, about 10,000 gallons of unleaded gasoline had leaked from a service station located approximately 600 ft upgradient from the city wellfield.

Confirmation of the hydrocarbon source at the station was made during an initial vapor survey. Inspection of the station by the State Fire Marshal found explosimeter readings of more than 100% under the concrete apron on the east side of the station. Based on these findings, four monitor wells were installed in the area. The well, MW-2, located nearest to and downgradient of the station, contained 1/4 inch of gasoline-free product. An analysis of the groundwater from well MW-2 found 33,090 μg/L of benzene, ethyl benzene, toluene, and total xylenes (BTX). Because groundwater quality was of primary concern, delineation and cleanup of the residual hydrocarbon was critical. After assessing remedial alternatives for the site, vacuum extraction was selected for demonstration, for without source control of the residual hydrocarbons, the pumping and treating of groundwater would be futile.

SUBSURFACE CONDITIONS

The gasoline contamination site is located in Belleview, Florida, as shown on Figure 1. The region is noted for its karst topography.

The surficial geology at the site consists of four generalized units. A surficial unit consisting of Pleistocene clayey sands is underlain by an intermediate confining unit consisting of Miocene clays. Clays grade to silty sands with depth. The underlying Eocene and older limestones which compose the Floridan aquifer, is the sole source of groundwater.

The formation is compromised of a highly karstic limestone in which dissolution has formed an extremely porous and permeable aquifer system. Monitor wells drilled into the limestone have encountered voids and large sand-filled cavities. The contaminated Belleview municipal wells are developed into the Floridan aquifer. Water supply wells in the area are generally less than 200 ft deep.

The surficial sands contain very little water. Where a shallow water table is encountered, it exists as a perched zone of limited areal extent situated above relatively impermeable clays at depths of about 20 ft. The clays beneath the site are interbedded with sand, causing them to exhibit relatively high permeabilities locally. Above this plastic clay, perched groundwater is observed; however, it appears to be variable in thickness and more permeable beyond the site boundaries. Beneath the clay a relatively dry unsaturated zone exists, above variable weathered limestone. Groundwater in the limestone aquifer is unconfined and flows rapidly in cavities and solution channels.

Regional potentiometric surface maps of the upper Floridan aquifer indicate that the downgradient direction of groundwater flow in the aquifer is northwest,

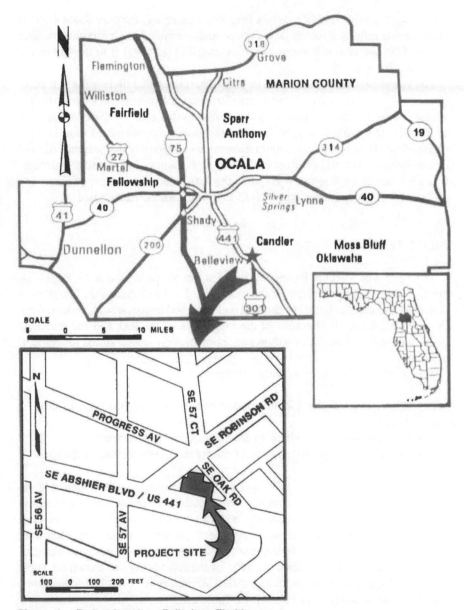

Figure 1. Project location—Belleview, Florida.

directly from the gas station site toward the municipal well field. Water level measurements from the monitor wells at the site have shown the downgradient direction to vary from northwest to northeast. Water level contours prior to remediation (April 1987) indicate that the groundwater gradient was toward the northwest. The Belleview municipal wells have not been operational, since they were taken out of service in 1982. When they were pumping, they no doubt exerted considerable influence on the flow regime and controlled the hydraulic gradient within their radius of influence, which extended at least 600 ft to the site of the gasoline leak.

The assessment work performed in 1984 determined that there were still components of gasoline in the Floridan aquifer downgradient of the station. It also verified the existence of a relatively thick unsaturated zone, at least for the state of Florida, averaging nearly 50 ft. The assessment recommended that the next phase of the site investigation should determine and quantify the presence of, and quantify the amount of, gasoline in the soil at the station.[2] If residual hydrocarbon were still present seven years after the leak, the effectiveness of removing the source of groundwater contamination could then be determined.

PILOT TEST

The Terra Vac Vacuum Extraction Process was demonstrated at the Belleview site to evaluate its effectiveness for the treatment of soil contamination. The process is designed to recover both free product and adsorbed hydrocarbons, in situ, from the contaminated soils. Because of the magnitude of the leak and the thickness of the unsaturated zone, the vacuum extraction process is well suited for remediation at the site.

The objectives of the pilot study were to:

- delineate the extent and magnitude of residual hydrocarbons in the subsoils
- quantify the rate at which hydrocarbons can be extracted
- develop a conceptual design and time frame for the cleanup of the site

Work began in December 1986 when four soil borings were drilled at locations shown in Figure 2. Soil samples were obtained from each boring to assess the subsurface conditions and quantify the hydrocarbons present. Soils were classified by a geologist and analyzed onsite for gasoline components. Benzene, toluene, and xylenes (BTX) were quantified in soil samples by gas chromatography using the headspace method. Results of the soil testing are shown on Table 1. Selected samples were sent to a certified laboratory for correlation purposes. Results of these analysis are shown in Table 2.

Soil samples were obtained from three locations in the tank pit area. A hand auger was used to collect samples for onsite hydrocarbon analysis. Selected samples were sent to the laboratory for analysis.

At each of the test boring locations a vacuum extraction monitoring well was installed. Two wells, VE-1 and VE-2, were installed primarily for vacuum extraction of subsurface hydrocarbons from above and below the confining clays

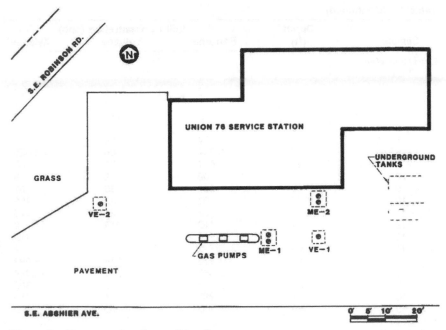

Figure 2. Vacuum extraction well locations.

Table 1. Soil Concentrations By Onsite Headspace Analysis.

Sample	Depth (ft)	Soil Concentration (ppb) Benzene	Toluene	Xylenes
Extraction Well VE-1				
1	1	N/A	1,103	2,486
2	3	232	606	511
3	6	10,900	28,180	16,730
4	10	1,877	4,264	1,586
5	12	29,345	38,518	29,068
6	15	730	1,701	4,597
7	17	526	596	252
8	18	1,160	1,140	1,090
9	21	466	518	686
10	23	270	205	256
11	26	113	92	122
12	28	70	54	114
13	31	28	28	22
14	33	60	47	107
15	35	26	40	65
16	37	43	53	55
17	40	75	103	136
18	44	175	428	558
19	46	166	470	527

Table 1. (Continued)

Sample	Depth (ft)	Benzene	Soil Concentration (ppb) Toluene	Xylenes
Extraction Well VE-2				
1	3	2	4	ND
2	7	4	3	1
4	12	ND	ND	113
3	10	4	2	2
5	15	304	98	72
6	17	2,744	3,266	4,035
7	20	877	121	33
8	22	708	9	9
9	25	230	80	37
10	27	93	33	103
11	35	36	24	25
12	44	ND	ND	2
13	40	14	4	ND
14	46	2	3	2
15	49	1,216	1,335	5,692
16	52	3,367	5,788	15,784
17	54	8	77	360
18	57	ND	21	ND
19	60	ND	ND	ND
Extraction Well ME-1				
1	3	158	35	ND
2	7	1,476	1,657	480
4	12	19,500	24,200	6,980
3	9	2,350	3,397	1,084
5	15	7,324	10,871	3,867
6	17	6,582	11,649	8,111
7	20	439	361	253
8	22	360	175	106
9	25	152	76	118
10	27	190	165	197
11	30	92	125	95
12	32	10	18	0
13	35	11	24	0
14	37	ND	ND	ND
15	40	ND	ND	ND
16	42	ND	22	ND
17	46	2,410	7,170	10,900
Extraction Well ME-2				
1	3	201	53	63
2	7	906	1,867	1,209
3	10	8,873	18,971	17,910
4	12	2,065	4,739	3,089
5	15	2,184	4,549	3,242
6	17	825	1,180	1,100
7	20	155	162	105
8	22	500	1,306	775

Table 1.   (Continued)

Sample	Depth (ft)	Soil Concentration (ppb)		
		Benzene	Toluene	Xylenes
9	25	538	699	591
10	27	229	171	353
11	30	112	45	72
12	32	61	72	78
13	35	176	281	269
14	37	107	213	217
15	40	55	95	140
16	42	15	30	55
18	47	237	1,063	1,499
19	50	925	2,860	1,780

Table 2.   Soil Concentrations By Offsite Laboratory.

Sample	Depth (ft)	Soil Concentration (ppb)		
		Benzene	Toluene	Xylenes
Extraction Well VE-1				
3	6	6,270	97,000	210,00
8	18	157	160	<134
10	23	<103	<103	<103
Extraction Well VE-2				
4	12	<112	<112	<112
17	54	<107	<107	<107
18	57	<115	<115	<115
Extraction Well ME-1				
4	12	12,000	69,000	127,000
8	22	<91	<91	<91
17	46	1,170	21,700	98,600
Extraction Well ME-2				
6	17	765	1.050	760
14	37	<107	<107	<107
19	50	<103	310	233

at 20 ft, respectively. At the other two boring locations, multilevel, dual-purpose wells were installed. These wells were designed for monitoring of subsurface vacuum and extraction of hydrocarbons from three different hydrogeologic zones.

In boring ME-1, three vacuum monitoring wells were installed: ME-1-13, ME-1-35, and ME-1-50, monitoring subsoils at 13, 35, and 50 ft deep, respectively. ME-1-13 and ME-1-50 are also capable of vacuum extraction of hydrocarbons from separate zones, above the clay layer and below, within the limestone formation. Similarly, in the borehole at ME-2, two monitoring/extraction wells, ME-2-16 and ME-2-58, were installed to monitor and extract hydrocarbons from clayey soils about 16 ft deep and from the limestone at 58 ft deep. A profile of these extraction wells and the subsurface stratigraphy is shown in Figure 3.

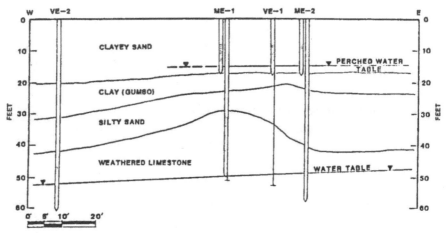

Figure 3. Subsurface profile.

Relative recovery rates of hydrocarbons from pilot test wells ranged from 450 to 2000 lb/day. Six wells (VE-1, VE-2, ME-1-13, ME-1-50, ME-2-16, and ME-2-58) were used simultaneously or individually for extraction of hydrocarbons. Quantification of subsurface vacuum, hydrocarbon recovery rates, and overall hydrocarbon concentrations were monitored on a daily basis.

Following the three-week pilot test, modifications were made to the system to include a vapor/water separator. Operations were continued over the next five months. Independent testing of subsoils was conducted by EPA, including subsurface vapor sampling, soil analysis, and limited operations.[3]

RESULTS AND DISCUSSION

Soil samples were analyzed from each of the four vacuum extraction well locations and three shallow borings in the tank pit area. The extent and magnitude of the subsurface contamination in vacuum extraction locations indicated high levels of residual hydrocarbons. Benzene concentrations observed in subsoils are summarized in Figure 4. Similar trends are apparent for other indicator parameters and total hydrocarbons.

The highest levels of benzene and total hydrocarbons (THC), up to 29,000 ppb benzene and 335,000 ppb THC, were observed beneath the product pipeline, the

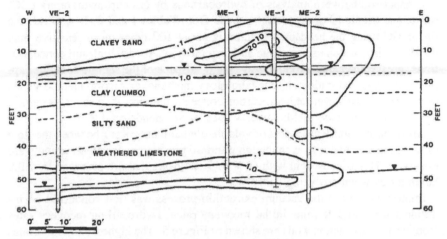

Figure 4. Initial benzene concentrations (ppm) December 1986.

suspected source of the subsurface contamination. The maximum level of benzene observed in the tank pit was 1,200 ppb at a depth of 10 ft. The relative magnitude of these residual concentrations confirmed the major source was at the leaky pipeline.

Concentrations generally decreased with depth and distance from the leaky pipeline until the perched groundwater was encountered just above the clay layer about 17 to 20 ft deep. Above the clay layer, benzene concentrations ranged from 1,100 to 6,600 ppb and extended to VE-2, about 50 ft away from the source. Similarly, total hydrocarbon concentrations were about the same beneath the leaky pipeline as those observed more than 50 ft away at VE-2.

Apparently, the clay layer at 17 to 20 ft has caused significant lateral migration of hydrocarbons from the source area beneath the leaky pipeline. Based on the magnitude of the original leak, an estimated 10,000 gallons, and the free product observed in downgradient groundwater monitoring wells, free product apparently migrated laterally across the top of the clay layer over much of the site where the clay exhibits low permeability. Since the clay layer is discontinuous at the monitoring well locations beyond the site boundaries, significant amounts of residual hydrocarbons remain in the soils above clay as a lingering source of groundwater contamination.

Beneath the clay layer at depths of about 30 ft, contaminant concentrations are about 10 times lower within the silty sand strata and the upper portion of the weathered limestone. However, limestone samples near the water table, about 50 ft deep, exhibited relatively high levels of benzene and total hydrocarbons, up to 3400 and 95,000 ppb, respectively. Even after several years of natural flushing of the weathered limestone by the rapidly moving groundwater flow, a "smear zone" of residual hydrocarbons near the water table still persists as an immediate threat to groundwater quality.

Comparison between analysis of hydrocarbons by gas chromatography (GC) onsite and offsite laboratory can be made from Tables 1 and 2. In general, the onsite GC using the headspace method is about 100 times more sensitive than the offsite lab, which used a purge and trap method. At very high soil concentrations, greater than about 1000 mg/kg, the headspace of the sample container approaches vapor saturation and the laboratory methanol extraction method yields higher results. However, at hydrocarbon concentrations less than about 10 mg/kg, the laboratory method yields lower values or are nondetectable.

Apparently, significant losses of volatile contaminants occur between the time of sampling and analysis, even though standard precautions for preservation were observed. This is consistent with 60% to 90% losses that are observed by EPA laboratories when analyzing soils.[3]

Implementation of the vacuum extraction process was first conducted on individual wells to determine initial recovery rates. Hydrocarbon recovery rates from vacuum extraction wells are shown in Figure 5. The highest extraction rates were observed in wells corresponding to the highest soil concentrations. This suggests a correlation between soil concentrations and vacuum extraction rates. Although lower flow rates were observed in the soils above the clay layer, the relatively high concentrations from these wells yielded recovery rates between 300 and 950 lb/day. Conversely, the concentration of hydrocarbon vapors in the limestone was relatively low. However, large vapor flow rates obtained from VE-2 and ME-1-50 yielded recovery rates of 1650 and 2000 lb/day, respectively.

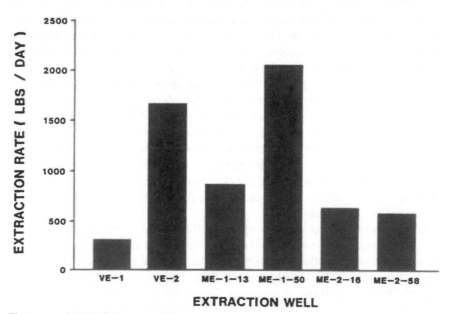

Figure 5.   Initial relative extraction rates.

A high recovery rate from just above the water table suggests that pockets of liquid product were hydraulically trapped within the limestone, yet effectively recovered by the vacuum extraction process.

About two weeks after the start of the vacuum extraction pilot test, recovery rates increased dramatically in response to another leak in the pipeline. Recovery rates of up to 1350 lb/day were observed from ME-1-13. After the leak was repaired it took about one month of vacuum extraction to reduced vapor concentrations to levels observed prior to the pipeline leak.

During the pilot test and subsequent vacuum extraction operations, about 22,000 lb of hydrocarbons were extracted from the subsoils at the site, corresponding to approximately 2750 gallons. As the subsoils were cleaned up by the vacuum process the extracted vapor concentration and the hydrocarbon recovery rates declined, with time.

Figure 6 illustrates hydrocarbon extraction rates at the end of the demonstration project. The reduction in extraction rates is due to the decline in concentrations that was observed at the site as the cleanup progressed. Other vacuum extraction wells indicated similar results. In addition, significant reductions in dissolved hydrocarbons (BTX) in downgradient monitoring wells have been observed as a result of the vacuum extraction process.

Soil sampling and soil gas sampling by EPA after the Terra Vac cleanup demonstrated significant reductions in soil concentrations of benzene, toluene, and total

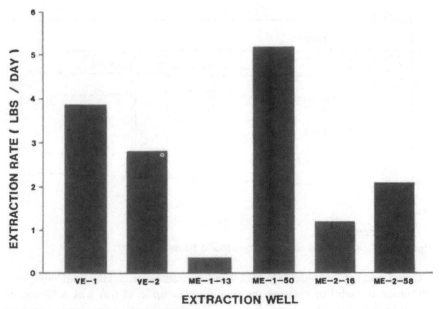

Figure 6. Final relative extraction rates.

xylenes. Results of soil borings taken next to extraction wells VE-1, VE-2, and ME-2 are shown in Table 3.

A benzene soil concentration profile measured after the cleanup period is illustrated in Figure 7, based on EPA sampling and analysis. Comparison with Figure 4 demonstrates the high degree of cleanup achieved in only six months of operations. Similarly, soil gas measurements at the end of the Terra Vac operations indicated low levels, nondetectable to 0.64 ppm, in soils throughout the gas station. After a short test period conducted by EPA, a subsequent soil gas survey indicated all samples were nondetectable for benzene, toluene, ethyl benzene, and xylenes.

Table 3. Results of Soil Borings Next to Three Extraction Wells Following Cleanup.

Well	Depth (ft)	Soil Concentration (ppm)		
		Benzene	Toluene	Xylenes
VE-1	5.5	ND	ND	ND
	9.5	ND	ND	ND
	12.0	0.13	0.25	0.48
VE-2	4.0	ND	ND	ND
	9.0	ND	ND	ND
	12.0	ND	ND	5.6
	19.0	0.34	0.14	0.20
ME-2	5.0	ND	ND	0.21
	9.5	ND	ND	ND

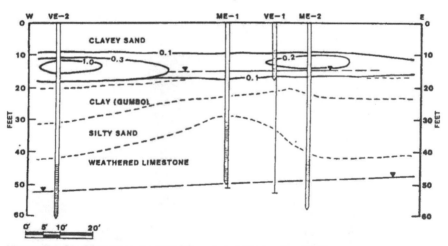

Figure 7.   Final benzene concentrations (ppm) September 1987.

Comparison of the initial extracted vapor concentration and the final vapor concentration is useful to evaluate the relative cleanup level that was achieved by the vacuum extraction process. Figure 8 indicates the relative cleanup achieved at VE-1 during the project. Similar results were observed in the other vacuum extraction wells. Table 4 compares the initial vapor concentrations with the final concentrations observed for each wellhead.

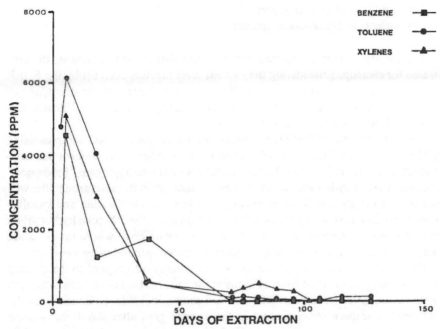

Figure 8. VE-1 wellhead vapor concentrations.

Table 4. Comparison of Initial and Final Vapor Concentrations at Wellheads.

| | Total Hydrocarbon Concentration at Wellhead (ppm) | | | | | |
	VE-1	VE-2	ME-1-13	ME-1-50	ME-2-16	ME-2-58
Initial	13,900	20,500	16,100	30,700	78,800	27,600
After EPA Testing	145	3	52	37	42	76
Relative Cleanup Achieved	99.0%	99.99%	99.7%	99.9%	99.95%	99.8%

The relative decline in extracted vapor concentrations is expected to be proportional to an aggregate soil concentration within the radius of influence of each vacuum extraction well. Essentially, the extracted vapors are in equilibrium with dissolved and adsorbed hydrocarbons within the soil matrix around the well. Overall, the relative degree of cleanup achieved by the vacuum extraction process during the demonstration period, based on initial and final wellhead concentrations of total hydrocarbons, ranged from 95.9% to 99.7%. Following the EPA test period, further reductions in concentrations were observed, ranging from 99.0% to 99.99% reductions.

CLEANUP CRITERIA

Two approaches for evaluating cleanup effectiveness were considered for this and other sites:

- residual soil concentration
- impact on groundwater quality

Extrapolation of the concentration versus time data is used to estimate the time frame for cleanup. Considering the current level that has been set by the FDER to define excess soil contamination, 500 ppm of hydrocarbons in headspace of soils, the relative time frame to reach this cleanup goal using the Vacuum Extraction Process can be evaluated.

Since the criteria of 500 ppm is based on a headspace of soil samples using a Flame Ionization Detector (FID) calibrated for methane, the concentration extracted from the wellhead may be considered essentially an aggregate "headspace" concentration of hydrocarbons (calibrated to gasoline) in the soils around the well screen. Assuming any disparity between methane standardization and gasoline standardization was negligible or at least quanitifiable, the 500 ppm hydrocarbon response would represent the upper limit of a cleanup goal, where lower values may be required at certain sites in order to protect groundwater resources.

Using this as a basis to evaluate the site conditions with regard to the cleanup goal of 500 ppm, the total hydrocarbons in the extracted vapor in each extraction well were analyzed using an FID, and are shown in Table 4. Considering the 500 ppm headspace concentration as the cleanup goal after the demonstration period, the deep wells VE-2 and ME-1-50 were far below the criteria, while ME-2-58 is slightly under the limit. Similarly, the shallow extraction wells VE-1 and ME-2 are just above the goal, while ME-1-13 is substantially below the cleanup goal. Following further extraction during EPA testing, all the wells exhibited concentrations well below the cleanup criteria.

An alternative criteria for considering the cleanup goal would be based on a drinking water standard for indicator parameters, such as benzene, toluene, ethyl benzene, and xylenes. Since benzene has the lowest Maximum Contaminant Level (MCL) in Florida (1 μg/L), the goals for cleanup of the vadose zone may be determined using the concentration of benzene in the soil water just above the water table and Henry's law to calculate the vapor concentration.

Based on Henry's law constant for benzene, the soil gas concentration in equilibrium with water at 1 μg/L would be about 0.18 μg/L or 0.05 ppm benzene in air. Assuming the extracted vapors from the wellheads are sufficiently close to equilibrium conditions, an area within the radius of influence of the well may be considered clean if the concentration of extracted benzene vapors is less than 0.05 ppm. This is analogous to a leaching model that is often applied at Superfund sites to determine acceptable MCLs for soils to prevent degradation of groundwater quality.

Applying this concept to evaluate a cleanup criteria for wells extracting benzene near the water table, the data presented in Table 5 would be used to evaluate the success of the cleanup.

This criteria is most applicable to wells extracting benzene from near the water table. Accordingly, VE-2, ME-1-50, and ME-2-58 have all surpassed this cleanup criteria. The increase observed in VE-1 may be related to residual hydrocarbons

Table 5.   Evaluation Criteria for Benzene Cleanup.

| | Benzene Concentration at Wellhead (ppm) | | | | | |
	VE-1	VE-2	ME-1-13	ME-1-50	ME-2-16	ME-2-58
Initial	4,600	490	730	740	910	360
After Demonstration	0.01	0.16	0.81	0.25	6.2	ND
After EPA Testing	0.20	ND	ND	ND	ND	ND

present above the clay layer that could slowly leach into the groundwater. Flow conditions in VE-1 during the EPA test period were also higher. This may have increased the radius of influence, causing recovery rates to increase compared to previous conditions during the demonstration period which indicated nondetectable levels at the end.

CONCLUSIONS

Based on the results of the pilot test, the Terra Vac Vacuum Extraction Process has been effective in cleaning up the residual hydrocarbons beneath the Belleview site. Testing of subsoils indicated elevated levels of gasoline hydrocarbons above a clay layer occurring at depths of 15 to 20 ft that are spread at least 50 ft from the source area of the leaky pipeline. In addition, high levels of BTX were observed near the water table. Rates of extraction of hydrocarbons from the subsoils using the vacuum extraction process reached a maximum of 2000 lb/day. More than 22,000 lb (2750 gal) of gasoline hydrocarbons were recovered during the pilot test and subsequent vacuum extraction operations. As a result of vacuum extraction operations, hydrocarbon concentrations in the groundwater aquifer have been reduced significantly.

The vacuum process operated a total of 150 days over a period of seven months and achieved significant reductions of hydrocarbons within the vadose zone. Soil concentrations were reduced from initial levels of about 30 ppm to levels ranging from less than 1 ppm benzene. Comparison with cleanup objectives for the site indicated that within the radius of influence of the vacuum system, soil decontamination had been achieved.

REFERENCES

1. M. P. Askarinec, L. H. Johnson, and S. K. Holliday. *Proceedings USEPA Symposium on Waste Testing and Quality Assurance, Washington D.C.*, July 1988. Vol II., Analytical Chemistry Division, Oak Ridge National Laboratory.
2. R. H. Patton. Florida Department of Environmental Regulation Memo, Groundwater—City of Belleview/Union 76 Station, Marion County—Volatile Organics Analysis (Gasoline)—File Reference T.15 A.1g.
3. Interim Report for Field Evaluation of Terra Vac Corrective Action Technology, for USEPA Research for Abatement of Leaks from Underground Storage Tanks Containing Hazardous Substances, CDM Federal Programs, December, 1987.

CHAPTER 25

Using Soil Vapor Contaminant Assessment at Hydrocarbon Contaminated Sites

H. James Reisinger, James M. Kerr, Jr., Robert E. Hinchee,
David R. Burris, Russell S. Dykes, and Garey L. Simpson

SOIL VAPOR SURVEY THEORY

When a nonaqueous phase liquid (NAPL) which is less dense than water leaks from an underground storage tank, the liquid migrates vertically downward under the influence of gravitational and capillary forces until it reaches an impermeable barrier or the groundwater table, where it begins to spread laterally as a function of the topographic or hydraulic gradient.

Upon reaching the water table, compounds in the NAPL will either remain in the NAPL phase, volatilize into the soil atmosphere, or dissolve into the groundwater. Most hydrocarbon liquids such as gasoline consist of a complex mixture of many compounds, each with its unique physicochemical properties; therefore, each compound behaves differently in the subsurface. Those with a high aqueous solubility and low vapor pressure (low Henry's law coefficient) have a greater tendency to solubilize in groundwater (Figure 1). Those with a high vapor pressure and low aqueous solubility (high Henry's law coefficient) have a greater tendency to volatilize. The result of these phenomena is that both vapor and dissolved phase plumes will form in the subsurface after a leak, and the chemical characteristics of these plumes will be distinctly different from each other and from the parent NAPL.

303

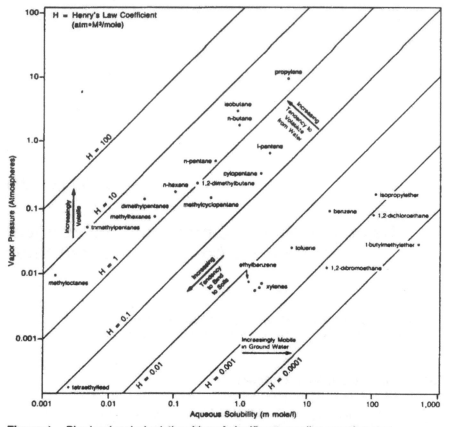

Figure 1. Physicochemical relationships of significant gasoline constituents.[4]

Hydrocarbons not remaining in the parent NAPL will partition into either the groundwater or the soil vapor and will then migrate in response to a variety of interacting forces. Equilibrium solute concentrations resulting from groundwater contact with a liquid hydrocarbon phase can be described by Henrys' law modified for component interactions in the hydrocarbon phase according to the following equation:

$$C_i = X_i \gamma_i C_i{}'$$

where:

C_i is the equilibrium aqueous phase concentration of component i.
X_i is the mole fraction of component i in the hydrocarbon phase.
γ_i is the activity coefficient of component i in the hydrocarbon phase.
$C_i{}'$ is the solubility of component i at the temperature of interest of any specific component.

As can be seen, the groundwater concentration of any given component is a function of both solubility and concentration in the parent NAPL.

As the dissolved contaminants pass through soil or rock, organic and mineral constituents in the medium interact with the contaminants, and some are adsorbed or bound to particulate surfaces. The result is a net retardation in the velocity of movement of those compounds relative to that of the groundwater in which they are dissolved.

The process can be described mathematically as follows beginning with the one-dimensional advective equation of Bear[1]:

$$\frac{\partial C}{\partial t} = D \frac{\partial^2 C}{\partial x^2} - v \frac{\partial C}{\partial x} + p/n \frac{\partial S}{\partial t}$$

where:

C = concentration of compound in water
t = time
D = Dispersion coefficient
v = groundwater velocity
p = density
x = distance of travel
n = porosity
S = concentration of compound adsorbed to soil

The relationship between the adsorbed concentration and the free concentration can be expressed as a linear partition equation:

$$S = K_d C$$

where:

K_d = distribution coefficient

this may be differentiated with respect to time:

$$\frac{\partial S}{\partial t} = K_d \frac{\partial C}{\partial t}$$

and solved with substitution as

$$\frac{\partial C}{\partial t} = D/Rf \frac{\partial^2 C}{\partial x^2} - v/Rf \frac{\partial C}{\partial x}$$

where:

Rf (the retardation factor) = $1 + p/nK_d$

For selected gasoline constituents, the relative mobility may be illustrated by the following continuum:

MOST MOBILE   *LEAST MOBILE*
Benzene > Toluene > Ethylbenzene > Xylenes

In many groundwater systems, the velocity term dominates transport. It has been found for organic compounds in an aqueous system flowing through soil that K_d can generally be related to the soil organic content (expressed as a total organic carbon) and to the compound's hydrophobicity.[2] The relative migratory velocity of a compound in groundwater can be estimated from either the aqueous solubility or the octanol:water partition coefficient. Differential compound solubility and affinity for the soil or rock matrix in which a compound is moving, as described above, explain how spill constituents will often be observed in groundwater in ratios that differ as a function of distance from the spill site.

In addition to dissolution and transport in groundwater after a hydrocarbon spill, volatilization begins. The degree to which a compound volatilizes is primarily a function of its vapor pressure. Compounds with higher vapor pressure are more volatile and will have a greater tendency to migrate in the vadose atmosphere. Equilibrium vapor phase concentrations resulting from a hydrocarbon phase at a given temperature can be described by Raoult's law modified to account for component interactions in the hydrocarbon phase:

$$P_i = X_i \, \gamma_i \, P_i'$$

where:

P_i = the equilibrium partial pressure of component i.
X_i = the mole fraction of component i in the fuel.
γ_i = the activity coefficient of component i in the hydrocarbon phase.
P_i' = the vapor pressure of component i at the temperature of interest.

Activity coefficients for gasoline and other hydrocarbon mixtures and the compounds that make them up tend to be relatively close to unity but can be as high as 1.3 or more for aromatic compounds. In the absence of active vapor movement (advection), the contaminant molecules will migrate along concentration gradients, causing the contaminated vapor to distribute radially around the source. The rate of vapor transport relates to the vadose soil or rock characteristics, the magnitude of the concentration gradient, the soil moisture content, the soil bulk density, the soil tortuosity, physical properties of the compound (i.e. diffusivity and hydrophobicity), and the physicochemical interaction between the compound and the geological matrix. Vapor phase transport can be expressed by the same one-dimensional transport equation utilized for groundwater except that the dispersion term must now incorporate diffusion.

The presence of a significant vapor phase in the vadose above NAPL containing volatile organic compounds and groundwater contaminated with them makes soil vapor surveying a useful assessment tool. As described in the theory outlined above, a detectable vapor plume is usually present in the vadose zone above volatile NAPL. The primary vapor plume will be most concentrated in the immediate vicinity of the source and vapor migration will be manifested as declining concentrations in all directions as a function of distance. However, the distribution may be influenced by volatilization from groundwater and therefore a trend in the direction of groundwater flow may often be observed. Other site-specific features (i.e., areal and vertical soil heterogeneity) may also influence plume configuration. These phenomena form the basis for the soil vapor assessment technique.

SOIL VAPOR SURVEY METHODOLOGY

The first step in conducting a soil vapor survey is to determine the appropriateness of the compounds spilled and the site for assessment by the vapor technique. In determining the degree to which the nature of a subsurface spill lends itself to vapor surveying, it is important to consider the physicochemical characteristics of the compounds spilled, and the concentrations that may be present. The compounds must be adequately volatile and present in adequately high concentrations to be detected. Volatility can be judged by examining the Henry's law coefficient. Concentration can be estimated using Henry's law for aqueous systems and Raoult's law for free-phase and soil systems. The estimated vapor concentration must exceed the analytical detection limit. Appropriateness of the site for vapor surveying rests in the logistical nature of the site, the character of the soil, the depth to bedrock, the water content of the soil, and the soil vertical heterogeneity. If, after making these determinations the site is judged to be a good candidate, target compounds are selected. At this point it is necessary to decide whether an active or passive approach is to be used. Both approaches have merit and applicability. The selection is generally made based on site characteristics and the history of the contamination episode.

The active technique employs installation of probes on a predetermined sampling grid and extraction of soil vapor with a vacuum system. As the samples are withdrawn, negative pressure through the sampling system may be recorded and used in data interpretation and remedial design. Negative pressure is an indication of the effective vapor permeability of the soil and is an important consideration in developing fate assessments and remedial approaches. In addition to areal coverage, the active system can be used to develop a three-dimensional perspective of the contamination on the site. This is accomplished by collecting and analyzing vapor samples at specific depths at a number of locations on the site. Depth profiling is an important aspect of vapor surveying as it provides not only a means of determining the vertical distribution of contamination, but also an indication of vertical heterogeneity in the soil profile. This aspect of the investigation can be of critical importance in interpretation of the data generated and in developing the most effective remedial approach. Reisinger et al. reported that vertical heterogeneity in the soil profile can impact soil vapor data.[3] Highly compacted or dense strata with low effective porosity can slow the migration of vapors, thereby producing negatively biased data, if not considered in the interpretation. Perched water may also be present in this type of scenario, and can slow and effectively preclude vapor migration. The result of this situation is the potential for an active survey conducted at a single shallow depth failing to detect the presence of contamination.

In the active approach, soil vapor samples are generally collected using one of two approaches. The first is to collect the sample and analyze in the field. The samples are generally analyzed in one of two ways. The first is gross contaminant analysis in the field. Gross contaminant field analysis is conducted by a gross total reading instrument such as a photoionization or hydrogen flame ionization meter. This approach measures total hydrocarbons and does not have the capability to determine the concentration of individual compounds. An additional field analytical approach commonly used is analysis with a gas liquid chromatograph, with a selective detector chosen based on the compounds of interest. This approach provides the capability to determine the concentrations of individual hydrocarbons. In the second active approach, samples are collected in gas bottles or plastic bags and transported to a laboratory for analysis by gas chromatography with selective detectors or mass spectrometry.

The passive approach employs deployment of hydrocarbon adsorber traps in flux chambers. The traps often consist of activated carbon; however, other materials can be used as site needs dictate. The chambers are installed in holes in the soil dug to a predetermined depth. The samplers are deployed in a grid pattern in much the same manner as the sampling points used in the active approach. The samplers are left in place to collect vapors for a predetermined length of time. At the end of the exposure period, the samplers are retrieved and transported to a laboratory for analysis. The analytical system commonly used in the passive system involves thermal or solvent desorption of the adsorbed constituents and analysis by an appropriate technique. The commonly used analytical methods are curie point mass spectrometry and gas liquid chromatography using a selective detector.

CASE HISTORY METHODOLOGY

The following methodologies were used to gather soil vapor data at the case history sites.

Soil vapor was sampled on a grid approximately 50 ft by 50 ft. Samples of vadose vapors were collected by drilling small diameter holes through the asphalt pavement at the sites. Small diameter stainless steel tubing was then driven into undisturbed material beneath the pavement. Samples were collected at specific depths through slotted sections of the tubing by applying negative pressure (vacuum) at the distal end of the sampling string. Negative pressures were measured during sampling by an integral vacuum gauge. These negative pressure readings were recorded at each location. The readings were later utilized by the sampling team in determining the permeability of the shallow soil. This analysis of permeability was later utilized in the design of a portion of the remedial measures used at the sites. In addition to the driven sampling points, samples of head space gases were collected from any existing monitoring wells to supplement the data.

Analysis of vapor samples was performed on site through the use of a gas chromatograph. The gasoline constituents benzene and toluene were used as analytical standards. These compounds were selected as they are major gasoline constituents. The concentrations of the aliphatic compounds that eluted prior to benzene were summed and also used in the analysis. These compounds in sum are termed pre-benzenes. Analytical results were thus available to the sampling team within minutes of collection, which allowed the grid to be modified as the sampling progressed during the day.

Case History Site 1

Site 1, situated in the Valley and Ridge Physiographic Province of the southeastern U.S., was the scene of a subsurface fuel spill of approximately 5,000 gallons. Gasoline was discovered in a storm drainage ditch in front of the site, (Figure 2). In response to the situation, the facility inventory records were examined and the system was pressure tested. Results of the test indicated a leak in a distribution line. The leak was repaired, the system was retested, and was restored to service, after successful completion of the tests.

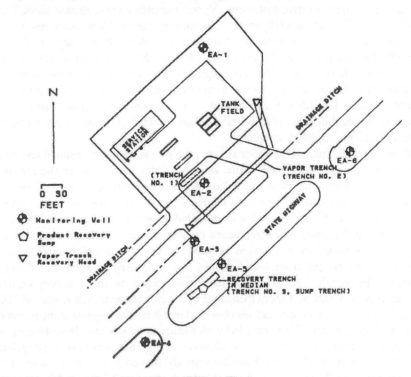

Figure 2. Site configuration at Case History Site 1.

The initial response included placement of hydrocarbon-adsorbent booms in the drainage ditch to intercept free-phase gasoline and preclude contamination of surface waters adjacent to the site. The storm ditch and a small stream south of the site were inspected, and the ditch was found to contain small areas of floating gasoline contamination near the adsorbent booms. However, the ditch was dry for the last 200–300 ft prior to its confluence with the stream. There was no visual evidence that gasoline had entered the stream, nor were there indications of environmental stress. Observations of water levels in the stream and the site topography suggested that groundwater might be encountered approximately 12–15 ft below surface. Because groundwater was expected to be shallow, a trench was installed downgradient and adjacent to the suspected source. The purpose of the trench was to determine the depth to groundwater and the presence and magnitude of gasoline on the shallow water table. The trench was dug and apparently fresh gasoline was encountered at about 10 ft. A volume of 30–50 gallons of gasoline was recovered from the trench. After this initial recovery, no additional gasoline or groundwater were observed in the trench and, therefore, the trench was backfilled.

As a result of observations made during initial trenching, it was determined that a soil vapor survey would be an expedient means of more completely determining the extent and distribution of gasoline on the site. The vapor survey was conducted using the active technique. Vapor samples were collected along lines 50 ft apart on 50-ft centers (Figure 3). The concentrations of low molecular weight gasoline constituents (pre-benzene), benzene, and toluene were determined in the field using a portable gas chromatograph with an SE-30 analytical column and a photoionization detector. The sampling lines were extended until two consecutive uncontaminated data points were observed. Results of the vapor survey are given in Table 1 and Figures 3, 4, and 5. The vapor survey data indicated the presence of gasoline-associated contamination adjacent to the leak source and extending into the highway median.

Based on the results of the vapor survey, the presence of soil residual and free-phase gasoline was suggested in the vicinity of the source and in the median. These observations and those made during trenching suggested that soil venting adjacent to the suspected source and a recovery trench in the median would remediate the problem. A vapor trench was installed adjacent to the source. The decision to use soil venting was made based on the contaminant distribution and the high vapor permeability observed in the vapor survey. The sampler gauge vacuum measurements ranged from 1 to 5 inches Hg, indicating a rather free-flowing system. The trench depth was also determined based on vapor survey vertical profile results. Gasoline vapors measured in the completed trench exceeded 1000 ppm. The high vapor concentrations observed in the median suggested the presence of free-phase gasoline. This, coupled with the apparent shallow depth to groundwater, suggested that a recovery trench and sump system would be appropriate. The system was installed and gasoline initially flowed into it at the rate of 10 gallons per hour. Flow continued for about three weeks and resulted in recovery of about 400 gallons of gasoline.

Table 1.   Results of Soil Vapor Contaminant Assessment Conducted at Site 1.

Location	Sum of Peaks Eluting Prior to Benzene (ppm as Toluene)	Benzene (ppm)	Toluene (ppm)
VP-1	<1	<1	<1
VP-2	511	<1	6
VP-3	1,300	6	62
VP-4	11,300	3	134
VP-5	6,000	<1	150
VP-6	85	1	14
VP-7	<1	<1	<1
VP-8	<1	<1	<1
VP-9	<1	<1	<1
VP-10	<1	<1	<1
VP-11	<1	<1	<1
VP-12	<1	<1	<1
VP-13	<1	<1	<1
VP-14	9,000	2	210
VP-15	75	<1	4
VP-16	64	<1	3
VP-17	<1	<1	<1
VP-18	<1	<1	<1

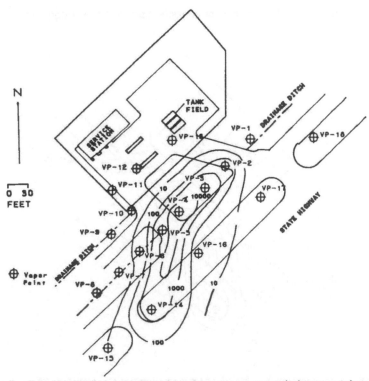

Figure 3.   Isoconcentration contours of pre-benzene compounds (ppm as toluene) at Case History Site 1.

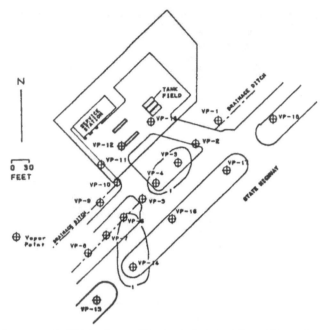

Figure 4.   Isoconcentration contours of benzene (ppm) at Case History Site 1.

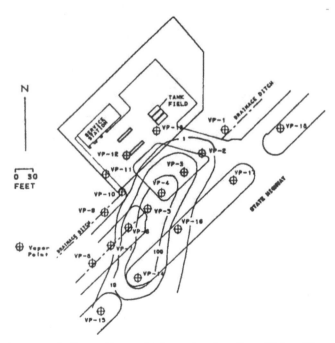

Figure 5.   Isoconcentration contours of toluene (ppm) at Case History Site 1.

Case History Site 2

A second case study in the use of soil vapor techniques to locate and design investigative and remedial measures at a fuel-contaminated site occurred at a service station in the northeast United States Piedmont Physiographic Province. The station is located in an urban area, surrounded by commercial and residential land uses. A fuel leak of approximately 10,000 gallons was discovered through a discrepancy in inventory records. The leak occurred in a product-dispensing line which was unearthed and repaired soon after the leak was discovered. At the time of the repair work, four groundwater monitoring wells were installed on the property in an attempt to define the location and direction of movement of the contaminant plume. Observations made in the monitoring wells indicated the presence of free-phase gasoline in two of the four which were converted to recovery wells (Figure 6). Due to the volume of fuel lost, a rapid determination of the location of the bulk of the product lost was necessary to prevent environmental impairment. A vapor survey was performed to make this determination.

The vapor survey team responded and within 24 hours was on site. Soil vapor was sampled on a grid approximately 50 ft by 50 ft. Samples were collected and analyzed utilizing the active method. In addition to the driven sampling points, samples of headspace gases were collected from the four existing monitoring wells.

Concentrations of gasoline constituents in soil vapor measured at the site are displayed in Table 2 and Figures 6 through 8. Examination of these figures reveals several facts. First, it is apparent that the extent of the vapor phase contaminant plume was largely confined within the property lines of the service station at the

Table 2.   Results of Soil Vapor Contaminant Assessment Conducted at Site 2.

Location	Sum of Peaks Eluting Prior to Benzene (ppm as Toluene)	Benzene (ppm)	Toluene (ppm)
VP-1	550	1	<1
VP-2	10,000	9	15
VP-3	<1	<1	<1
VP-4	470	11	32
VP-5	<1	<1	<1
VP-6	2,500	<1	7
VP-7	<1	<1	<1
VP-8	<1	<1	2
VP-9	<1	<1	4
VP-10	19,000	130	210
VP-11	<1	<1	<1
VP-12	23	1	2
VP-13	<1	<1	<1
MW-1	4	<1	10
MW-2	250	44	400
MW-3	640	20	740
MW-4	290	16	230

time of the vapor sampling. Second, the vapor concentrations displayed indicate that gasoline soil residual was present in the area running in an arc between VP-10 and VP-2/MW-3. Third, the orientation of the soil vapor contours indicated vapor phase transport through the soil in a northwest direction (along the axis MW-4 to MW-3) from the area of MW-3 and in a northeast direction (along the axis MW-2 to VP-10) from the area of VP-10.

A design for a remediation system consisting of soil venting and product recovery wells was prepared utilizing the results of the vapor survey. Figure 9 shows the location of the various components of the remediation system. The rationale

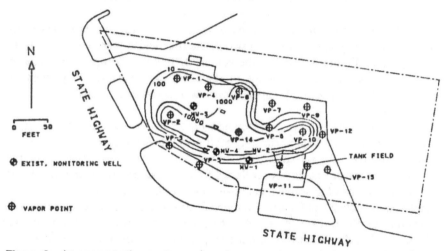

Figure 6.   Isoconcentration contours of pre-benzene compounds (ppm as toluene) at Case History Site 2.

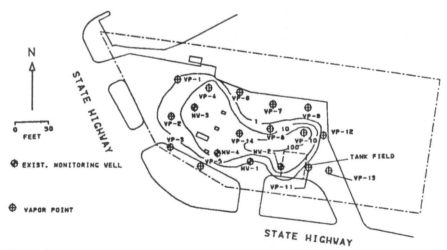

Figure 7.   Isoconcentration contours of benzene (ppm) at Case History Site 2.

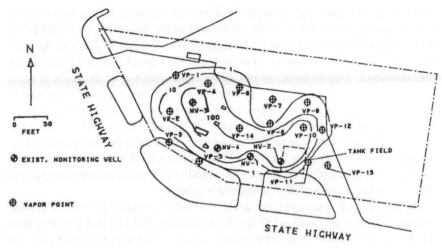

Figure 8. Isoconcentration contours of toluene (ppm) at Case History Site 2.

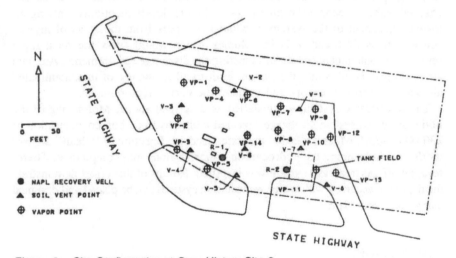

Figure 9. Site Configuration at Case History Site 2.

for placement of the vent points and recovery wells followed from the locations of the "hot spots" and boundaries of the contaminant plume revealed by the vapor survey results. Slight negative pressure readings obtained during the sampling effort indicated that the soil below the pavement had a relatively high vapor permeability and that point vents would be effective. The locations of vent points were designed to contain the contaminant plume and prevent its migration offsite, and to recover free gasoline. Comparison of Figure 9 with the vapor contours shown in Figures 6–8 reveal that V-1 through V-8 were placed to prevent migration of the contaminant plume offsite. V-6 and V-7 were placed in the midst of the area of highest vapor contamination. The vent points were manifolded together and individually valved to provide for control of the flow from each vent point. Thus, higher negative pressures could be applied to V-6 and V-7 than to the other peripheral vent points in order to induce gas flow toward the areas of highest contamination. This situation would allow rapid removal of the highest vapor concentrations while limiting the amount of ambient air drawn into the system through the peripheral vent points, while still preventing the movement of the contaminant plume away from the source of the leak. (It should be noted that both V-6 and V-7 were vadose well points and both contained free gasoline after completion, confirming the location of the "hot spot" determined by the vapor survey.) Existing monitoring wells MW-1 and MW-4 were converted to product recovery wells as part of the remediation system.

The remediation system was designed and installed rapidly in order to prevent further migration of gasoline offsite. Free gasoline (NAPL) was pumped from the two recovery wells to a holding tank on site. Vapors were drawn from the soil through the venting system and discharged to the atmosphere. A state air discharge permit limited the amount of total hydrocarbons which could be discharged without treatment of the off gas. The previously mentioned valving allowed adjustment of the system to withdraw vapors from the areas of highest concentrations (V-6 and V-7) in volumes which would allow the most rapid removal of contaminants while still meeting the discharge requirements. Ambient air could also be bled into the system both to allow mixing of uncontaminated air with the gasoline vapors and to meet the permit requirements.

The soil venting system was installed as described above. Monitoring of the initial untreated system discharge revealed a total hydrocarbon concentration of 230,000 mg/m^3, which exceeded the state standard. Injection of "clean" air was ineffective in bringing the hydrocarbon concentrations into compliance. Therefore, an off-gas treatment system was installed. Results of this effort demonstrate the degree of accuracy with which remediation systems can be placed and designed using soil vapor data.

CONCLUSIONS

Soil vapor surveying has been widely applied as a useful tool in assessing sites contaminated with volatile organic compounds. It is typically utilized only as an initial assessment tool to screen for site contamination and to aid in monitoring well placement. Careful planning and an understanding of vapor fate and transport theory can extend the utility of the data generated. In many instances, when rapid response is required, data generated via a soil vapor survey can form the basis for design and implementation of remediation systems. Prudent utilization of soil vapor survey data can reduce the time between discovery of contamination and implementation of a remedial system.

REFERENCES

1. Bear, J. *Dynamics of Fluids in Porous Media* (New York: American Elsevier, 1972).
2. Mackay, D. M., P. V. Roberts, and J. A. Cherry. "Transport of Organic Constituents in Ground Water," *Environ. Sci. Techn.*, 19(5), pp. 384–392.
3. Reisinger, H. J., D. R. Burris, L. R. Cessar, and G. D. McCleary. "Factors Affecting the Utility of Soil Vapor Assessment Data," in *Proceedings of the First National Outdoor Action Conference on Aquifer Restoration, Ground Water Monitoring and Geophysical Methods.* May 18–21, Las Vegas, Nevada, 1987.
4. Lyman, W. J., W. F. Reehl, and D. H. Rosenblatt. *Handbook of Chemical Property Estimation Methods: Environmental Behavior of Organic Compounds.* McGraw-Hill, New York, 1982.

PART V

Case Histories

CHAPTER **26**

Application of Quantitative Risk Assessment Evaluation of Underground Storage Tanks to Insurance, Banking, and Real Estate Transactions

John Casana and Gina Dixon

In the 1984 Resource Conservation and Recovery Act (RCRA) amendments, Congress required the U.S. Environmental Protection Agency (EPA) to impose a minimum coverage of $1,000,000 per underground storage tank (UST) release. This requirement, as well as a minimum annual aggregate coverage of up to $6,000,000 per tank owner, is included in the proposed UST rules. Currently, there is a dearth of UST insurance available, and that which is available is generally restricted. The key to providing coverage, we believe, is a concerted risk management program with sound technical assessment as the cornerstone for underwriting. Such a program also provides a basis for sound property transactions involving underground tanks.

Versar, Inc., has developed a simple yet effective risk assessment procedure as part of our overall risk management program for underground storage tanks. We believe the risk rating procedure is highly applicable to assessing exposures for insurance and property transfer purposes, and we have compiled environmental data bases and specialized software to allow us to perform the risk rating in a highly efficient manner. This chapter will briefly describe the risk rating procedures and illustrate their use in an insurance and property transfer application.

Versar has drawn from its key role in supporting the technical development of the proposed rules, from its onsite risk evaluation of over 1,500 facilities, and from its computerized analysis of over 13,000 underground tank releases to develop an economical yet sound risk rating and prioritization system. The system is based on tank information derived from the UST notification data form (required by all state UST agencies) and from local environmental characteristics derived from Versar's comprehensive in-house data network. This information is applied in a consistent manner to determine where the tank owner should focus his resources to minimize overall risk and to ensure compliance with all federal, state, and local regulations. For example, an older, unprotected tank located in the recharge zone of a major public drinking water supply will warrant expedient replacement with a double-walled system. On the other hand, a new tank in a less environmentally sensitive area may be economically retrofitted at a later time to meet minimum requirements.

Versar uses two sets of criteria to quantify risks and prioritize tanks. The first set consists of release potential characteristics (e.g., frequency of release), such as tank age, corrosion protection controls, leak detection controls, tank design, and soil corrosivity. The second set of criteria consists of site vulnerability factors (e.g., severity of release), such as depth to water table, soil hydraulic conductivity, surrounding population, aquifer use, and proximity to surface waters.

Tank system characteristics are provided by the tank owner, and primarily consist of information already collected for the EPA's tank notification program. The only additional information involves existing leak detection practices and secondary containment systems. In the absence of specific information, we conservatively assume that no leak detection program is currently in place and that the tank system has no secondary containment.

Site vulnerability characteristics are determined using Versar's comprehensive in-house data bases and information sources that cover the nation. These data bases bring together information compiled by a variety of government agencies and others. For example, we have centralized access to municipal drinking water well and potable surface water supply locations with supporting data, such as the population served by each potable water supply source. The information allows us to determine the relative likelihood that a potable water supply will be contaminated, which directly affects the cleanup costs. Similarly, we have computerized access to soil characteristics on a county-by-county basis throughout the United States. This information has been compiled by the U.S. Soil Conservation Service through field surveys that have been conducted over several decades. The information allows us to assess the relative likelihood that a release will travel quickly to impact a large area, which will directly affect the severity and associated costs of the release. As another example, our in-house population data base allows us to readily determine the population surrounding a tank. In this way, we can determine the relative likelihood of fumes in basements and other impacts resulting from a release. Finally, with our centralized access of depth to groundwater information, Versar can determine the relative potential for groundwater contamination in the event of a release, which significantly affects the severity and the associated costs. In this way. we can efficiently compile the basic information needed for each tank and site.

Once this basic information is compiled, Versar uses specific criteria that we have developed based on analysis of our in-house data base on over 13,000 underground storage tank releases and based on our experience in risk evaluation and site remediation. For example, we have found that the mean age to leak for an unprotected steel tank is 17 years, based on over 1000 releases for which tank age is reported. These criteria are used to assign each tank a score of zero to ten for release potential and damage potential, respectively. The specific factors used to determine these scores are summarized in Table 1. Each factor is given a score of zero to two, resulting in a release potential score and a damage potential score of zero to ten, respectively.

Table 1. Risk Rating Criteria.

	Low Risk (0 points)	Moderate Risk (1 point)	High Risk (2 points)
Release Potential Factor			
•Tank system age (years)	0 to 10	11 to 20	21 or more
•Corrosion protection	tank and/or piping protected	either tank or piping protected	none
•Leak detection	continuous	intermittent	none
•Soil corrosivity	nonacidic to weakly acidic (pH > 6.5)	moderately acidic (pH 5.5 to 6.5)	strongly acidic (pH < 5.5)
Damage Potential Factor			
•Depth to water table (ft)	more than 50	15 to 50	less than 15
•Soil hydraulic conductivity (gal/day/ft^2)	less than 1 (e.g., glacial till, clay)	1 to 10^3 (e.g., sand, silty sand)	10^3 to 10^6 (e.g., gravel, limestone)
•Local population served by public groundwater supply	less than 1,000	1,000 to 10,000	more than 10,000
•Distance to surface water (ft)	more than 1,000	1,000 or less (other than public water supply)	1,000 or less (public surface water supply)
•Population within 2 miles	less than 100	100 to 1,000	more than 1,000

Once the damage potential and the release potential scores are determined for each tank, the values are placed on a risk rating chart, as illustrated in Figure 1. The tanks that fall in the upper right portion of the chart are of highest priority since they represent a high risk of release in an environmentally vulnerable area. Those in the upper left to lower right portions of the chart are of moderate risk and, therefore, moderate priority, while those falling in the lower left are of lowest priority. This consistent ranking scheme allows us to efficiently make an initial prioritization of tanks based on their relative risk.

Underwriters and tank owners can use the risk rating results summarized in Figure 1 to assist in the underwriting determination as to which tanks are insurable

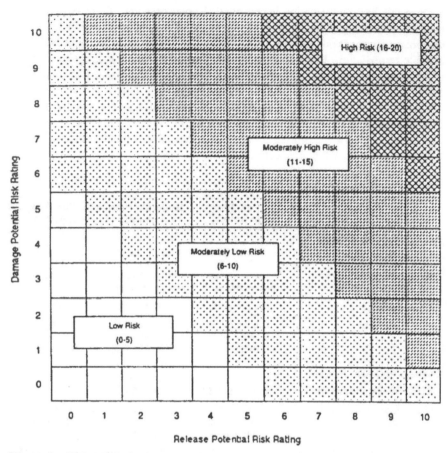

Figure 1. Risk rating chart.

(e.g., the tanks falling in the lower left portion of the chart are the most insurable). The chart may also be used to provide tank owners with options to bring their tanks into the insurable range. For example, the damage potential rating is based entirely on site characteristics that cannot be readily improved. However, the release potential characteristics are based primarily on tank system characteristics that can be improved by upgrading or replacing the tank. In this way, exposure is minimized and the tank owner's resources are focused where they are needed most.

Case A—Property Transfer Application

In this case study, the risk rating system is applied to a property transfer involving a facility with 28 tanks. A summary sheet is prepared for each tank, as illustrated in Table 2. This tank summary provides identification information, specific results of the risk rating analysis, areas of compliance/noncompliance, and recommended action.

Since the release potential is relatively high (7), and the damage potential is relatively low (2), the recommended actions are:

1. Perform tank integrity testing to verify that the tank is not leaking, and
2. Retrofit the existing tank with cathodic protection and implement a regular tank detection program.

These relatively low cost actions will ensure that a leak will be discovered quickly if it occurs, thereby minimizing overall costs.

Table 2. Tank Summary Sheet.

TANK NUMBER:	313
TANK LOCATION:	Tank Facility 18, Bethesda, MD
TANK CAPACITY (Gallons):	20,000
MATERIAL STORED:	Unleaded gasoline

RELEASE POTENTIAL

Factor	Value	Score (points)
Tank system age (years)	16	1
Corrosion protection	none	2
Leak detection	inventory control	1
Secondary containment	none	2
Soil acidity	moderately acidic	1
	Total	7

DAMAGE POTENTIAL

Factor	Value	Score (points)
Depth to water table (feet)	15 to 50	1
Soil hydraulic conductivity	glacial till	0
Local population served by public groundwater supply	less than 1,000	0
Distance to surface water (feet)	more than 1,000	0
Population within 2 miles	685	1
	Total	2

Areas of noncompliance with existing regulations:
•Tank integrity test required by Maryland state regulations

Areas of noncompliance with proposed regulations:
•Lack of corrosion protection system and adequate leak detection program

Recommended short term action:
•Perform tank integrity test

Recommended long term action:
•Retrofit with cathodic protection and test annually; perform annual tightness testing; continue inventory monitoring.

Table 3 presents a risk rating summary and compliance summary for all 28 tanks at this location. This facility summary, together with the summary sheet for each tank (e.g., Table 2), provide the buyer, seller, and lender with a framework to ensure that the risk associated with the sale is managed rationally, at minimum cost to all parties.

Table 3. Facility Summary Sheet.

FACILITY NAME/LOCATION:	Tank Facility, Bethesda, MD
NUMBER OF TANKS:	28
RELEASE POTENTIAL RATINGS:	Min: 5 ; Mean: 65 ; Max: 8
DAMAGE POTENTIAL RATINGS:	Min: 1 ; Mean: 1.7 ; Max: 3
TANK TESTING RESULTS:	Number Tested: 1 Number Found Leaking: 0

Federal EPA

Tank Status	Regulatory Requirement	Regulatory Status	Compliance Status
Existing	Upgrade all tanks with corrosion protection within 10 years. Test cathodic protection systems annually.	Proposed	24 tanks would not meet this requirement
Existing	Implement leak detection system for all tanks within 3 to 5 years. Options include: 1) semiannual tank integrity testing plus inventory control; 2) monthly testing using external or automated systems; 3) continuous monitoring systems.	Proposed	27 tanks would not meet this requirement
New	Corrosion protection and leak detection for all tanks installed since May, 1985.	In force	One tank in violation

State of Maryland

Tank Status	Regulatory Requirement	Regulatory Status	Compliance Status
Existing	Tank integrity testing for tanks over 15 years old at 5 year intervals.	In force	14 tanks in violation
New	Corrosion protection for all tanks.	In force	One tank in violation
New	Two monitoring wells for each installation.	In force	In compliance
All	Inventory monitoring.	In force	In compliance

Montgomery County

Tank Status	Regulatory Requirement	Regulatory Status	Compliance Status
All	Permits required to remove, install or abandon in place.	In force	One tank in violation

Case B—Insurance Application

This case involves evaluating the insurability of a retail gasoline station chain with 634 tanks at various locations. The results of the risk rating analysis are summarized in Figure 2.

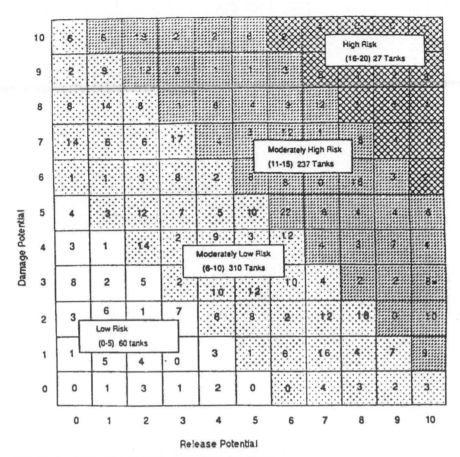

Figure 2. Risk rating chart for retail gasoline chain.

The results provide a framework for the tank owner and the insurance company to negotiate a fair insurance policy. For example, the insurance company may wish to exclude all 27 tanks in the high risk category from the policy until specific actions are taken to reduce those risks. On the other hand, the 237 tanks falling in the moderately high risk range may be insurable at a higher premium than those in the lower risk ranges. This premium differential could provide the tank owner with a monetary incentive to upgrade/replace his higher risk tanks to achieve a better premium.

Over time, the risk rating system, as applied in this case, would greatly reduce the tank owner's expenditures, by allowing him to focus his resources where they are most needed. Similarly, the insurance company would be able to provide competitive rates by being able to efficiently identify and address the highest risk tanks first. Ultimately, the significant savings in tank upgrading/insurance costs and in reduced claims will be passed to the general public through lower retail prices.

CHAPTER 27

Assessment and Remediation of Residential Properties Contaminated with Home Heating Oil: Case Studies

Robert W. Wilhelm, II, and Robert J. Bouchard

INTRODUCTION

During the past decade, as consultants to a variety of potentially responsible parties, we have noted an increase in reported releases involving fuel oil at residential properties.

When considering the abundance of residential properties with tanks, the potential for releases is high—generally not as high as gasoline stations, but the impacts can be just as serious. These releases are frequently sudden in nature, as opposed to the more routine leakage over time from underground storage tanks. As a result, we have been retained by homeowners, local oil distributors, and most frequently by insurance companies representing potentially responsible parties who have become frustrated with conventional oil recovery and hazardous waste contractors, and often with state and local regulatory agencies. Recently, banks and mortgage companies have taken greater interest in the remediation process since property values are at risk, not only human health. Therefore, the potential for liability is increased, coupled with losses in revenues, the threat of priority liens, and the possibility of total loss of the property.

This chapter presents three case studies which exhibit some of the complexities encountered when faced with the restoration of a residential property following a release of home heating oil. Our intent is not merely to discuss the common exploration and remediation techniques employed at these sites, which have become fairly routine in the industry, but to highlight some of the problems associated with these activities in a residential environment. These problems have included in general:

1. Physical Constraints—relatively small work areas often limited by the living space itself
2. Costs of Investigations and Remedial Work—completely borne by the homeowner, and limited through property value or limits set by the insurance policy
3. Time—the time needed to temporarily relocate families while restoration takes place, and for dealing with adverse weather conditions
4. Regulatory Involvement—the imposition of standards and policies designed for larger commercial and industrial sites.

Another problem which must be dealt with is the homeowner's insurance policy. The typical policy excludes environmental impairment, and is restricted to replacement value of the structure and contents, and not the resultant contaminated soil or groundwater beneath it.

Most individual family homes with oil heat have a conventional 275-gallon tank located in a basement, attached garage, or often buried underground near the foundation. As a result of a release in these typical settings, the oil will become trapped in the granular fill material beneath the foundation until it seeps through cracks or migrates along buried utility lines.

In the three case studies reviewed, each residence possessed an aboveground tank. The potential for leaks and spills in this situation is usually due to sudden and accidental causes, as opposed to the gradual loss of product typically associated with underground storage tanks. In fact, the most common causes of spills at residential sites is human error. We have investigated cases where oil distributors have attached a fill nozzle to a disconnected fill pipe, a monitoring well, made a delivery to the wrong house, and two deliveries to the same house in a one day period. As a result, overfilling of the tank causes it to rupture, spilling oil into basement and living areas.

Other incidents have involved leaking fuel lines which, although the tanks are aboveground, have gone undetected because the lines are often buried beneath the floor itself. The loss is actually not detected until fuel consumption is noticeably higher, or seepage and odors occur.

In the following case studies, preferred engineering options for the restoration of the residential property versus the undesirable alternative of demolition, are reviewed relative to the magnitude of the fuel oil released under the structure. The subsurface fuel oil containment and removal actions at a site that is abutted by a wetland and which utilized an onsite drinking water supply well are reviewed with regard to these sensitive receptors. The assessment techniques and remedial actions presented were performed at residential properties located in contrasting hydrogeologic and urban settings in southeastern Massachusetts.

The fact that these case histories are residential is problematic when considering typical remedial investigations and cleanups are commercial or industrial in nature. For instance, human exposure standards are often set for occupational environments (8-hour Time Weighted Average) limits and not the potential 24-hour exposure, long-term risks associated with living areas.

CASE HISTORY #1

Site Description

The first site is situated in a densely populated residential neighborhood and consisted of a one-story, three bedroom, single family house. The house was constructed as a slab-on-grade with all underground utilities and radiant heat.

General Conditions

A release was first suspected when above-average fuel consumption was noted and the residents smelled fuel in their kitchen and living room areas. The release occurred due to leaks in the fuel distribution line. The fuel line extended from a 275-gallon aboveground fuel oil tank located inside the attached garage, under the cement slab floor, to the furnace located in the kitchen, as shown on Figure 1. Just prior to the detection of odors in the house, the residents had noticed an unusually quick drop in the level of fuel oil in their tank. Unsuspecting, and due to the unusually cold weather, they promptly had the tank refilled, thereby increasing the problem. It was later estimated that a minimum of 300 to 400 gallons of #2 fuel oil was released at the site as a result.

The residents immediately abandoned the property following the detection of the release, under order of the local Board of Health. The owners' insurance carrier accepted the initial responsibility for the cleanup, since the heating system was covered under the homeowners' policy. Coverage included replacement of the leaking oil supply line from underneath the cement slab floor with a new line, and repairs to the floor and radiant heating system. This was necessary so that the heating system could remain on, to keep water pipes from freezing and bursting.

Since the leak constituted a "release" to the environment, the Massachusetts Department of Environmental Quality Engineering (DEQE) was notified. As a result, the owner of the site was subsequently served with a Notice of Responsibility which identified the owner as a Potentially Responsible Party (PRP) with liability under Massachusetts General Law (M.G.L.), Chapter 21E, the state superfund law. Under Ch. 21E, if a PRP does not accept responsibility for actions deemed necessary to respond to a release, then the PRP's liability for any costs incurred by the DEQE for actions they take to remediate the release, constitutes a debt to the Commonwealth. In addition to a foreclosure remedy provided in the priority lien, the Attorney General of the Commonwealth may recover the

Site plan

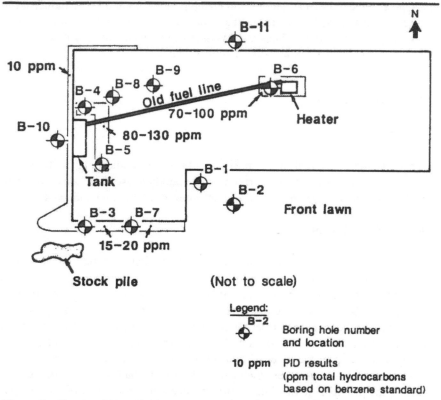

Figure 1. Results of 11 test borings ranging in depth from 1 to 3.5 feet below ground surface.

debt (treble damages) or any part of it in action against the PRP. The PRP may also be liable under Ch. 21E for additional fines or penalties for damages pursuant to other statutes of common law.

An initial site visit to assess the site conditions was conducted by all parties involved, including a representative of the Massachusetts DEQE. Fuel oil staining was identified in sections of the kitchen floor and outside along cracks in the concrete walkway and asphalt paved driveway. Extensive snow cover prohibited direct observation of ground surface conditions during the site visit.

It was recommended at that time that a licensed oil recovery contractor be retained to initiate cleanup actions inside the house. Cleanup of exterior areas would be delayed and performed under favorable weather conditions. Cleanup actions recommended at that time included the removal of any free product, contaminated soils, rock and/or concrete from excavations performed within the garage and kitchen inside the house. The extent of the oil migration would also be estimated during this initial cleanup effort.

Due to the limitations in the homeowners' insurance plan covering the structure, heating system, and the actual contents of the house (e.g., concrete, flooring, furniture, etc.), the insurance carrier became apprehensive about assuming total liability for the costs associated with contaminated soil removal, which conceivably could be very high. The insurance carrier agreed to accept the cleanup costs up to the limits of the policy coverage ($60,000). The insurance adjuster then retained a consultant to develop a remediation plan and coordinate site restoration activities with the various parties involved. It was decided at that time that a series of test holes would be performed within the kitchen, bedroom, garage, and outside along the sidewalk and the perimeter of the house to preliminarily assess the extent of the contaminated soil.

Test holes were conducted in the areas outlined above with a pneumatic chisel to approximately 2 feet below the ground surface. The soils and concrete were removed with a shovel and stockpiled on polyethylene, outside along the driveway. Groundwater was noted to be approximately six inches below the concrete slab in the excavations. No free product was observed on the water table and volatile organic vapor readings detected with a photoionization detector (PID) indicated levels for the soils ranging between 0–20 ppm. PID levels were observed to be the highest in the soil beneath the kitchen, near the heater, and beneath the garage around the fuel tank. It was thought at the time that the cleanup could be adequately performed by limited soil excavation and removal.

Cleanup actions were subsequently initiated by a licensed hazardous waste contractor. This consisted of drumming the previously stockpiled soil and excavating and removing a limited amount of contaminated soil from within the kitchen and garage.

It became apparent with the changing weather conditions and during the excavation of the soils that the contamination was more extensive than was detected in the initial assessment. Oil stains, dead grass and shrubbery were now obvious along the entire perimeter of the garage, along the front sidewalk, and around the outside perimeter of the house. The PID levels recorded during the soil excavations were also higher than those previously detected. PID readings ranged from 70–100 ppm for the soils excavated inside the house, whereas the levels ranged between 10-50 ppm for the soils excavated outside. Groundwater was not encountered in the excavations. It became apparent after reviewing the data collected during this initial cleanup effort that the site would require further assessment and cleanup measures.

In order to define the extent of the contamination and the site's hydrogeologic characteristics, a series of hand-augered test borings were performed. A total of 11 test borings were performed, as shown on Figure 1, and ranged in depth between 1 to 3.5 ft below ground surface. Test borings were advanced to refusal at a very dense glacial till layer consisting of a gray silty fine sand, with medium to fine gravel and few cobbles. The soils encountered above the glacial till layer consisted of a light brown, coarse to medium sand and gravel fill. The fill material had been placed as a pad for the concrete floor when the house was constructed.

PID levels were recorded for the soil samples collected at selected depth intervals and from the open space during the advancement of each test boring. The results of this screening indicated generally elevated PID levels for the soils collected from the bottom of each boring. A summary of the boring depths, soil conditions, and PID levels is provided in Table 1. At the completion of the test borings, six standpipes were installed with the screened section (Schedule 40, 2-inch PVC, 10-slot) positioned 1–3 ft below the ground surface. The standpipes were installed to monitor headspace volatile organic levels in the soil. Headspace readings were taken periodically in each standpipe to evaluate volatile organic levels over time. This analysis indicated a significant drop in the PID levels in each standpipe within a two week period between readings.

Table 1. Case History #1
Summary of Boring Depths, Soil Conditions, and PID Levels.

Boring	Depth (ft)	Soil Type	PID (ppm)
B-1	0–1	fill	15–25
	1–1.6	fill	50–75
	1.6–2.0	glacial till	50–85
B-2	0–1	fill	0–4
	1–2	glacial till	20–25
B-3	0–2	fill	80–100
B-4	0–1.5	fill	70–80
B-5	0–1.7	fill	100–120
B-6	0–1	fill	30–50
	1–2.6	glacial till	100–105
B-7	0–1	fill	0–10
	1–3.5	fill	0–25
B-8	0–2.3	glacial till	40–45
B-9	0–1	fill	0–25
B-10	0–2	fill	0–25
B-11	0–1	fill	0–10
	1–1.3	fill	0–4

Note: ppm = parts-per-million
   PID = Photoionization Detector Reading
   Background Levels = 0–5 ppm

The following interpretations were arrived at based on the results of the test borings and PID screening:

- Due to the presence of a shallow groundwater table (6 in. below the ground surface) observed during the initial investigation in the late winter and early spring and the absence of a water table during the summer, water table conditions were assumed to be perched. Therefore, groundwater elevations were expected to fluctuate seasonally, dependent upon groundwater recharge from precipitation.
- The dense glacial till layer encountered beneath the site would facilitate a perched water table environment. An inspection of a U.S.

Geological Survey surficial soils and bedrock map for the site area indicated that the glacial till layer was consistent with the soils mapped for the region. The depth to bedrock was also assumed to be relatively shallow at the site due to the presence of bedrock outcrops mapped in the immediate site vicinity.

- Soil conditions encountered indicated that the released fuel oil probably migrated fairly rapidly through the sand and gravel fill material and the upper weathered surface of the glacial till layer when the leak initially occurred. During the spring thaw and subsequent above average precipitation events, migration of the fuel oil was thought to have occurred both horizontally and vertically as it was entrained and "pushed" by the recharging groundwater.

- The results of the volatile organic vapor screening recorded during advancement of the test borings indicated that the fuel oil had migrated and became concentrated in the topographically low areas of the glacial till surface. The PID levels also indicated a general contaminant migration direction toward the south and southeast.

- The PID screening results for the headspace in standpipes indicated that the volatile organic vapor levels had decreased over a two week time period. The drop in levels was attributed to venting of the soils when the excavations and test borings were conducted, which aided in volatilizing the aromatic fuel oil compounds.

Based upon the findings of this phase of our assessment, it was recommended that the excavations inside the house be continued to further assess the extent of the contamination. Efforts were concentrated within the kitchen and garage areas, the suspected focal points of the release.

Initially, the cleanup effort was concentrated within the kitchen, where the highest levels of volatile organic compounds were previously detected. Cleanup activities consisted of removing the concrete slab floor and loosening the underlying soils with an air-driven hammer and chisel. Once loosened, the contaminated soils were excavated with an air-flow "Vactor" system which loaded the contaminated soil directly into a truck for offsite disposal. The removal process was performed in 1-ft depth increments in an attempt to reach acceptable PID levels (0–10 ppm) for the underlying soils. Excavations were continued to a depth of approximately 5 ft below the floor surface inside the kitchen by this method. PID levels for the soil at this depth were consistent between 50–70 ppm at the bottom of the excavation and 10–20 ppm for the headspace of the open hole. At this point, further excavation of the soil inside the kitchen was determined inappropriate without endangering the structural stability of the house. Approximately 20 cubic yards of contaminated soil had been removed from the excavation located inside the kitchen.

The concrete slab floor was removed in additional areas of the house, and contaminated soil was observed under the living room and bedroom. The soils exhibited PID levels in the range of 50–70 ppm at a depth of approximately 3.0 ft below the floor surface. Localized "hot spots" with PID levels of 100–130 ppm were observed in low areas of the glacial till surface.

At this stage, it was determined that the required actions necessary to remove all the contaminated soil and to preserve the structural integrity of the house could not be performed. Operations were ceased until a new plan of action could be developed. As the limits of the homeowner insurance policy were rapidly being approached by the labor and disposal costs, a review of the conditions relative to fuel oil contamination, and the feasibility of restoring the site to a usable and habitable status within the financial constraints was performed.

The following conclusions were determined:

1. The fuel oil contamination was contained within the soils onsite;
2. No free product was observed in any of the excavations;
3. The site exhibited seasonally perched water table conditions; and
4. No sensitive environmental receptors or public water supplies were identified downgradient of the site.

After reviewing the site conditions and following discussions with the DEQE on current policies and guidelines, restoration of the house to a habitable status without significant demolition/restoration seemed a viable option. Schematic plans and specifications for a subsurface soil ventilation system which would remove accumulated petroleum vapors trapped beneath the floor were prepared and submitted to the insurance company and the DEQE for review. This proposal included replacing the radiant heating system with baseboard heat.

After installation of the ventilation system, a new concrete slab would be poured and the utilities would be hooked up. The new baseboard heating system would be installed and the house would be cleaned of any materials that may have absorbed petroleum odors, (i.e., wallpaper, carpeting, wood, etc.).

At completion, the heating system would be turned on and a determination of the ambient air quality within the house would be performed. If the air quality was determined to conform to a predetermined DEQE and local Board of Health air quality criteria, the house could be restored to a habitable status. A monitoring program would then be established to verify air quality within the house on a regular basis. Particular concerns would include monitoring during the spring months when perched water table conditions typically occurred and during the winter heating season when the house received reduced air exchanges.

This proposal was rejected by the owner's public adjuster as less than adequate. Only complete removal of the contaminated material would be acceptable. In order to achieve this level of remediation, only demolition of the house would allow such large scale excavation. The proposed costs for such activities exceeded the mortgage value of the property as well as the limits of the insurance policy

($60,000). As a result, no further action occurred at the site for over a year while the parties involved pursued litigation.

Site Status

In the absence of any assumed liability for this case, the final remediation was unresolved. The owners remain living at an offsite location. The home is currently unheated and approximately 15 drums of oil-laden soil and debris blemish the neighborhood.

CASE HISTORY #2

Site Description

Our second case history is also located in southeastern Massachusetts, in a thickly settled area. The property is a residential lot developed with a 2-story, wood-framed, single family house. The 275 gallon fuel oil tank was located in the basement on a concrete slab floor. The house is founded on bedrock and abutted by residential properties.

General Conditions

An accidental release occurred at this property when an oil distributor misidentified the address to which a scheduled oil delivery was to be made, a rather common occurrence. Since the tank was at 3/4 capacity to start, the oil soon began spurting out the vent pipe, saturating the exterior walls and surrounding ground surface. This discharge could not sustain the rate of fill and eventually the tank ruptured, spilling at least 300 gallons into the basement laundry and recreation areas. The owners were absent at the time and the operator left an invoice for the oil delivery, adding insult to injury. After returning, the owners discovered the problem and immediately notified the fire department and heating contractor. The fire chief assessed the situation and recommended notification of the DEQE. The DEQE responded immediately and ordered a contractor to undertake initial remedial measures. However, by this time, the oil had drained into an old abandoned sump. The following day, a multitude of interested parties converged on the site including:

- an attorney for the homeowner and an environmental consultant;
- the heating contractor;
- the Board of Health and DEQE;
- the Fire Department;

- the owner's insurance company;
- the oil company, their attorney, insurance company, claims adjusters, and their consultant.

Once again, no resolution of the problem was foreseeable without intervention. The oil distributor's insurance company engaged another consultant to direct remedial activities. However, the process was hampered by the owner's attorney who, in anticipation of eventual litigation, would not consent to any remedial activity considered to be less than total restoration of preexisting conditions. Not wishing to set a precedent in the absence of remedial guidelines for residential properties, the DEQE required that a solution be agreed to by consultants for both parties.

Removal of the concrete floor in the basement was undertaken in selected areas to assess the contamination. The house had been originally built where bedrock had been previously blasted on the site. Large blocks of bedrock had been used as backfill under the foundation. The foundation consisted of granite blocks with mortar. The concrete floor had cracked along form joints, and the fuel oil had migrated through the cracks into the soils below.

Cleanup activities included removal of the ruptured oil tank, heating system with asbestos-covered piping, concrete flooring, 3 ft of soil, oil-saturated materials such as wall coverings, vinyl flooring, stairway support columns, and miscellaneous personal items. Two trailer-loads of soil were removed from the site. No further excavation was possible due to the placement of large boulders beneath the floors. Total Petroleum Hydrocarbon (TPH) levels in remaining soils were in the hundred ppm range.

A passive soil venting system was proposed to address the residual contamination. However, a risk assessment by another consultant indicated the potential for induction of fumes through cracks in the slab during winter heating periods. The consultant recommended a series of elaborate air exchange units as compensation. This equipment included air induction units for the new furnace, addition of range-top ventilation hoods, and a tamperproof active ventilation system for the subfloor. The final decision was left to the homeowner to accept this remediation or, alternatively, a cash settlement for the property value.

After initial excavation and reinspection, counsel for the owner indicated that rehabitation was not feasible under any circumstances in the event that the risk assessment was in error or that other contaminants went undetected which may have adverse effects.

It was the owner's decision not to accept any further remediation. Therefore, all proposals were withdrawn, and the owner's bank entered the dispute in order to ensure coverage for their interests.

Site Status

In spite of pressure from many factions, no clear resolution to this problem has been determined. The lack of clearly defined remedial goals caused unnecessary delays in resolving the problem. The owner is currently living in a mobile trailer parked on a relative's property at the expense of the homeowner's insurance.

CASE HISTORY #3

Site Description

The third case study offers, in contrast, the positive resolution of a similar situation. The property is a residential lot with a seasonal cottage that has been converted for year-round use located on Cape Cod, Massachusetts. The site abuts a pond to the south, a small stream and wetland bog that flows into the pond to the west, and residential properties to the north and east. The existing cottage and abutting cottage to the east both utilize private wells for their drinking water supply, as shown on Figure 2.

Site plan
GW contours

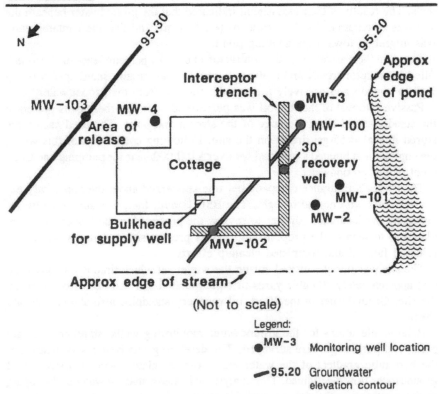

(Not to scale)

Legend:

● MW-3   Monitoring well location

━ 95.20   Groundwater
          elevation contour

Figure 2.   Location of groundwater monitoring wells installed at the completion of each test boring.

General Conditions

A release of #2 home heating oil occurred from an aboveground 300-gallon fuel oil tank located outside the cottage. The release was suspected during the winter months when the tenants complained of odors inside the cottage, but this was attributed to the furnace located in the living room. A release was eventually identified in the spring when a fuel line ruptured under the house and oil was observed seeping into the bog and the crawl space for the supply well bulkhead. Initially, cleanup and containment efforts were performed by an oil recovery contractor. This included removal of the fuel line and relocating the fuel storage tank. A containment boom and sand bags were also placed along the edge of the pond and stream to contain fuel oil breakout.

A Notice of Responsibility (NOR) was issued for the site by the DEQE, which outlined specific site actions to be completed to assess the property with regard to soil, groundwater, and surface water contamination. The oil and hazardous material contractor who performed the emergency containment actions was retained by the owner to investigate the release.

Their assessment included the installation of three monitoring wells, groundwater sampling, and sampling of two water supply wells, the owners and an abutter's. The results of their assessment indicated that the groundwater beneath the site was contaminated with petroleum (#2 fuel oil) and that the contamination was migrating toward the abutting pond.

Based on their assessment, the contractor installed a polyethylene-lined, stoned-filled interceptor trench and a 30-in. corrugated aluminum standpipe/recovery well to contain and passively remove the fuel oil from the groundwater.

Passive removal of the fuel oil was performed with absorbent pads placed in the standpipe and along the edge of the stream and pond. The used pads were stored in sealed 55-gal drums on the site. Following completion of this work, the contractor submitted a proposal for over $100,000/year for pumping and treatment of the groundwater.

The owner's insurance company became concerned about the apparently excessive costs with no end in sight. The BSC Group, Inc., was subsequently retained by the insurance adjuster to review the proposed groundwater treatment plan and to inspect the property and assess present site conditions relative to released fuel oil and completed cleanup efforts.

The site was initially visited during the winter, at which time it was observed that approximately 20 cubic yards of contaminated soil had been stockpiled on the site. Groundwater in the 30-in. oil recovery standpipe also showed an oily sheen.

Relative elevations for the groundwater monitoring wells, standpipe, stream bank, and pond edge were surveyed. The depth to groundwater was measured, the hydraulic gradient of the water table was calculated, and the direction of groundwater flow determined. This analysis indicated a shallow water table sloping to the west across the site, toward the pond.

General observations noted during the visit indicated that one of the monitoring wells was located in close proximity to the onsite septic system. Therefore, groundwater elevations in this well would be affected by mounding of the water table. Also, no upgradient wells were installed, and thus background water quality at the site could not be determined.

Absorbent pads were present in the recovery well and along the bank of the pond and stream. A containment boom was also located along the pond and stream outlet. The stream and pond were frozen over at the time of the visit, thereby limiting the effectiveness of the floating absorbent pads. Free product was not observed in the stream or pond, due to the ice coverage.

A soil sample was collected from the stockpiled material and analyzed for disposal parameters. The test results indicated that the soils exceeded the DEQE criteria for onsite containment, met the appropriate criteria for disposal at an out-of-state landfill location, and may have been acceptable for disposal at an approved Massachusetts landfill.

The stockpiled soil was also screened at selected locations with a photoionization detector (PID) for the presence of volatile organic vapors. No readings were recorded above normal background levels of 2 parts per million (ppm). It was believed that the unexpectedly low readings may have been affected by the cold ambient air temperature inherent at the time of screening.

Based on the findings of the initial site visit, it was determined that the site would have to be reevaluated in the spring when there would be no ice coverage and the water table would be typically elevated.

The site was visited in the spring to evaluate the effectiveness of the absorbent pads placed in the surface water bodies, and to check for the presence or absence of free product in the wells and the standpipe during elevated water table conditions.

The absorbent pads located in the pond and stream were noted as clean, and free product was not observed. A surface film believed to be produced by iron oxidizing bacteria was observed on the surface of the stream. The monitoring wells and recovery well were sampled with a 1.5-in. clear acrylic bailer, and free product was not detected. The cover to the bulkhead of the onsite supply well was removed and minor surface oil stains were observed on the ground around the top of the well head and on the cinder blocks that lined the bulkhead.

The existing monitoring wells, the recovery well, the onsite supply well, the supply well located on the adjacent property, the stream, and the pond were sampled for analysis of Total Petroleum Hydrocarbons (TPH). The results of the TPH analyses indicated that residual fuel oil contamination was present in the groundwater beneath the site. The distribution of the concentrations detected indicated the contamination was migrating offsite, possibly into the abutting pond. Table 2 provides a summary of the TPH analyses.

Water samples from the onsite supply well were also analyzed for the following drinking water quality parameters: RCRA metals, sodium, nitrate, fluoride,

Table 2.   Case History #3
Summary of Total Petroleum Hydrocarbon (TPH) Analyses

Sample Location	TPH (ppm)
MW-2	430
MW-3	6.9
MW-4	130
Recovery Well	210
Pond	ND
Stream	ND
Residential Well (onsite)	ND
Residential Well (neighbors)	ND

Note: ppm = parts per million
ND = below minimum detectable level (MDL)

hardness, pH, pesticides, and herbicides. The results indicated generally acceptable drinking water quality conditions. A slightly elevated water hardness supported the observation of the iron bacteria in the stream, as elevated water hardness is principally caused by elevated calcium and manganese levels. Iron bacteria oxidize and precipitate dissolved iron and manganese. In addition, samples from the two water supply wells were analyzed for volatile organic compounds (EPA Method 624) and none were detected.

The following conclusions were determined upon completion of this phase of the evaluation:

1. Free-floating product was not observed on the groundwater or surface water at the site;
2. Minor fuel oil surface stains were observed on the ground surface around the top of the supply well;
3. The results of the chemical analyses detected residual concentrations of petroleum hydrocarbons in the groundwater beneath the site;
4. No petroleum hydrocarbons were detected in samples from the on-site supply well, neighbor's supply well, stream, or pond.

Based upon the findings of this phase of our investigation, it was recommended that additional investigation was required. The actual extent of the soil and groundwater contamination was unknown and a prediction of the migration potential could not be determined in the absence of any defined hydraulic gradient.

Therefore, it was recommended that additional monitoring wells, one upgradient well and three downgradient wells, be installed. The wells would be installed using DEQE-approved well construction and development procedures. Soil samples collected during the test borings would be screened with a PID for volatile organic vapors, and the wells would be stadia-surveyed in order to construct a localized groundwater contour map.

Four additional test borings were performed at the site with a truck-mounted drill rig, using 3-in. hollow-stem augers. The augers and drilling tools were decontaminated prior to the start of drilling by steam cleaning. The tools were washed with tri-sodium phosphate soap between borings to prevent cross-contamination.

Soil samples were collected at 5-ft depth intervals and/or at each change in the stratigraphy with a split-spoon sampler. The results of the soil sampling indicated that the site is predominantly underlain by glacial outwash, sand, and gravel deposits. Additional soil samples were collected from each boring at the depth when the water table was first encountered and screened with a PID. The samples were collected in glass jars, covered with a double layer of aluminum foil, tightly sealed with a metal lid, placed in a cooler, and transported to an in-house laboratory for headspace screening. The results of this screening procedure indicated elevated volatile organic vapor levels of 12.5 ppm and 17.5 ppm at test borings B101 and B102, respectively.

Groundwater monitoring wells were installed at the completion of each test boring, and were denoted as MW-100, 101, 102, and 103, as shown on Figure 2. Monitoring wells MW-100 and MW-101 were located downgradient of the interceptor trench. Monitoring well MW-102 was located within the interceptor trench and adjacent to the onsite supply well. MW-103 was placed upgradient of the release area to provide background water quality conditions. The depth to groundwater measured at the time of the monitoring well installations ranged from 2.0 to 4.5 ft below the ground surface.

The monitoring wells and the recovery standpipe installed by the previous contractor were removed with a backhoe, as they were no longer useful. Fuel oil stains were observed on the screened sections and silt was accumulated in the screen slots and in the bottom of the wells when they were removed.

The excavated area created by the removal of the standpipe was backfilled with clean sand and gravel. The absorbent pads located in the stream and pond, and several drums containing contaminated absorbent material, were collected and loaded onto a roll-off dumpster for disposal. A containment boom and several absorbent pads were left in place at the mouth of the stream as a precautionary measure against future oil breakout, if it were to occur.

Contaminated soil and debris loaded at the site were disposed of at an approved out-of-state landfill location. An out-of-state location was chosen for disposal due to time constraints and the complexities of the DEQE approval process for in-state disposal of contaminated material.

The remaining monitoring wells and a stake set in the stream were stadia-surveyed to determine the hydraulic gradient and therefore a potential oil migration direction. The groundwater elevations, analyzed using a 3 point problem, indicated a shallow groundwater table gradually sloping from east to west (non-pumping) across the site, as shown on Figure 2. The streamflow was elevated, and assumed to have been influenced by a rain event that occurred the morning prior to the survey. The stream was presumably a groundwater discharge boundary, as evidenced by the observed fuel oil breakout along the stream bank when the release initially occurred.

A limited duration pump test was conducted on the supply well to determine if petroleum hydrocarbons would be drawn into the well under abnormally prolonged pumping conditions. Static and dynamic water levels were taken over a 5-hr period at each monitoring well location and at the gauging stake surveyed in the stream. A discharge rate of approximately 10 gal per minute was established for the test. The supply well was not accessible during pumping, therefore drawdown could not be measured.

Water level measurements recorded during the first 4 hr of the pump test indicated that water levels in the monitoring wells and stream did not change appreciably from the static water levels recorded prior to the start of the test. The minor fluctuations observed between readings at each location were within the degree of accuracy for reading the electronic water level indicator used to gauge the wells.

Groundwater samples were collected from each monitoring well prior to the start of the pump test, and a water sample was collected from the tap located on the supply well at the completion of the test. The samples were analyzed for total petroleum hydrocarbons, and relatively low levels of hydrocarbons were detected in the monitoring wells, and nondetectable levels were indicated from the supply well. Table 3 presents a summary of the chemical analyses.

Table 3.   Case History #3
Summary of Total Petroleum Hydrocarbon (TPH) Analyses

Sample Location	TPH (ppm)
MW-100	0.1
MW-101	48
MW-102	17
MW-103	ND
Residential Well (onsite)	ND

Note: ppm = parts per million
ND = below minimum detectable level (0.1 ppm)

Constant head (discharge) permeability tests were conducted at each monitoring well location to determine the hydraulic conductivity of the aquifer, thereby evaluating the migration potential of the residual fuel oil. The horizontal hydraulic conductivities determined by this analysis ranged between 226 to 56 ft per day. The results of this analysis indicated that the soils at the site generally exhibited high permeabilities typical of sand and gravel hydrogeologic environments.

To assess the ambient air quality at the site, the interior of the cottage was screened with a PID and no volatile organic vapors were detected above normal indoor background concentrations of 0–1 ppm. Surface soils at the site were also screened and no volatile organic vapors were detected above normal background concentrations of 0–2 ppm.

The following interpretations and conclusions were arrived at based upon the investigations at the site:

- The results of the chemical analyses of the groundwater indicated relatively low levels of residual hydrocarbons in the vicinity of monitoring wells MW-100, 101, and 103. The TPH levels detected in the groundwater samples collected in our first sampling round, from the wells installed by the oil recovery contractor were higher than the levels detected in the monitoring wells installed in our evaluation. The higher TPH levels for the earlier samples were attributed to the accumulated silt inside the monitoring wells and the fuel oil smeared on the outside of the well screens which were acting as sources of contamination each time the wells were sampled.
- The samples collected from our wells were expected to more accurately reflect the concentrations occurring in the groundwater due to preferred well construction and well development techniques. The highest concentration of 48 ppm TPH, detected in the groundwater from MW-101, correlated with the low hydraulic conductivity of 56 ft/day determined for the soils at this location. It was also noted that MW-101 is located downgradient of the interceptor trench, indicating that the contamination had migrated past the interceptor trench.
- No petroleum hydrocarbons were detected in the sample collected from the supply well at the completion of the pump test. This indicated that the screened section of the supply well is positioned deep enough into the aquifer so that any residual contamination in the aquifer above would not be drawn into the well during normal (intermittent short-term) pumping conditions.
- The results of the groundwater gradient and flow direction analysis indicated an essentially flat groundwater table sloping from east to west across the site. Groundwater was expected to discharge into the pond and/or stream, presumably natural groundwater discharge boundaries. The interceptor trench was not believed to be influencing groundwater flow patterns at the site since the trench excavation was only approximately 1 ft deep into the water table, the aquifer materials were shown to have a high hydraulic conductivity and the water table was determined to be essentially flat.
- Due to relatively high hydraulic conductivities at the site, the low level residual contamination in the groundwater was expected to migrate fairly rapidly offsite, into the abutting pond. Since petroleum hydrocarbons were not detected in the water samples collected from the pond or stream, these sensitive aquatic environments were not believed to have been adversely impacted by the released fuel oil. Any fuel oil that may have reached these surface water bodies was presumed to have undergone chemical and biological degradation effects.
- Furthermore, the results of the pump test indicated that petroleum hydrocarbons were not drawn into the onsite water supply well after a prolonged pumping period of 5 hr.

Based upon our evaluation of the site, it was determined that the site was restored to a usable and habitable status and that no further soil removals or groundwater recovery and treatment operations were warranted.

Site Status

The successful assessment and remediation of this site led to a significant cost reduction for our clients, the insurance company, and owner. Occupancy of the cottage was restored with a recommendation that the onsite water supply well and septic system be relocated upgradient, to minimize potential impacts from any residual fuel oil. No further odors were detected during the past winter heating season and fuel oil breakout has not occurred in the surface water bodies.

PART VI

Risk Assessment and Risk Management

CHAPTER 28

Overview: Risk Assessment/Risk Management

John A. Del Pup

From the time we get up in the morning until we retire at night, each of us is involved in a series of risk assessment/risk management decisions. Many of these decisions are made by us consciously, or have become part of our way of life, while others have been made for us by others such as regulatory agencies, spouses, and colleagues.

For example, the choice of whether to smoke a cigarette or forego bacon and eggs for oat-bran was probably a conscious decision for some of us, but part of our habits for others.

We rely on regulatory agencies and industry to provide:

- safe—health based limits for the quality of water we use to brush our teeth
- safe—airplanes we ride and roads we drive on
- safe—products we use at work
- safe—food we eat

RISK ASSESSMENT

Toxicologists, as well as other health and technical professionals in industry, have been conducting health assessments for decades to assure safe products for consumers as well as a safe environment for employees. These assessments, however, tend to be more qualitative than quantitative because of the varied conditions under which the products are used—that is, there is a lack of site specificity and defined population potentially at risk.

Over the past 4 or 5 years there has been an increasing focus on the use of quantitative risk assessment in addressing environmental issues with special attention being paid to chronic human health effects. For example, prior to initiation of any remediation at National Priority List (NPL) hazardous waste sites, a risk assessment is required pursuant to the Comprehensive Environmental Response, Compensation, and Liability Act (CERCLA) and Resource Conservation and Recovery Act (RCRA).[1] The basic approach used in risk assessments is the one formalized by the U.S. National Research Council.[2]

It consists of four steps based upon:

Hazard Identification: to ascertain if there is a causal relationship between chemical(s) found at the site and health effects in man and/or animals. It may be necessary, owing to cost as well as lack of biological data on all chemicals identified or predicted to occur, to select a surrogate list of compounds for further evaluation in the risk assessment. This may be done on the basis of known physical/chemical properties such as solubility in water, biodegradation, and toxicity.[3] It is also important to identify the chemical form of the substance. For example, inorganic mercury compounds affect organs such as the kidney, whereas methyl mercury affects the nervous system.

Dose—Response Assessment: to ascertain whether known and/or predicted levels of chemicals can cause the effect based upon epidemiological and laboratory bioassays. Although the focus of the risk assessment is human health, few epidemiological data, either workplace or community, exist for many chemicals. In addition,where health effects data are available for laboratory animals they are often for only a single route of exposure and more often than not acute rather than chronic data. These, as well as the crucial differences between man and animals in terms of doses, effects, and target organ responses, result in a great deal of uncertainty and consequential conservativism in estimating the dose and target organ response in man.

Exposure Assessment: to ascertain the potential routes of exposure and levels of human exposure. A pathway analysis may include a review of the source of emissions; air, water, soil and the potential routes of human exposure; inhalation, soil ingestion and contact, water ingestion and contact as well as through the food chain. Exposure assessment is probably the most critical step because if there is no real exposure there is no real risk, and a lot of unwarranted concern on the part of the public can arise by an overzealous attempt to force the association.

Risk Characterization: describing the nature of the risk and the uncertainty. This step in the risk assessment process should address what the estimated risks are for a given health effect and route of exposure as well as the attendant uncertainty in this estimation. In addition, it should also place the risk(s) in perspective with other known risks in order to properly communicate the relative risk to the public and others who may be part of the decision-making process to manage the risks.

RISK MANAGEMENT

Risk management is the selection of option(s) to address the identified risks. Depending upon the purpose of the risk assessment, such as hazardous waste site cleanup, property acquisition and/or sale, as well as the location and end use of the site, participants' expertise may include legal, technical, business and, regulatory.

In summation, the goal of the risk assessment/risk management process should be to identify risks, place the risks in perspective, and manage the risks in a cost-effective manner to benefit man and the environment.

REFERENCES

1. The Environmental Protection Agency. "The Endangerment Assessment Handbook." Prepared by PRC Engineering. Chicago, IL, 1985.
2. National Research Council. "Risk Assessment in the Federal Government: Managing the Process," National Academy Press, Washington, D.C., 1983.
3. Ibbotson, B.G., et al. "A Site Specific Approach for the Development of Soil Clean Up Guidelines for Trace Organic Compound. Presented at Second Conference on Environmental and Public Health Effects of Soils Contaminated with Petroleum Hydrocarbons, University of Massachusetts, Amherst, MA, 1987.

CHAPTER 29

Creative Approaches in the Study of Complex Mixtures: Evaluating Comparative Potencies

Rita S. Schoeny

INTRODUCTION

An ongoing problem for agencies such as the U.S. Environmental Protection Agency (EPA) is that of providing guidance as to health risk assessment of mixtures from a variety of sources such as wastewaters, hazardous waste sites, air particulates, or materials spilled in waters or soils. In the absence of data on the mixture in question and its particular matrix, some approach using surrogates must be taken.

One such approach, which is discussed in the U.S. EPA 1986 Guidelines for Health Risk Assessment of Chemical Mixtures,[1] is to obtain an index of toxicity by summing the results of the risk assessments for individual components that have been identified as part of the mixture, after consideration of the potential for interaction among those components. There are generally considerable difficulties with this method. In the case of complex mixtures, such as organic extracts of drinking water or combustion products, the majority of compounds may not have been identified, much less quantified. If identified, the biological activity may not be known for a particular agent. And lastly, the current state of knowledge regarding interactions is such that, with few exceptions, nothing is known as to the type or extent of interactions that could be predicted to occur in any given combination of compounds. Currently, the recommendation in the Guidelines is for the development of a toxicity index that can assist in assigning priorities for action, but does not give a likelihood of adverse effect or an individual risk attributable to mixture exposure.

In the case of a homolog mixture for which the components are known or can be reasonably conjectured, the U.S. EPA Risk Assessment Forum has recommended that a series of toxicity equivalency factors (TEF), be developed. This has been proposed for estimating risks associated with chlorinated dibenzo-dioxins (CDDs) and dibenzofurans (CDFs) (Bellin and Barnes[2]), and is also under consideration for a set of polycyclic aromatic hydrocarbons (PAH). In this risk estimation procedure, information is obtained on the concentrations of the various congeners or homologs in the mixture of interest. Next, available toxicological data and structural activity relationship judgments are used to estimate the potential human hazard associated with exposure to each mixture component. For the CDD and CDF, this potential is expressed in terms of "equivalent amounts of 2,3,7,8-tetrachlorodibenzo-p-dioxin," (TCDD), an isomer for which a reasonable health effects data base exists. In other words, exposures are translated into a common scale from which risk can be derived. The assessment outcome is, in this case, an estimate of risk rather than a priority ranking. For the CDD and CDF, TCDD was assigned a value of 1 and the other CDDs and CDFs ranked according to the following criteria. In the absence of human carcinogenicity data, results from long-term animal tests for carcinogenic activity were used. In the absence of any carcinogenicity data, data on reproductive effects were considered. Lacking such data, information from in vitro assays was used, in particular the ability of a specific congener to bind to a cellular receptor which controls induction of aryl hydrocarbon hydroxylase enzymes. This last activity is associated with expression of several manifestations of TCDD toxicity (e.g., cleft palate in the mouse, and thymic involution and body weight loss). Thus, based on available data, the various CDD and CDF components were assigned a toxicity equivalency factor in relation to a signature, well-studied compound. For the PAH, benzo[a]pyrene serves as the signature compound, and the others are ranked according to their activities in tests such as skin painting bioassays. For a given mixture, the estimated risks for each component are summed, and information on exposure incorporated to determine the potential human hazard.

For mixtures of components that are not readily identified or separated, another method, that of comparative potency determination, has been suggested, both in the recent (1988) NRC report on *Methods for In Vivo Toxicity Testing of Complex Mixtures*,[3] as well as in a draft *Technical Support Document for the Risk Assessment of Mixtures*. Both the TEF and comparative potency approaches utilize types of data not generally considered to be suitable for risk assessment. One point of difference is that the TEF method assumes that either doses or responses can be added (so-called dose or response additivity). This assumption is most suitably applied to agents that share a mode of action or are believed not to interact in producing a biological effect. Comparative potency judgments, by contrast, are validated by comparison to in vivo data on a mixture presumed to be similar to the one in question. Mixture is compared to mixture; the possibility of accounting for interactions is, therefore, not excluded.

DESCRIPTION OF COMPARATIVE POTENCY APPROACH

In its simplest form a comparative potency estimate is done on a mixture for which data are incomplete by carrying out a comparison to a similar mixture on which a human risk assessment has been done. For example, if there are human dose-response data on mixture A and rodent skin painting bioassays for both mixtures A and B, then a comparative potency approach may be used to determine a human risk estimate for mixture B. The underlying hypothesis is that in different assays the potency of the agents relative to one another is the same; if mixture A is twice as potent a human carcinogen as B, then it will be twice as potent a rodent skin carcinogen, twice as active in a *Salmonella* mutagenicity assay, and so on. This can be expressed mathematically as:

$$RP_1 = k(RP_2)$$

where RP_1 and RP_2 are the relative potencies of the agent in assays 1 and 2, and k is a constant.

Albert et al.[4] and Lewtas[5] applied this method in the estimation of human lung cancer risks from a number of combustion products including coke oven emissions, roofing tar, cigarette smoke condensate (CSC), and particulates from diesel- and gasoline-fueled engines. Epidemiologic data were available for the first three materials; namely, dose-related increases in the observation of lung cancer in exposed humans. These data permitted calculation of the lung cancer unit risks presented in Table 1. If the most potent of these three mixtures, coke oven emissions, is given the relative potency of 1, then the RPs for roofing tar and CSC for human lung cancer are those in column two of Table 2. There were no epidemiologic data suitable for deviation of a unit risk for Nissan diesel emissions. An extract of a particulate sample from this engine type, however, had been tested in vivo in a mouse skin initiation/promotion protocol, as well as in vitro for genotoxicity. Likewise, samples of the other three combustion products had been assayed in a similar fashion. The relative potencies of all four agents based on these short-term bioassays are presented in columns 3-5 of Table 2. Examination of the comparative potencies shows good agreement among the assays, with the exception of the relatively high activity of CSC and Nissan diesel in the *Salmonella* mutagenicity assay. Having thus shown that for at least some assays the relative potencies of these mixtures were constant, a determination of Nissan diesel unit risk for humans was made by Albert et al.[4] and Lewtas,[5] by using the relative potency based on the mouse initiation/promotion data (Table 3). The unit risks thus derived (column 4, Table 3) were within a factor of 2 of one another; the final estimate was prepared by taking a simple mean.

Table 1. Human Lung Cancer Risks For Combustion Products.[a]

Combustion Product	Unit Risk/μg/m^3
Coke oven emission	$9.3 \times 10^{-\frac{1}{2}}$
Roofing tar	$3.6 \times 10^{-\frac{1}{2}}$
Cigarette smoke condensate	$2.2 \times 10^{-\frac{3}{4}}$

[a]Values from Albert et al.[4]

Table 2. Relative Potencies of Emission Extracts In Bioassay Systems.[a]

	Relative Potency			
Combustion Product	Human Lung Cancer	Mouse Skin Tumor Initiation, Promotion Assay[c]	Mutation for Mouse Lymphoma Cells[c]	Mutation for Salmonella Strain TA98[c]
Coke oven emission	1.0	1.0	1.0	1.0
Roofing tar	0.39	0.20	1.4	0.78
CSC[b]	0.0024	0.0011	0.066	0.52
Nissan diesel	—	0.28	0.24	12.0

[a]Values from Albert et al.[4]
[b]CSC = Cigarette smoke condensate.
[c]In vitro tests employed hepatic homogenates (S9) to provide metabolic activation.

Table 3. Derivation of Nissan Unit Risk.[a]

Combustion Product	Comparative Potency Mouse Skin Tumor Initiation[b] A	Unit Risk Human Lung Cancer/μg/m^3 B	Nissan Diesel Unit Risk /μg/m^3 A x B
Coke oven	0.28/1	9.3×10^{-4}	2.6×10^{-4}
Roofing tar	0.28/0.20	3.6×10^{-4}	5.2×10^{-4}
CSC	0.28/0.0011	2.2×10^{-6}	5.4×10^{-4}
		Average	4.4×10^{-4}

[a]Values are from Albert et al.[4]
[b]Nissan relative potency based on mouse skin tumor data + combustion mixture relative potency based on mouse skin tumor data.

This method was further applied to an extended set of diesel emission samples for which there were data in mouse skin tumor and short-term tests for genotoxicity (Table 4) but no epidemiology data. With the exception of the Nissan sample, the engine emission samples were only weakly active in the mouse skin tumor initiation/promotion assay. The correlations between relative potency in that assay and those from the genotoxicity assays were, however, very strong, as can be seen from Table 4. Unit risks for the diesel sample were derived by using the Nissan unit risk and the average relative potencies from Salmonella mutagenicity data, mouse lymphoma cell (L5178Y) mutagenicity, and sister chromatid exchanges in CHO cells (Table 5). In a later publication, Lewtas[6] reported an average unit lung cancer risk for the diesel engines to be $2.6 \times 10^{-5}/\mu$g/m^3

particulate matter. The author notes that this compares very well with a unit risk of $1.2 \times 10^{-5}/\mu g/m^3$ derived from data on incidence of adenomas and adenocarcinomas observed in several studies of rats exposed to diesel particulate for their lifetimes. Lewtas[5,6] has published unit risks based on relative potency determination for several emission extracts from several automotive engines and for combustion sources including woodstoves, oil furnaces, and utility power plants. Overall, the author reported a range of only about 150-fold in cancer unit risks calculated using comparative potency based on mutagenicity and tumor initiation data for particulate material obtained from these combustion processes.

Table 4. Relative Potencies of Diesel Emission Extracts Based on Four Bioassays.

	Relative Potencies[a]				
	Mouse Skin Tumor Initiation	Mutation in L5178Y Cells[b]	Sister Chromatid Exchange in CHO Cells[b]	Mutation for Salmonella Strain TA98[b]	Average
Nissan	1.0	1.0	1.0	1.0	1.0
Volkswagen	0.41	0.25	0.42	0.23	0.30
Oldsmobile	0.53	0.45	0.24	0.11	0.27
Caterpillar	NEG[c]	0.022	NEG	0.023	0.015
Correlation with mouse skin tumor initiation	(r^2)	0.95	0.83	0.72	

[a]Values from Albert et al.[4]
[b]In vitro tests incorporated rat hepatic extracts (S9) to provide metabolic activation.
[c]NEG = Negative, no increase in response by comparison to controls on assay.

Table 5. Unit Lung Cancer Risk Estimates for Diesel Emissions.[a]

	Unit Risk Estimates (Lifetime Risk/$\mu g/m^3$)			
	Comparative Potency Average	Organics[b]	Organic Extractables %	Particulates[c]
Nissan	1	$4.4 \times 10^{-\frac{1}{2}}$	8	$3.5 \times 10^{-\frac{5}{6}}$
Volkswagen	0.30	$1.3 \times 10^{-\frac{1}{2}}$	18	$2.3 \times 10^{-\frac{5}{6}}$
Oldsmobile	0.27	$1.2 \times 10^{-\frac{1}{2}}$	17	$2.0 \times 10^{-\frac{5}{6}}$
Caterpillar	0.015	$6.6 \times 10^{-\frac{1}{2}}$	27	$1.8 \times 10^{-\frac{3}{4}}$

[a]Values from Albert et al.[4]
[b]Unit risk estimates were obtained by multiplying the Nissan unit risk by the average comparative potency.
[c]These values were based on percent organic extractables in preceding column.

ASSUMPTIONS, USES, AND LIMITATIONS OF COMPARATIVE POTENCY APPROACHES

The example provided by the comparative potency and cancer risk estimation for combustion products, especially the diesel particulates, indicates that this method could be used to predict the potential human risk of similar mixtures based on short-term in vivo or in vitro tests. If such a method were validated, considerable resources could be saved that otherwise would be used to conduct animal bioassays. This promise of the relative potency approach, however, is tempered by a consideration of the limitations of this approach and an investigation of its underlying assumptions.

Models and Methods of Validation

Although there are references to the use of comparative potency to predict human health risks for endpoints other than cancer, the greatest use of the approach has been for cancer risk prediction. Specifically, methods of deriving the estimates have been tied to use of a linear, nonthreshold model for the low dose region of the dose-response curve for carcinogens. This model is based on the concept of a carcinogen as an agent that produces an irreversible change in a critical site, which initiates the process of neoplasia. Often a carcinogen is thought to produce a mutation (or mutations) as a first step in carcinogenesis. Thus, the tie to in vitro tests that measure mutagenicity or changes (such as damage or repair) in DNA and which generally show low-dose linearity of response, is on a reasonable conceptual basis. Chemical and biological characterization of the combustion materials indicates that these mixtures do contain components which have initiating activity. There are certainly promoting agents present as well, and some of the materials known to be initiators may also serve as promoters for other components of the whole mixture. The contribution of these to the overall carcinogenic potential of the mixture may not be accounted for in a comparative potency judgment based on in vitro genotoxicity, or one based on an initiation/promotion assay. Recently more attention has been turned to the subject of nongenotoxic carcinogens. Short-term assays designed specifically for their detection are now being developed; however, these are not as well characterized or validated as the better-established mutagenicity assays and are not readily applicable to comparative potency determinations at this time.

It seems reasonable to assume that greater confidence can be attached to a comparative potency approach which relies at some point on human data. The agents, both mixtures and single compounds for which such dose-response data exist, are limited. Albert[7] listed 13 human carcinogens considered suitable for comparative potency derivation and suggested animal models that reflect the carcinogenic specificity of these agents. Compounds for which there are no quantitative human data could be used in the process if they are known to have a well-characterized response in one of the animal models.

Another method of preparing estimates of potency for purposes of comparison has been presented by Peto et al.[8] They propose a numerical measure of carcinogenic potency (analogous to an LD_{50}) which they term a TD_{50}, or tumorigenic dose rate for 50% of the test animals. In other words "for any particular sex, strain, species, and set of experimental conditions, the TD_{50} is the dose rate (in mg/kg body weight/day) that if administered chronically for a standard period—the 'standard lifespan' of the species—will halve the mortality-corrected estimate of the probability of remaining tumorless throughout that period." The TD_{50} as an index of carcinogenic potency has been used as the basis for comparison for a Carcinogenic Potency Database (Gold et al.[9]) comprising data from about 3000 long-term animal bioassays on some 770 compounds. One advantage of this method for use in comparisons is that the actual experimental range of doses tested will generally bracket the TD_{50}; this allows for accurate estimation and eliminates the necessity of modeling the data. A disadvantage is the requirement for many interim sacrifices to identify nonlethal tumors.

Decision Points

At many points in the course of preparing a comparative potency determination there are important choices to be made, some of which may not be obvious. The units in which carcinogenic potency will be compared is one such choice. Unit risk can be compared on the basis of risk/μg/m³ for materials emitted to the air, risk/μg/L of water, risk/mg ingested material/kg body weight, and so on. For particulate samples, one could evaluate the risk per unit of extracted organic material from collected particles (which is generally what is tested in vitro or by skin painting), or risk per unextracted particle (which is the material tested in long-term inhalation studies and which represents the likely exposure to humans). Lewtas[5] suggests normalizing data to responses per kg fuel, km traveled (for engine emissions), or megajoule of heat produced. For spilled materials it will be particularly important to consider the matrix wherein environmental exposure takes place. For example, unit risk might be expressed in terms of risk/kg soil.

Another series of choices particularly germane to comparative potencies of mixtures is the form, source, and preparation of the environmental mixture sample. Optimally one wishes to use data on a sample most relevant to human exposure. This is not always practical in terms of in vitro or short-term testing. For example, diesel particulate is not amenable to assay in *Salmonella typhimurium*. The bacteria have no capacity for phagocytosis, and exposure to potentially mutagenic nonwater-soluble materials associated with particles embedded in agar would be minimal. Likewise, particles would be an inappropriate vehicle for a mouse skin tumor induction assay. When the sample has been modified to facilitate testing (by extraction, fractionation, or in some other way) the validity of its use in the context of potential human exposure must be assured. This is also true for locating points of sampling (e.g., samples taken from the top of a coke oven battery vs samples that could be obtained in greater quantity from within the coke oven battery). An example of this decision-making process and the validation of such decisions is found in the Albert et al.[4] paper.

The type of assay to be used for comparison with the human data is another point of decision. It is clear that no single assay will be suitable for risk estimation of every sort of mixture; thus, those materials that have undergone a battery of assays will be most amenable to comparative potency determination. An instructive illustration of this can be seen in Table 2. The relative potencies of both CSC and the Nissan diesel particulate extract relative to coke oven emissions were a great deal higher when based on the *Salmonella* mutagenicity data than when based on data from the other assays. Chemical composition data for diesel particulates show the presence of nitropyrenes as well as other nitroaromatics. Due to the presence of an endogenous nitroreductase enzyme system, *Salmonella* is particularly sensitive to the mutagenic activity of nitropyrenes (Mermelstein et al.[10]). It was shown that for one diesel sample, a total of 23 nitroaromatics was responsible for about 40% of the TA98 mutagenicity of the complete sample. Use of a congenic *Salmonella* strain which lacked the nitroreductase enzyme system resulted in a decrease of about 50% of the mutagenic activity of the sample.[6] Thus, it would seem that use of the *Salmonella* assay as a basis for comparative potency determinations of nitropyrene-containing materials would have resulted in over-estimation of potential human risk.

The use of a tiered approach to testing agents for potential carcinogenicity has been recommended on many occasions (e.g., NRC[3]) as a means of prioritizing testing and conserving resources. The U.S. EPA Health Effects Research Laboratory has proposed a three-tiered system. The first tier consists of screening bioassays determined to produce a minimum of false negatives (nonpositive agents which are, in fact, carcinogens); this tier is comprised of *Salmonella* mutagenicity and mammalian cytogenicity tests. Tier 3 consists of lifetime bioassays in rodents. The intermediate tier (also called the carcinogen testing matrix or CTM) would be made up of short-term in vivo assays designed to eliminate false positives and to predict the relative carcinogenic potency of agents. The skin tumor initiation/promotion assay in SENCAR mice, the lung adenoma assay in Strain A mice, and the rat liver altered enzyme focus bioassay, have been evaluated as to their ability to identify as positive 18 compounds of varying structure and target organ specificity which are known to be carcinogenic in long-term animal studies.

The short-term in vivo tests were also assessed for their ability to predict the relative potency of carcinogenic compounds or mixtures. Statistical methods for use with the CTM are being developed and validated in the following areas: (1) robust sensitive statistical methods to evaluate each of the three tests for a positive dose-related response; (2) statistical methods for estimation of carcinogenic quantitative risk from the long-term animal tests; (3) statistical methods for estimation of dose-response rate from the short-term tests; (4) and finally, methods for evaluation of the ability of the CTM to predict relative potency in a reliable fashion.

Taken as a battery of three tests, the CTM was able to identify 17 of 18 animal carcinogens, using the criterion of a positive response in at least one of the tests. In this phase of the validation, the lung adenoma assay was most reliable, and the liver focus assay did not identify any compound as positive that was not also detected by the other two.

Relative potencies were also done using data from a standard long-term and from the CTM short-term assays on the basis of TD_{50} determinations. In terms of prediction of relative potency for the long-term cancer bioassays, it is noteworthy that the initiation/promotion test provided significantly poorer predictions than either the liver focus assay or the lung adenoma assays. Future work with the CTM will probably encompass tests with additional chemicals to assist in ascertaining which assays will be most useful for particular chemical types, both for identification of potential carcinogens and for derivation of comparative potencies. The data from these efforts should provide not only some useful bases for relative potency development, but also guidance as to the types of short-term tests most applicable to various classes of agents and ultimately to various types of mixtures.

Many short-term tests have more than one endpoint available for measurement and evaluation. For example, the CTM evaluation effort determined that the most sensitive response measure for the skin tumor assay was time-to-tumor; for lung adenoma, the number of tumors; and for the liver focus assay, the number of foci per unit area. The selection of test endpoint or measurement is, thus, another decision to be made in the choice of assay data for comparative potency. Choice of measurement and the way in which data were handled is discussed at some length in one published description of use of mouse skin tumor data in the calculation of relative potencies of coke oven emission, roofing tar, and CSC (Albert et al.[4]).

CONCLUSIONS

Two requirements for application of comparative potency are (1) information on the composition of the mixture, and (2) data on a mixture which is sufficiently similar, both as to types of components present and as to the biological activity of those components. Thus, this procedure would seem to be less useful in the evaluation of highly complex, very changeable mixtures, such as those associated with toxic waste sites. It should be noted that developing test designs to generate data for comparative potency determinations may involve controlling factors not necessary in other experimental designs. These (as described in Reference #3) include the following: simultaneous testing of all agents used for comparison in one experiment for those bioassays characterized by large variability between experiments; use of identical measures of dose or overlapping dose ranges to facilitate statistical analysis; and a thorough consideration of the specific objectives of the comparative potency study and expected use of the data.

Despite its reliance on some untested assumptions, the comparative potency method offers great promise as a means of evaluating the hazard potential of complex environmental mixtures. It offers the advantage of treating the mixture as such, rather than as a sum of its known components. As in vitro and short-term bioassay data on various mixtures become increasingly available, this procedure should become a focus for research on improving the methods employed and validation of its hypothesis.

REFERENCES

1. U.S. Environmental Protection Agency. "The Risk Assessment Guidelines of 1986," EPA/600/8-87/045, 1986.
2. Bellin, J., D. G. Barnes. "Interim Procedures for Estimating Risks Associated with Exposures to Mixtures of Chlorinated Dibenzo-p-Dioxins and -Dibenzofurans (CDDs and CDFs)," EPA Risk Assessment Forum, EPA/625/3-87/012, March 1987.
3. National Research Council. *Complex Mixtures. Methods for In Vivo Toxicity Testing* (Washington DC: National Academy Press, 1988).
4. Albert, R. E., J. Lewtas, S. Nesnow, T. Thorslund, and E. Anderson. "Comparative Potency Method for Cancer Risk Assessment: Application to Diesel Particulate Emissions," *Risk. Anal.* 3:101-117 (1983).
5. Lewtas, J. "Development of a Comparative Potency Method for Cancer Risk Assessment of Complex Mixtures Using Short-term In Vivo and In Vitro Bioassays," *Toxicol and Indus. Health.* 1:193-203 (1985).
6. Lewtas, J. "Genotoxicity of Complex Mixtures: Strategies for the Identification and Comparative Assessment of Airborne Mutagens and Carcinogens from Combustion Sources," *Fund. Appl. Toxicol.* 10:571-589 (1988).
7. Albert, R. E. "The Comparative Potency Method: An Approach to Quantitive Cancer Risk Assessment," in *Methods for Estimating Risk of Chemical Injury: Humans and Non-human Biota and Ecosystems*, V. B. Vouk, G. C. Butler, D. G. Hoel, and D. B. Peakall, Eds., pp. 281-287. SCOPE. 26, SGOMSEC Conference, 1985.
8. Peto, R., M. C. Pike, L. Bernstein, L. S. Gold, and B. N. Ames. "The TD_{50}: A Proposed General Convention for the Numerical Description of the Carcinogenic Potency of Chemicals in Chronic-Exposure Animal Experiments," *Environ. Health Persp.* 58:1-8 (1984).
9. Gold, L. S., C. B. Sawyer, R. Magaw, G. M. Backman, M. De Veciana, R. Levinson, N. K. Hooper, W. R. Havender, L. Bernstein, R. Peto, M. C. Pike, and B. N. Ames. "A Carcinogenic Potency Database of the Standardized Results of Animal Bioassays," *Environ. Health Perspect.* 58:9-319 (1984).
10. Mermelstein, R., D. K. Kiriazides, M. Butler, E. McCoy, and H. S. Rosenkranz. "The Extraordinary Mutagenicity of Nitropyrenes in Bacteria," *Mutat. Res.* 89:187-196 (1981).

CHAPTER 30

How Much Soil Do Young Children Ingest: An Epidemiologic Study

Edward J. Calabrese, Harris Pastides, Ramon Barnes, Carolyn Edwards, Paul T. Kostecki, Edward J. Stanek, III, Peter Veneman, and Charles E. Gilbert

INTRODUCTION

It has long been recognized that contaminated soil may present a potential public health concern because of groundwater contamination since groundwater is a significant drinking water source. Recently, regulatory and public health agencies have become concerned that consumption of contaminated soil by children may present a significant public health problem. For example, soil levels of lead that range from 2500–7500 $\mu g/g$ in certain sections of Boston are suspected of contributing to elevated blood lead levels as a result of soil ingestion.[1] Also widely discussed has been the dioxin contamination of Times Beach, Missouri. The Centers for Disease Control (CDC) derived a theoretical cancer risk associated with levels of dioxin in soil.[2] A major concern was the assumed consumption by children of soil containing dioxin, a contaminant that is relatively tightly bound to soil.

While initial qualitative attempts at estimating childhood soil ingestion have been made by several groups,[3-5] these attempts lack sufficient quantitative evaluation to allow confident estimation of actual soil ingestion. In addition, scientists at Centers for Disease Control (CDC) developed an estimation for specific age groups based on unpublished behavior observations of children aged 1.5 to 3.5 yr. These children were estimated to ingest 10 g of soil/day.[2]

The first attempt to estimate human soil ingestion quantitatively was presented by Binder et al.[6] using aluminum (Al), silicon (Si), and titanium (Ti) as soil tracer elements. These elements were selected since their concentration is high in soil but low in food products, and their gastrointestinal tract absorption is low. The amount of soil ingested was calculated based on the fecal and soil concentrations of the tracer elements and the amount of fecal output. In their study involving 59 diapered children aged 1–3 years, the calculated mean soil ingestion estimates for the tracers Al and Si were 181 and 184 mg/day, and 10 times higher (1834 mg/day) if Ti was used. The authors were unable to resolve the apparent conflict between estimates based on the various tracers.

Clausing et al.[7] have estimated the amount of soil consumed by children following a method similar to Binder. They reported mean soil ingestion values ranging from 127 mg/day to 1084 mg/day, depending on the marker, with Ti yielding the highest estimate. The Binder report and the Clausing study are indirect attempts to estimate soil ingestion, and could be improved by measuring the concentration of tracers in food and other ingested products (e.g., medicines), as well as the presence of tracers in diapers and other materials that contact the feces. These factors combined with reported average adult dietary Al, Si, and Ti intakes of 15 mg/day, 7 mg/day, and 0.3 mg/day, respectively,[8,9] may lead to an overestimation of soil consumption.

Based on the collectively cited studies, it is evident that the amount of soil consumed by humans, especially children, is uncertain. The present study was designed to extend the previously cited reports by assessing the amount of soil ingested in children aged 1–4 years, using a mass-balance methodology.

An experimental adult soil ingestion study was also conducted to validate the methodology used in the children's study by assessing the extent to which soil tracer levels contribute to the fecal levels, and inferring the comparative bioavailability of the tracers in soil versus foods. Since the adults served to validate the children's study, we present the results of the adult study first.

ADULT SOIL INGESTION VALIDATION METHODOLOGY

Participants in the study were six healthy volunteer adults (three males, three females), 25–41 years old, who did not have chronic illness such as diabetes, heart disease, or gastrointestinal disorders. Each of the volunteers ingested: (1) one empty gelatin capsule at breakfast and one with dinner Monday, Tuesday, and Wednesday during the first week of the study; (2) 50 mg of sterilized soil within a gelatin capsule at breakfast and dinner for a total of 100 mg of sterilized soil per day for the three days during week two; and (3) 250 mg of sterilized soil in a gelatine capsule at breakfast and dinner during week three, for a total of 800 mg of soil per day over the three days. Duplicate food samples were collected on Monday through Wednesday in each of the study weeks, and total excretion was collected from Monday through Friday for the three study weeks.

Soil Preparation and Security

Soil for the adult ingestion study was selected from a soil library maintained by the Department of Plant and Soil Sciences at the University of Massachusetts. The soil was selected because it was previously characterized as noncontaminated and the tracer elements (aluminum, barium, manganese, silicon, titanium, vanadium, yttrium, and zirconium) were of sufficient concentration to be detected during analysis of the volunteer's excretory materials. A formal security system was employed following the selection of this soil to prevent tampering with the soil and soil-filled capsules.

The soil was sterilized by autoclaving, oven-dried, and evaluated microbiologically to detect the presence of anaerobic and aerobic organisms and fungi.

Human subjects review required the soil to be chemically analyzed for lead and the U.S. Environmental Protection Agency's (EPA's) extractable priority pollutants, consisting of approximately 100 compounds, including chlorinated hydrocarbon insecticides, PCBs, numerous polynuclear aromatic hydrocarbons, and phenolic pollutants. The concentrations of these compounds in the soil were less than the detection limits: all compounds except polychlorinated biphenyls, 1 microgram per gram; polychlorinated biphenyls, 5 micrograms per gram.

Duplicate Food

Duplicate meal samples collected from the six adults included all food and beverage ingested from breakfast Monday through the evening meal Wednesday during each of the three weeks, to reflect total dietary intake. All medications and vitamins ingested by the adults were included in the samples or were addressed during chemical analysis. Each study subject prepared two meals each, containing the same quantities and types of food. One meal was eaten by the subject; the second meal was used to prepare the duplicate meal sample. All uneaten food on the subject's plate was consolidated and compared to the duplicate meal. The food portions in the duplicate meal were separated to represent ingested food and uneaten food. The estimated amount of eaten food was placed in the storage container for analysis. Duplicate beverages were collected in a similar manner.

Total excretion, feces and urine, was collected from Monday noon through Friday midnight (five days) over three consecutive weeks. The collection, transport, and storage of dietary intake and excretory output followed the same procedures as the children's study described below.

CHILDREN SOIL INGESTION METHODOLOGY

Sixty-five healthy children between the ages of 1.0 and 4.0 years were identified through two University of Massachusetts affiliated day-care facilities and through referred recruitment in the greater Amherst, Massachusetts area. Once enrolled in the study, duplicate meal samples were collected for all children on Monday through Wednesday on two consecutive weeks, while the total fecal and urine was collected on Monday through Friday morning in the corresponding weeks. Soil and dust samples were also collected in each child's home and play areas in the study period. The parents were trained in study protocol and their children were enrolled in the study after obtaining informed parental consent. Data collected from the parents included: dates of birth, education level, occupation, marital status, and other demographic information. Parents also reported any missing or lost samples of food, medicine, or fecal material, along with their children's outdoor and indoor activities and health status changes daily during the data collection period.

Duplicate Meal Collection Storage and Transport

Duplicate meals were collected for all children by the parents and day-care staff, all of whom had been trained in standard procedures. The duplicate meals represented all food and beverages ingested from breakfast Monday through the evening meal Wednesday over each of the two consecutive study weeks. Over-the-counter medications and vitamins ingested by the children were included in the duplicate meals.

The children in the study were supplied with toothpaste that contained non-detectable levels of the tracer elements, with the exception of silica, which was present in trace quantities.

The children's duplicate meals representing the three daily meals were placed in clean, one-liter polyethylene containers, then enclosed in ziplock bags, and transported in an insulated cooler bag with a reusable refrigerant.

Diaper and Commode Collection, Storage, and Transport

Baby cornstarch, diaper rash cream, and soap which had been evaluated for and found to be low in tracer content, were provided for infant hygiene. The parents and day-care staff were supplied with cotton cloth diapers and waterproof pants and advised to use a standardized procedure to reduce stool and urine contamination and sample variability. The parents and day-care staff were instructed to collect diapers and commodes starting Monday at noon and ending with the first void on Friday morning. A blank control with distilled water was run on every 10th diaper and commode used at the day-care and some homes. Wipes and toilet paper used on the child's bottom were not included in the diaper or the commode, and not collected by the research team. Used diapers and commodes were enclosed in ziplock plastic bags and placed in the insulated bags containing portable refrigerants. Diaper specimens were collected every morning in the insulated storage bag.

Soil and Dust Sampling

Soil and dust samples were taken from each child's home, play areas, and from each day-care facility. The parent identified the three areas (i.e., sites) outside the house where their child played most often during the study. Each soil sample site was located on a plot map and its distance from the house measured, using the two closest corners of the house as reference points. Parents estimated the percentage of the total time spent outdoors that the child played at each site. To simplify the procedure, the parents were instructed to assume that the three play areas would correspond to 100% total time outside.

Once the site was located, a composite sample was drawn using the eight major compass points. A stainless steel soil sampling corer was placed on the ground at each indicated sampling compass point and driven three inches into the soil.

Dust samples were collected using a specially modified commercial vacuum cleaner from those indoor floor surfaces identified by the parents and day-care staff as where the children crawled, walked, or played during the study.

Analytical Chemistry Methodology

Inductively Coupled Plasma-Atomic Emission and Mass Spectrometry

Two sequential ICP systems, a Perkin-Elmer Plasma 40 and a Perkin-Elmer Plasma II, were used to analyze food, urine, fecal, soil, and dust samples by atomic emission spectrometry for Al, Ba, Mn, Si, Ti, V, Y, and Zr.

An ICP-MS system (Sciex Elan Model 250) was coupled with flow injection for V, Y, and Zr measurements.

Processing Equipment

A flow injection analyzer was used to introduce dust and soil samples into the ICP-MS. A Spex mixer/mill (Spex Industries, Metuchen, NJ, Model 5100), food processor, programmable ashing furnace (Fisher Scientific Model 495), and commercial microwave digestion system (MDS-81D, CEM Corp., Indian Trail, NC) were used for soil and dust sample grinding, food sample ashing, fusion, wet digestion, and other sample treatment.

Reagents

Reagent-grade chemicals (hydrogen peroxide, boric acid, and nitric acid) and standard solutions (1000 μg/mg of Co, Al, Mn, Ba, Y, V, and In) were used to prepare the multielement standard solution. Canadian reference soil samples (Regosolic clay, CANMET SO-1; Podzolic soil, CANMET SO-2; Calcareous soil, CANMET SO-3; and Chenozemic soil, CANMET SO-4 from Canada Center for Mineral and Energy Technology, Ottawa, Ontario K1A0G1), NBS mixed diets (NBS RM 8431), and USDA mixed diets were prepared for the quality control of the analysis.

Sample Preparation and Handling

Food Samples

All of the tracer elements were below the detection limits of the instrument after overnight leaching of the polyethylene containers with 2M HNO_3, and therefore, no sample contamination was believed to occur from the polyethylene containers.

The duplicate food samples were collected daily, frozen, freeze-dried in a commercial bulk freeze dryer at 2.05 to 2.15 Torr pressure $-12°$ to $-7°C$ for 21 days, weighed, and then stored frozen until analysis. Food samples for all duplicate meals in one day for each child were homogenized in a food processor with stainless steel blade. Ten grams of the homogenized food samples were weighed in a platinum or quartz crucible (100 mL) and ashed in a programmable furnace.

The ashed samples were transferred into a 60 mL Teflon digestion vessel (CEM Corp.) for dissolution in 2 mL of concentrated HNO_3 and 1 IL of H_2O_2 at room temperature. Heat (100°C) was used to enhance predigestion if the decomposition of organic material was incomplete (black residue could be seen) during the muffle furnace ashing. The solution was cooled to room temperature, diluted to 10 mL, and 1 mL of HF was added. The vessel was heated in the microwave digestion system (MDS–81D) at 80% power for 1 min and 25% power for 5 min. The resulting solution was cooled to room temperature, transferred to a 25 mL volumetric flask, and diluted to volume.

Urine and Fecal Samples

Fecal and urine samples also were freeze-dried and stored frozen until analysis. The frozen commode samples were pooled into a large quartz dish (250 mL) manually in a clean room (class 100). Specimens over a 24-hour period, noon of the first day to that of the next day, were pooled into one sample and ashed in a programmable oven. The ash percent of feces was measured and calculated against the freeze-dried weight in the range of 10% ±4%. The ashed samples were then ground for a few minutes and fused with lithium metaborate in a programmable furnace.

After cooling, the fused solid was dissolved in 5% nitric acid, and transferred into a 100 mL volumetric flask, 2 mL of 1000 $\mu g/mL$ Co was added as an internal reference element, and the solution was diluted to volume.

Dust Samples

The house dust samples were completely ashed, sieved (stainless steel sieves 200-65 mesh), and particles larger than 65 mesh were collected and ground in a mixer-mill for 10 minutes. The final dust sample size was less than 65 mesh. The mean loss, mainly of organic materials, of four representative dust samples after ashing was 39%. Final dust samples were dissolved with the lithium metaborate fusion method. A further 100-fold dilution was necessary for final analysis because of the high concentrations of Si, Al, and Ti in the dust samples (final dilution factor of 4000).

Soil Samples

Samples were air-dried and thoroughly mixed by tumbling for one minute. The soil sample was separated into particles larger than 2 mm and those less than 2 mm, because the investigators assume that children are unlikely to ingest particles of this size or larger. This premise was based on examination of the size of objects in a random sample of 20 stool specimens from the total collection of the children's stool output. The sample was split by coning, and half the sample was passed through a 2 mm nylon sieve. An approximate 30 g subsample was obtained from the thoroughly mixed nylon sieved sample and ground for 5 minutes on a Spex Shatterbox®, producing a powder passing a 200-mesh sieve. The lithium metaborate fusion method was used for soil sample dissolution. The final dilution factor for the soil samples was 5000 to 1.

Quality Assurance of the Analysis

Quality control of the analyses were carried out with standard solutions and reference materials (NBS diets 8431: Canadian soil SO-1, SO-2, SO-3, SO-4, SO-5, and SO-7), and were performed every 6–10 samples.

The fecal sample analyses were evaluated with an artificial dirty diaper containing a known amount of Canadian reference soil (SO-2), since there are no known fecal reference materials. The artificial dirty diaper was made with 150.0 g tomato sauce mixed with 1.000 g of SO-2 on a new clean diaper, and freeze-dried for 48 hours. Tomato sauce (150.0 g) without soil was placed on a clean diaper as a control and freeze-dried for 48 hours.

Urine Analysis

Extremely low concentrations of the original eight tracer elements were found in urine samples except for Al and Si. Silicon was found to constitute a substantial fraction of adult urine excretion. Except for the low recovery of Zr, the recoveries for the rest of the elements were satisfactory (i.e., 88.1 to 100.2%). The loss of Zr was studied along with the recovery of residue of the dirty diaper (ashed all residue), the residue on the filter paper, and the recovery of Zr from the addition of 1001 μg/ml Zr standard solution into the tomato sauce prior to the freeze-drying. However, the recoveries were still very low after these additional treatments. Further investigation of Zr loss was carried replacing tomato sauce with white flour paste. Homogeneous flour paste 100.0 g was mixed. with 1000 g of SO-2 prior to freeze-drying. All the elements in 1000 g of SO-2 were found, even though the recovery for Zr was still lower (70%). It is possible that Zr may form some unknown complexes which were lost during ashing or the Zr may not be completely dissolved in one fusion process. Finally, it was discovered that Zr was very easily hydrolyzed; therefore, if the samples were not immediately analyzed after sample preparation, the loss of Zr may occur.

RESULTS

Adult Validation Study

The adult validation phase of the study was designed to determine if the analytical methodology could adequately detect the impact of the soil tracer elements in the feces at levels of postulated daily soil ingestion by children (i.e., 100 mg/day to 1000 mg/day). Consequently, this study selected levels of daily soil ingestion of 100 mg/day and 500 mg/day for evaluation. The use of children of the same age to validate the soil ingestion epidemiological study would be ideal. However, since young children are considered vulnerable subjects and cannot legally consent to treatment, ethical concerns precluded their participation in this phase of the study. In addition, it was unlikely that children aged 1–4 years would have been able to fulfill the needs of the project with high compliance by ingesting the capsules of soil. Consequently, adult volunteers were selected for the validation study.

Baseline evaluations revealed considerable variation in the concentrations of tracer elements in the duplicate meals. This variation was observed both from week to week for each volunteer and among individuals.

The ingestion of a total 300 mg of soil during the second week of the study, i.e., 100 mg/day for 3 days, resulted in a marked increase in fecal excretion of the tracer elements that could not be accounted for solely by food. Total recovery values for the tracer elements during the second week of the study indicated that of the eight tracer elements, Al, Si, Y, and Zr most closely approached 100% recovery values, while Ba and Mn grossly exceeded the 100% recovery values (Table 32). The variation in tracer recovery values for the participants was extremely large. Barium and Mn had standard deviations (SD) of 5433 and 1341, respectively. In contrast, Y and Zr had the lowest SD of 42.4 and 43.7, respectively.

During the third week a total of 1500 mg of soil was ingested, i.e., 500 mg/day for 3 days. The tracers most closely approaching the 100% recovery values were Al, Si, and Y, with Ba and Mn again displaying values which grossly exceeded 100% recovery (Table 1). The standard deviation of recovery of Si, Al, and Y was lowest and percent recovery was closest to 100%. Zirconium was more variable as a tracer, both between volunteers and elements. Titanium was not a reliable tracer at either the 300 mg/week or 1500 mg/week levels. The results of this adult validation study suggest that the most reliable tracer elements for the study protocol are Al, Si, and especially Y.

Table 1. Percent Recovery of Eight Tracer Elements in Adult Validation Study.

Sample	Ba	Mn	Si	Al	Ti	V	Y	Zr
0.3 g soil ingested								
0966	361	1150	88.9	86.0	105	693	108	28.4
0967	0	0	0	70.0	0	99.1	94.4	35.5
0968	11500	0	101	96.1	522	260	118	89.6
0969	1373	3320	431	301	762	527	205	146
0970	592	2190	133	281	120	67.9	92.6	101
0971	0	403	81.7	82.8	0	423	105	83.0
mean	2304.3	1177.2	139.3	152.8	251.5	345	120.5	80.6
SD	4533	1341	149.6	107.5	316	247	42.4	43.7
1.5 g soil ingested								
0966	179	187	84.8	92.9	337	245	99.4	63.2
0967	168	201	81.7	91.2	25.3	178	87.2	32.2
0968	95.7	234	94.6	108	139	116	84.2	58.6
0969	258	613	120	107	92.8	179	102	6.6
0970	139	120	97.6	96.5	93.8	114	66.7	61.1
0971	59.3	135	72.0	65.4	1030	53.8	85.5	106
mean	149.8	248.3	91.8	93.5	286.3	147.6	87.5	54.6
SD	69.5	183.6	16.6	15.5	380	66.8	12.6	33.4

Insufficient evidence exists to explain conclusively the reasons for the performance of these three tracer elements. However, in general, as the ratio of tracer concentration in the soil vs food rises, the ability of the tracer to estimate accurate soil ingestion is increased. Thus, the third week of the study in which the subjects ingested 1500 mg (500 mg/day) of soil yielded more consistent and reliable recovery values than did the values for the second week of the study.

The design of this three-week study with 3–5 days per week involvement of the subjects did not appear to allow the subjects to achieve a mass-balance equilibrium for any of the tracer elements. While a continuous noninterrupted mass-balance study would have been ideal, it was not employed because of a major concern over participant compliance. However, the intent of this adult validation study was to determine if the analytical methodology would detect the impact of soil tracers on fecal tracer ratios. These collective findings indicate that ingestion of soil at 100 mg/day and 500 mg/day and will significantly increase fecal tracer ratios and with greater than 85% recovery occurring for three of the tracers (i.e., Al, Si, and Y) by this method.

Given the limitations of the mass-balance methodology employed in the present study, these findings revealed with a high degree of confidence that 500 mg of soil ingestion per day was reliably detected. While ingestion of 100 mg of soil per day was also reliably detected, these data reflected higher variability of recovery and less intertracer consistency than the 500 mg/d exposure.

The reliability of the soil tracers in estimating soil ingestion is likely caused in large part by the concentration of tracer levels in soil compared to food and by the duration of the mass-balance methodology. As the soil tracer to food tracer ratio decreases in conjunction with short duration mass-balance study, the variability of recovery will increase. Thus, the third study week with higher soil ingestion provided more stable and reliable recovery values. Nevertheless, the study was able to verify that the methodology was adequate to estimate soil ingestion in adults in the range studied.

Children Soil Ingestion Study

Comparison of UMass Study Results with Binder Results

As a starting point for summarizing the results from the University of Massachusetts (UMass) study, the results were compared with those reported by Binder et al. This comparison is made by constructing tables from the UMass study that are similar in format to those reported by Binder et al.

Comparison of Age and Sex

There were 65 subjects initially enrolled in the UMass study. One subject dropped out; results are reported on the remaining 64 subjects. The age and sex distribution of these subjects is given in Table 2a, with the comparable age and sex distribution reported in the Binder study given in Table 2b. Mean age in the UMass study was 2 years, 4 months, while the mean age in the Binder study was 1 year, 7 months. There was no statistically significant difference in age by sex in either study. A wide range of demographic information was collected on the study population including family size, birth order, mother and father's education, marriage status, distribution of age difference in years between subjects and their next oldest sibling, and is provided in the Appendix.

Next, we compared the mean daily fecal freeze-dried mass (in grams) for males and females between the UMass and Binder study. Two measures of the mean daily fecal mass were computed based on the UMass data. The first measure is directly comparable to Binder's reported values, and is computed by dividing the total fecal mass over 8 study days (4 per week) and dividing by 8 to get a daily fecal output. The second measure is formed by averaging fecal estimates only over days when fecal output was reported. Although 8 days were included for fecal collection in the study protocol, not all subjects reported a fecal output on each day (see Table 3.)

Table 2.   Age and Sex Distribution of Study Participants.

Age (Yrs)	Sex		Total Number	Percent
	Male (= 1)	Female (= 2)		
a. UMass Study				
1.0–1.49	5	7	12	18.7%
1.5–1.99	6	3	9	14.1%
2.0–2.49	7	7	14	21.9%
2.5–2.99	2	11	13	20.3%
3.0–3.49	4	3	7	10.9%
3.5–3.99	6	3	9	14.1%
Total	30	34	64	
Percent	46.9%	53.1%		
b. Binder Study				
1.0–1.49	24	11	35	53%
1.5–1.99	7	7	14	22%
2.0–2.49	9	5	14	22%
2.5–2.99	2	0	2	3%
3.0–3.49	0	0	0	0%
3.5–3.99	0	0	0	0%
Total	42	23	65	
Percent	65%	35%		

Table 3.   Distribution of Number of Days with Fecal Output Reported.

No. of Days	No. of Subjects	Percent
3	1	1.6%
4	12	18.8%
5	12	18.8%
6	9	14.1%
7	14	21.9%
8	16	25.0%

The distribution of average daily fecal output was skewed to the right for the sample of 64 subjects, and not normally distributed. However, the distribution of the natural logarithm of daily fecal output was normally distributed. A summary of UMass study results is given in Table 4a, with comparable results for Binder's study in Table 4b.

Since the age of subjects in the UMass study was older than in the Binder study, the large fecal outputs are not directly comparable to those reported by Binder. A simple linear regression of age on log (fecal output) was significant ($p < 0.001$) in the UMass study, resulting in regression equations:

$$\text{Log (fecal mass/B)} = 1.02 + 0.0463 \text{ (age in mths)}$$

and   Log (ave daily

$$\text{fecal mass)} = 1.71 + 0.0326 \text{ (age in mths)}$$

Since the average age in the Binder study was 19 months, the comparable esti-mates from the UMass study (adjusting for age) are 6.7 g/day (based on 8 day average) and 10.3 g/day (based on daily averages where fecal mass was reported). These results are similar to those obtained by Binder.

Fecal sample concentrations were multiplied by reported fecal dry weight to obtain the total fecal mass output for each element in the UMass study. These total fecal masses were then divided by the total freeze-dried fecal mass to obtain the average concentration, so that results could be compared directly with those reported by Binder. The results are expressed in terms of the distribution of ele-ment concentrations in stool samples for the 64 children in the UMass study (Ta-ble 5a) and the Binder study (Table 5b). Additional elements were measured in the UMass study, and their concentrations in fecal samples are summarized in Table 6.

Table 4.   Mean Daily Fecal Mass for Males and Females, Based on Total Mass
(g freeze-dried weight/d)

Sex	n	Mean	Median	Geometric Mean	Std	Min–Max
a. From the UMass Study (8 days)						
Male	30	12.9	9.8	10.7	8.4	3.8–37.4
Female	34	12.4	8.7	10.1	7.9	1.7–33.1
Total	64	12.6	9.3	10.4	8.1	1.7–37.4
b. From the Binder Study (3 days)						
Male	42	7.5	7.1	6.9	3.0	1.9–15.7
Female	23	6.9	6.0	6.1	3.4	1.8–17.2
Total	65	7.3	6.7	6.6	3.2	1.8–17.2

Table 5.   Fecal Al, Si and Ti Concentrations (mg/g).

Element	Mean	Median	Geometric Mean	Std	Min–Max
a. UMass Study (64 Subjects)					
Al	0.56	0.29	0.30	1.16	DT– 9.1
Si	4.53	2.49	2.67	8.43	0.29 –59.7
Ti	0.14	0.07	0.06	0.22	0.006– 1.16
b. Binder Study (65 Subjects)					
Al	0.83	0.53	0.60	0.82	0.11 – 5.10
Si	3.65	2.56	2.56	3.45	0.66 –14.88
Ti	0.34	0.12	0.08	0.54	<0.01 – 2.83

Table 6. Fecal Ba, Mn, V, Y, and Zr Concentrations for 64 Subjects in UMass Study (μg/g).

Element	Mean	Median	Geometric Mean	Std	Min–Max
Ba	21.5	18.3	18.7	13.4	2.4 – 97.
Mn	113.4	92.9	93.7	72.0	3.7 –403.
V	2.91	1.41	1.56	7.3	0.2 – 58.
Y	0.21	0.13	0.13	0.52	0.03– 4.2
Zr	0.75	0.64	0.52	0.86	0.04– 5.9

Unlike the Binder study, the total weight of the fecal output was used in calculating the mass of each element excreted. This mass was calculated in two ways, using the average (over 8 days) of fecal output, or based on an average including only days with fecal output reported. These average daily fecal outputs are summarized in Table 7, with the corresponding average and standard deviation reported by Binder (assuming 15 g fecal weight). Soil and dust samples were collected on all study subjects. In Binder's study, soil and dust samples were only available on 42 of the 65 subjects. The soil and dust sample estimates are based on a time-weighted average of time at school and at home (Table 8a and 8b). Concentrations of other elements reported in the UMass study are summarized in Table 9.

Table 7. Fecal Al, Ti, Si, Ba, Mn, V, Y, and Zr Output for 64 Subjects in UMass Study (μg/day).

	Mean			SD			Median		
Element	UMass (Subset)[1]	UMass (8 Days)	Binder	UMass (Subset)[1]	UMass (8 Days)	Binder	UMass (Subset)[1]	UMass (8 Days)	Binder
Al[2]	10.5	8.9	12.5	37.4	37.4	12.3	3.8	2.8	na
Si[2]	78.3	67.5	54.8	245.	245.	51.8	36.0	24.6	na
Ti[2]	2.04	1.61	5.1	3.3	2.9	8.1	.86	.58	na
Ba	339.	265.	na	393.	393.	na	273.	199.	na
Mn	1607.	1219.	na	951.	810.	na	1453.	975.	na
V	52.9	38.5	na	157.	85.	na	17.1	12.6	na
Y	4.16	3.71	na	17.4	17.4	na	1.58	1.12	na
Zr	12.1	10.1	na	24.3	24.4	na	8.1	5.7	na

[1]Using only days with fecal output reported.
[2]Estimate in mg.

The Binder study did not include measures of food and medicine intake during the study period. The UMass study included food and medicine intake measures for the first three days of each study week. These intake measures were averaged for each week, and then averaged for the two weeks in the study. The intake measures are summarized in Table 10 for the eight elements.

Table 8.   Soil and Dust Concentrations of Al, Si, and Ti (mg/g).

Element		Mean	Median	Geometric Mean	Std	Min–Max
Soil (n=62), dust (n=64), and total (n=62) concentrations in UMass study						
Soil	Al	54.0	55.1	53.2	7.8	14.3– 76.6
Dust	Al	47.2	48.9	39.6	15.1	DT– 68.8
Total	Al	47.6	49.	43.1	13.4	1.6– 67.8
Soil	Si	307.	316.	302.	42.	91. –367.
Dust	Si	234.	260.	204.	92.	36. –389.
Total	Si	243.	268.	221.	84.	54. –388.
Soil	Ti	3.41	3.43	3.30	0.8	1.3– 5.2
Dust	Ti	5.76	4.23	4.71	5.8	2.0– 45.0
Total	Ti	5.47	4.25	4.59	5.1	2.0– 40.4
Soil (n=59), and dust (n=45) Al, Si and Ti concentrations in Binder study						
Soil	Al	66.6	67.0	66.0	9.2	45.1–111.4
Dust	Al	33.6	34.1	29.9	13.8	5.4– 64.2
Soil	Si	302.	302.	302.	18.	243. –332.
Dust	Si	202.	203.	199.	39.	116. –319.
Soil	Ti	2.98	2.93	2.95	0.4	2.4– 4.0
Dust	Ti	2.72	2.77	2.67	0.5	1.2– 3.8

Table 9.   Soil (n=62), Dust (n=64), and Total (n=62) Ba, Mn, V, Y, and Zr Concentrations in UMass Study (μg/g).

Element		Mean	Median	Geometric Mean	Std	Min–Max
Soil	Ba	356.	331.	337.	124.	90. – 705.
Dust	Ba	624.	503.	466.	942.	32. –7821.
Total	Ba	594.	494.	458.	843.	41. –7020.
Soil	Mn	731.	729.	706.	187.	249. –1209.
Dust	Mn	690.	628.	470.	401.	DT–1908.
Total	Mn	678.	634.	548.	328.	28. –1523.
Soil	V	86.6	64.6	68.6	111.	23. – 706.
Dust	V	59.6	54.4	55.3	27.	28. – 203.
Total	V	63.3	56.8	58.5	30.	31. – 205.
Soil	Y	24.0	23.5	23.8	3.7	15.8– 32.6
Dust	Y	17.2	16.5	16.5	4.9	8.4– 40.4
Total	Y	17.8	17.5	17.3	4.3	9.1– 39.1
Soil	Zr	196.	187.	179.	82.	55. – 451.
Dust	Zr	228.	168.	177.	247.	51. –1865.
Total	Zr	199.	164.	175.	119.	66. – 703.

DT = detection limit = 0.28 mg/g for Al and 12.0 μg/g for Mn.

The average estimates of soil ingestion were calculated based on several methods. The soil ingestion algorithm used by Binder et al. employing the UMass data is compared with the data presented by Binder et al. The Binder et al. method corresponds to taking a simple daily average of fecal output recorded over the

Table 10.   Average Daily Intake for 6 Days for Al, Si, Ti (in mg) and for Ba, Mn, V, Y, and Zr (in μg) in UMass Study (n = 64).

Element	Mean	Median	Geometric Mean	Std	Min–Max
Al	1.87	1.21	1.14	2.52	0.1– 15.6
Si	16.6	14.2	14.7	9.3	4.6– 59.1
Ti	1.01	0.20	0.24[1]	1.9	DT– 8.12
Ba	266.	224.	218.	341.	70.7–2867.
Mn	1452.	1288.	1283.	793.	412. –4287.
V	9.0	5.3	5.3[1]	11.6	DT– 60.1
Y	1.9	1.1	1.2[1]	4.6	DT– 36.0
Zr	6.7	2.5	4.3	18.1	DT– 112.

DT = detection limit.
[1]Values less than detection limit excluded in calculations.

entire 8-day study period, and using separate estimators of soil and dust concentrations. Binder's method does not adjust for ingestion of the element in food. Estimates of soil ingestion based on soil concentrations are given in Table 11a–11c (with comparable estimates from Binder), while estimates based on dust concentrations are given in Table 12.

For comparison, we also report the soil ingestion estimates after subtracting the average daily food intake (in Tables 13 and 14), and the soil ingestion estimates based on an estimate of fecal output averaged over days where fecal output was reported (Tables 15 and 16).

Finally, the various types of estimates for soil ingestion by element for the UMass study are summarized in Tables 17–24 and Figures 1–8. The results indicate that the range of median soil ingestion estimates based on excretory tracer levels minus the food tracer levels, ranged from a low of 9 mg/day (Y) to a high of 96 mig/day (V). The median estimates of combined soil/dust were similar to

Table 11.   Soil Ingestion Estimates for Children Based on Soil Concentrations (in mg/day).

Element	N	Mean	Median	Std	Min	P5th	P10th	P90th	P95th	Max
a. In UMass Study, Using Average Fecal Output Over 8 Days, and Ignoring Food Intake										
Al	64	187	59	850	1	12	20	190	243	6858
Si	64	211	81	692	6	21	25	310	402	5582
Ti	64	577	192	1220	10	12	25	1365	2353	6911
b. In Binder Study										
Al	59	181	121	203	25				584	1324
Si	59	184	136	175	31				578	799
Ti	59	1834	618	3091	4				9590	17076
c. In UMass Study, Using Average Fecal Output Over 8 Days, and Ignoring Food Intake										
Ba	64	785	593	937	50	208	270	1265	1599	7219
Mn	64	1807	1329	1412	29	343	591	3390	3929	9245
V	62	587	194	1056	5	41	58	1493	2220	6019
Y	62	175	45	864	15	18	22	148	167	6855
Zr	62	63	31	178	1	6	9	87	135	1412

Table 12.   Soil Ingestion Estimates for Children Based on Dust (mg/day).

Element	N	Mean	Median	Std	Min	P5th	P10th	P90th	P95th	Max
a. In UMass Study, Using Average Fecal Output Over 8 Days, and Ignoring Food Intake										
Al	64	446	63	1617	2	17	20	313	554	9267
Si	64	1077	103	6878	6	22	34	496	780	55198
Ti	64	360	154	742	7	12	13	713	1399	4670
b. In Binder Study										
Al	45	515	289	326	33					3474
Si	45	286	197	287	48					1265
Ti	45	2068	871	3006	5					13923
c. In UMass Study										
Ba	64	695	428	967	16	136	159	1421	2088	5841
Mn	64	10189	1706	28083	32	417	583	8612	71047	151213
V	64	616	282	1052	40	51	66	1937	2147	7193
Y	64	174	77	640	14	27	29	196	248	5186
Zr	64	60	33	109	1	5	6	109	161	801

Table 13.   Soil Ingestion Estimates for Children Based on Soil in UMass Study, Subtracting Average Ingestion and Using Average of 8 Day Fecal Output (in mg/day).

Element	N	Mean	Median	Std	Min	P5th	P10th	P90th	P95th	Max
Al	64	153	29	852	−75	−23	−3	138	223	6837
Ba	64	32	−37	1002	−3608	−429	−362	228	283	6773
Mn	64	−294	−261	1266	−3297	−2056	−1244	595	788	7281
Si	64	154	40	693	−53	−29	−17	219	276	5549
Ti	64	218	55	1150	−3069	−601	−381	702	1432	6707
V	62	459	96	1037	−650	−148	1	1366	1903	5676
Y	62	85	9	890	−1733	−84	−56	91	106	6736
Zr	62	21	16	209	−597	−70	−35	67	110	1391

Table 14.   Soil Ingestion Estimates for Children Based on Dust in UMass Study, Subtracting Average Ingestion and Using Average of 8 Day Fecal Output (in mg/day).

Element	N	Mean	Median	Std	Min	P5th	P10th	P90th	P95th	Max
Al	64	317	31	1272	−118	−27	−7	264	506	8462
Ba	64	31	−18	860	−3598	−335	−246	222	337	5480
Mn	64	−1289	−340	9087	−6E+4	−4449	−2339	717	2916	20575
Si	64	964	49	6848	−234	−34	−27	278	692	54870
Ti	64	163	28	659	−1024	−469	−320	448	1266	3354
V	64	453	127	1005	−595	−177	0	1700	1918	6782
Y	64	62	15	687	−1835	−163	−93	130	169	5096
Zr	64	27	12	133	−442	−54	−32	91	160	789

Table 15. Soil Ingestion Estimates for Children Based on Soil in UMass Study, Subtracting Average Ingestion and Using Average of Reported Days with Fecal Output (in mg/day).

Element	N	Mean	Median	Std	Min	P5th	P10th	P90th	P95th	Max
Al	64	183	54	849	−48	0	9	214	299	6837
Ba	64	258	160	1005	−3333	−271	−183	581	980	6773
Mn	64	254	125	1378	−2748	−1711	−931	1486	2260	7281
Si	64	190	70	695	−48	−21	−5	243	287	5549
Ti	64	351	99	1158	−2860	−601	−127	1118	1958	6707
V	62	633	194	1640	−650	−124	12	1579	1962	11695
Y	62	104	22	887	−1674	−62	−35	125	136	6736
Zr	62	34	28	211	−597	−50	−28	97	168	1391

Table 16. Soil Ingestion Estimates for Children Based on Dust in UMass Study, Subtracting Average Ingestion and Using Reported Days with Fecal Output (in mg/day).

Element	N	Mean	Median	Std	Min	P5th	P10th	P90th	P95th	Max
Al	64	461	53	1913	−113	−1	11	404	526	14022
Ba	64	229	100	965	−3324	−252	−108	480	831	5480
Mn	64	2415	120	16588	−3E+4	−2607	−1538	2916	4671	111303
Si	64	1031	106	6841	−111	−20	−6	450	992	54870
Ti	64	244	68	655	−710	−321	−127	689	1266	3354
V	64	656	225	1805	−595	−118	42	1771	1970	13974
Y	64	91	40	683	−1773	−150	−64	178	193	5096
Zr	64	42	23	136	−438	−50	−29	124	183	789

that for soil alone, ranging from a low of 11 mg/day (Y) to a high of 123 mg/day for V. Thus, these values reveal a high degree of consistency among themselves. In addition, the estimates based on soil and dust separately are in close agreement. Since there was a large degree of interindividual values for estimations of soil/dust ingestion, with one and/or two children being extreme outlyer(s), depending on the tracer, the mean values for these tracers were considerably higher than the median. Because of the large impact of a single individual on the mean, the median value is believed to be a better measure of central tendency.

DISCUSSION

The present study utilized eight different tracer elements. It retained Al, Si, and Ti, as were employed by Binder et al.[6] and Clausing et al.[7] (Al, Ti), while adding Ba, Mn, V, Y, and Zr. Additionally, a mass-balance appraisal of tracer intake from food and medicines and excretory output of the tracer elements was conducted. Consequently, in this investigation, we were able to ascertain the

Table 17. Soil Ingestion Estimates Based on Al for Children in UMass Study (in mg/day).

Name	N	Mean	Median	Std	Min	P5th	P10th	P90th	P95th	Max
(8 day fecal—ingest)/soil	64	153	29	852	-75	-23	-3	138	223	6837
(8 day fecal—ingest)/dust	64	317	31	1272	-118	-27	-7	264	506	8462
(8 day fecal—ingest)/combined	64	154	30	629	-114	-27	-6	247	478	4929
(Obs daily fecal—ingest)/soil	64	183	54	849	-48	0	9	214	299	6837
(Obs daily fecal—ingest(/dust	64	461	53	1913	-113	-1	11	404	526	14022
(Obs daily fecal—ingest)/combined	64	204	56	647	-97	-1	12	393	478	4929
(8 day fecal)/soil	64	187	59	850	1	12	20	190	243	6858
(8 day fecal)/dust	64	446	63	1617	2	17	20	313	554	9267
(8 day fecal)/combined	64	211	58	644	2	17	19	292	524	4944
(Obs daily fecal)/soil	64	217	81	849	2	20	29	262	338	6858
(Obs daily fecal)/dust	64	590	78	2222	2	24	32	501	733	14827
(Obs daily fecal)/combined	64	261	84	670	2	25	31	468	689	4944

Table 18. Soil Ingestion Estimates Based on Ba for Children in UMass Study (in mg/day).

Name	N	Mean	Median	Std	Min	P5th	P10th	P90th	P95th	Max
(8 day fecal—ingest)/soil	64	32	-37	1002	-3608	-429	-362	228	283	6773
(8 day fecal—ingest)/dust	64	31	-18	860	-3598	-335	-246	222	337	5480
(8 day fecal—ingest)/combined	64	29	-19	868	-3601	-323	-245	229	331	5626
(Obs daily fecal—ingest)/soil	64	258	160	1005	-3333	-271	-183	581	980	6773
(Obs daily fecal—ingest)/dust	64	229	100	965	-3324	-252	-108	480	831	5480
(Obs daily fecal—ingest)/combined	64	220	107	928	-3327	-245	-115	473	846	5626
(8 day fecal)/soil	64	785	593	937	50	208	270	1265	1599	7219
(8 day fecal)/dust	64	695	428	967	16	136	159	1421	2088	5841
(8 day fecal)/combined	64	675	448	902	18	140	162	1390	1787	5996
(Obs daily fecal)/soil	64	1011	842	964	57	297	353	1772	1830	7219
(Obs daily fecal)/dust	64	893	540	1170	26	167	205	1604	2636	7007
(Obs daily fecal)/combined	64	866	578	1035	29	180	213	1529	2550	5996

Table 19. Soil Ingestion Estimates Based on Mn for Children In UMass Study (In mg/day).

Name	N	Mean	Median	Std	Min	P5th	P10th	P90th	P95th	Max
(8 day fecal—ingest)/soil	64	−294	−261	1266	−3297	−2056	−1244	595	788	7281
(8 day fecal—ingest)/dust	64	−1289	−340	9087	−6E+4	−4449	−2339	717	2916	20575
(8 day fecal—ingest)/combined	64	−496	−340	1974	−9607	−3123	−1954	708	3174	4189
(Obs daily fecal—ingest)/soil	64	254	125	1378	−2748	−1711	−931	1486	2260	7281
(Obs daily fecal—ingest(/dust	64	2415	120	16588	−3E+4	−2607	−1538	2916	4671	111303
(Obs daily fecal—ingest)/combined	64	686	120	3498	−5105	−2588	−1381	2327	4467	18749
(8 day fecal)/soil	64	1807	1329	1412	29	343	591	3390	3929	9245
(8 day fecal)/dust	64	10189	1706	28083	32	417	583	8612	71047	151213
(8 day fecal)/combined	64	3573	1685	6051	31	457	639	7514	15277	36359
(Obs daily fecal)/soil	64	2355	2111	1587	33	547	931	3945	4948	9245
(Obs daily fecal)/dust	64	13894	2319	39717	36	556	793	10227	97318	241940
(Obs daily fecal)/combined	64	4754	2204	8026	36	551	822	8666	17459	41553

Table 20. Soil Ingestion Estimates Based on Si for Children In UMass Study (In mg/day).

Name	N	Mean	Median	Std	Min	P5th	P10th	P90th	P95th	Max
(8 day fecal—ingest)/soil	64	154	40	693	−53	−29	−17	219	276	5549
(8 day fecal—ingest)/dust	64	964	49	6848	−234	−34	−27	278	692	54870
(8 day fecal—ingest)/combined	64	483	49	3105	−150	−32	−27	263	653	24900
(Obs daily fecal—ingest)/soil	64	190	70	695	−48	−21	−5	243	287	5549
(Obs daily fecal—ingest(/dust	64	1031	106	6841	−111	−20	−6	450	992	54870
(Obs daily fecal—ingest)/combined	64	540	103	3100	−80	−20	−6	369	704	24900
(8 day fecal)/soil	64	211	81	692	6	21	25	310	402	5582
(8 day fecal)/dust	64	1077	103	6878	6	22	34	496	780	55198
(8 day fecal)/combined	64	575	101	3115	6	22	34	383	737	25049
(Obs daily fecal)/soil	64	248	116	695	7	28	51	318	402	5582
(Obs daily fecal)/dust	64	1143	159	6873	7	32	52	794	1247	55198
(Obs daily fecal)/combined	64	632	155	3111	7	32	51	626	891	25049

Table 21. Soil Ingestion Estimates Based on Tl for Children In UMass Study (in mg/day).

Name	N	Mean	Median	Std	Min	P5th	P10th	P90th	P95th	Max
(8 day fecal—ingest)/soil	64	218	55	1150	-3069	-601	-381	702	1432	6707
(8 day fecal—ingest)/dust	64	163	28	659	-1024	-469	-320	448	1266	3354
(8 day fecal—ingest)/combined	64	170	30	691	-1035	-470	-320	442	1059	3597
(Obs daily fecal—ingest)/soil	64	351	99	1158	-2860	-601	-127	1118	1958	6707
(Obs daily fecal—ingest)/dust	64	244	68	655	-710	-321	-127	689	1266	3354
(Obs daily fecal—ingest)/combined	64	253	68	688	-718	-348	-126	729	1059	3597
(8 day fecal)/soil	64	577	192	1220	10	12	25	1365	2353	6911
(8 day fecal)/dust	64	360	154	742	7	12	13	713	1399	4670
(8 day fecal)/combined	64	373	163	783	8	12	14	701	1119	4911
(Obs daily fecal)/soil	64	710	280	1327	13	21	33	1462	2729	6911
(Obs daily fecal)/dust	64	441	216	769	11	17	23	1141	1512	4670
(Obs daily fecal)/combined	64	458	211	810	12	17	23	1119	1561	4911

Table 22. Soil Ingestion Estimates Based on V for Children In UMass Study (in mg/day).

Name	N	Mean	Median	Std	Min	P5th	P10th	P90th	P95th	Max
(8 day fecal—ingest)/soil	62	459	96	1037	-650	-148	1	1366	1903	5676
(8 day fecal—ingest)/dust	64	453	127	1005	-595	-177	0	1700	1918	6782
(8 day fecal—ingest)/combined	62	456	123	1013	-617	-168	3	1701	1783	6736
(Obs daily fecal—ingest)/soil	62	633	194	1640	-650	-124	12	1579	1962	11695
(Obs daily fecal—ingest)/dust	64	656	225	1805	-595	-118	42	1771	1970	13974
(Obs daily fecal—ingest)/combined	62	661	250	1821	-617	-134	42	1735	1967	13881
(8 day fecal)/soil	62	587	194	1056	5	41	58	1493	2220	6019
(8 day fecal)/dust	64	616	282	1052	40	51	66	1937	2147	7193
(8 day fecal)]dust	62	603	254	1055	26	41	61	1861	2161	7144
(Obs daily fecal)/soil	62	761	309	1665	6	54	93	1706	2292	12039
(Obs daily fecal)/dust	64	819	352	1846	53	71	103	1937	2302	14385
(Obs daily fecal)/combined	62	808	353	1860	35	55	87	2105	2273	14289

Table 23. Soil Ingestion Estimates Based on Y for Children In UMass Study (in mg/day).

Name	N	Mean	Median	Std	Min	P5th	P10th	P90th	P95th	Max
(8 day fecal—ingest)/soil	62	85	9	890	-1733	-84	-56	91	106	6736
(8 day fecal—ingest)/dust	64	62	15	687	-1835	-163	-93	130	169	5096
(8 day fecal—ingest)/combined	62	65	11	717	-1805	-149	-88	118	159	5269
(Obs daily fecal—ingest)/soil	62	104	22	887	-1674	-62	-35	125	136	6736
(Obs daily fecal—ingest)/dust	64	91	40	683	-1773	-150	-64	178	193	5096
(Obs daily fecal—ingest)/combined	62	92	34	713	-1744	-137	-58	166	177	5269
(8 day fecal)/soil	62	175	45	864	15	18	22	148	167	6855
(8 day fecal)/dust	64	174	77	640	14	27	29	196	248	5186
(8 day fecal)/combined	62	174	70	673	14	27	29	185	249	5362
(Obs daily fecal)/soil	62	194	66	862	19	33	35	161	191	6855
(Obs daily fecal)/dust	64	204	101	638	18	42	46	284	321	5186
(Obs daily fecal)/combined	62	202	94	670	19	39	47	269	297	5362

Table 24. Soil Ingestion Estimates Based on Zr for Children In UMass Study (in mg/day).

Name	N	Mean	Median	Std	Min	P5th	P10th	P90th	P95th	Max
(8 day fecal—ingest)/soil	62	21	16	209	-597	-70	-35	67	110	1391
(8 day fecal—ingest)/dust	64	27	12	133	-442	-54	-32	91	160	789
(8 day fecal—ingest)/combined	62	23	11	138	-465	-55	-32	89	159	838
(Obs daily fecal—ingest)/soil	62	34	28	211	-597	-50	-28	97	168	1391
(Obs daily fecal—ingest)/dust	64	42	23	136	-438	-50	-29	124	183	789
(Obs daily fecal—ingest)/combined	62	37	26	140	-460	-46	-31	107	181	838
(8 day fecal)/soil	62	63	31	178	1	6	9	87	135	1412
(8 day fecal)/dust	64	60	33	109	1	5	6	109	161	801
(8 day fecal)/combined	62	57	33	111	1	5	8	94	159	851
(Obs daily fecal)/soil	62	77	40	178	1	9	12	105	174	1412
(Obs daily fecal)/dust	64	74	42	112	2	6	10	170	184	801
(Obs daily fecal)/combined	62	71	43	112	2	7	10	146	181	851

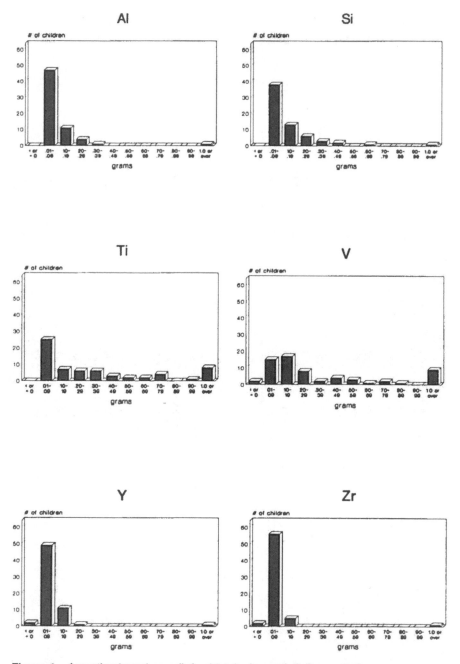

Figure 1. Ingestion based on soil: food intake ignored, 8 day average.

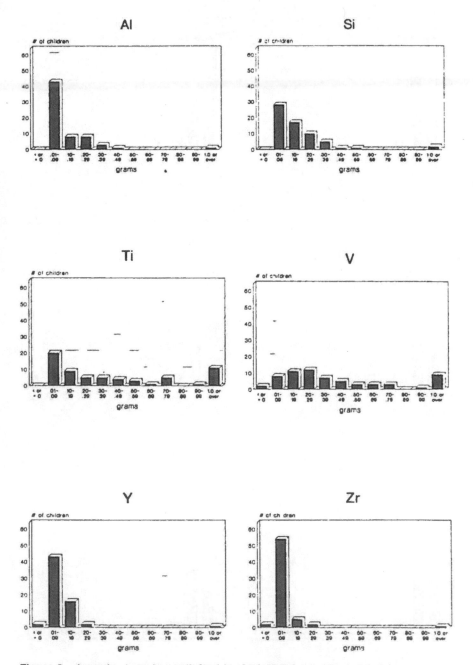

Figure 2. Ingestion based on soil: food intake ignored, average number days.

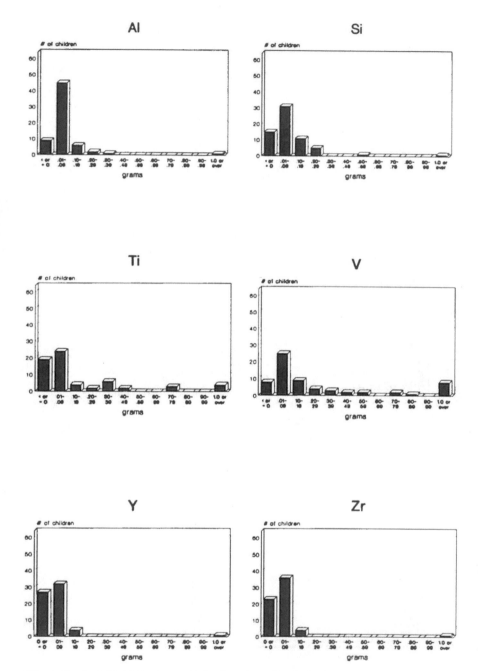

Figure 3. Ingestion based on soil: food intake subtracted, 8 day average.

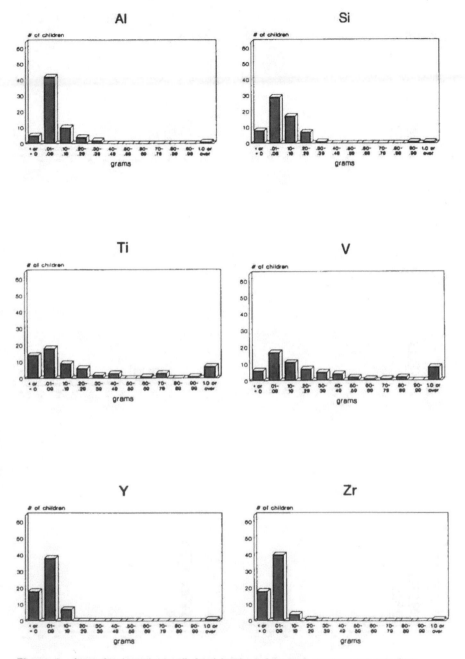

Figure 4.   Ingestion based on soil: food intake subtracted, average number days.

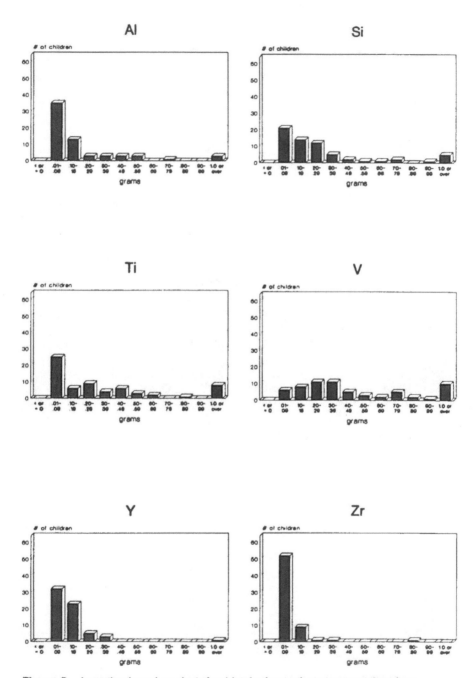

Figure 5.   Ingestion based on dust: food intake ignored, average number days.

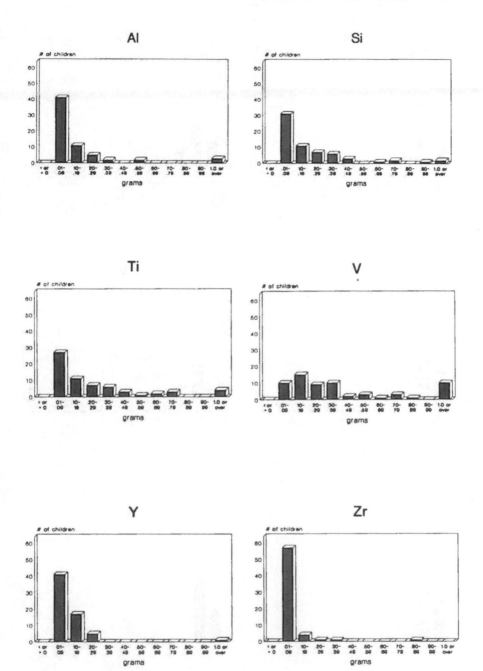

Figure 6. Ingestion based on dust: food intake ignored, 8 day average.

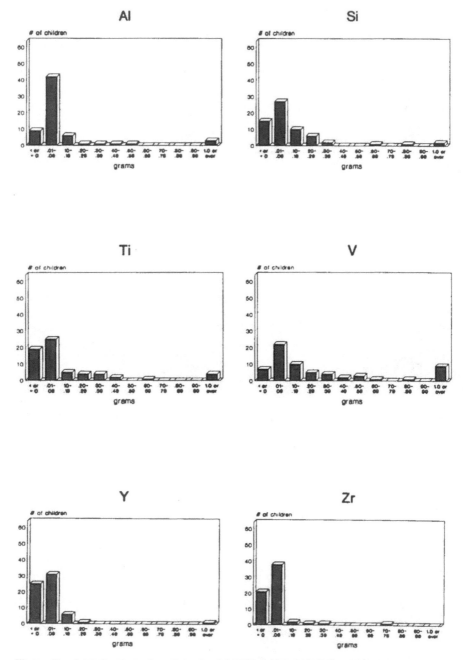

Figure 7. Ingestion based on dust: food intake subtracted, 8 day average.

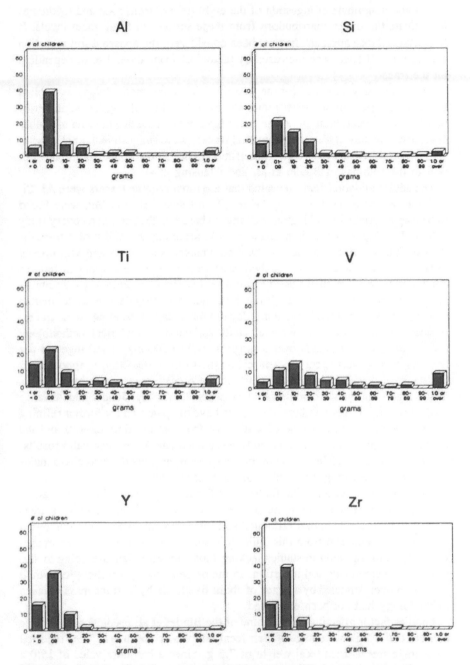

Figure 8. Ingestion based on dust: food intake subtracted, average number days.

extent and magnitude of ingestion of the eight tracers from food and medicines to evaluate the tracer contributions from these sources in fecal tracer levels. If this mass-balance approach had not been considered, the apparent soil ingestion estimate would have been increased by factors of from about 2 to 8, depending on the tracer.

The adult study validated for the first time the tracer methodology as applied to human subjects in previous studies for the estimation of soil ingestion. Prior to this adult validation study, a case could have been made that the data of Binder et al. and Clausing et al. using elemental tracers were suggestive but not conclusive that children ingested soil. These findings have now added additional support to the findings of Binder et al. and Clausing et al.

The adult validation study revealed that the most reliable tracers were Al, Si, and Y based on percent recovery values. Two tracers, Ba and Mn, were found to be inappropriate for soil ingestion estimates based on the percent recovery study values. It appears that an important variable affecting the utility of a tracer is the extent to which it is present in the diet. Tracers, such as Ba and Mn, proved to be quite unreliable because of their relatively high concentrations in food as compared to soil, and their possible variable absorption, retention, gastrointestinal exfoliation, or transit rates. While the children's study data revealed median soil ingestion estimates of from 9 mg/day to 96 mg/day, depending on the tracer, it should be emphasized that the adult study validated the analytical methodology used in the children's study over a range of 100 to 500 mg of soil ingested per day. As with all analytical procedures, as the limit of detection is approached, the standard deviation increases. Thus, in the present study, with lower levels of soil ingestion, the standard deviations increased. Although longer mass-balance evaluations and larger numbers of subjects have the potential for further refining results obtained with this type of study, we feel that the time periods and the numbers of children used in this study were adequate to produce valid results. The exceptionally high level of compliance for sample collection was also a major factor in substantiating the validity of the present study.

This study has extended the findings of Binder et al. and Clausing et al. in several important areas. Both the Binder et al. and the Clausing et al. studies revealed significantly higher estimated soil ingestion values for Ti as compared to Al and Si. The data from this children's study suggest that the Ti estimated values for soil ingestion in studies lacking food ingestion data are going to significantly overestimate soil ingestion. In the present study, soil ingestion would have been overestimated by a factor of about five to six by Ti if the mass-balance methodology had not been employed.

Another factor affecting the outcome of the Binder et al. predictions was their concern that a high percentage of their fecal samples had been lost. They reported an average freeze-dried fecal weight of 7.5 g. Since a literature value of 15.0 g dry weight was reported for young children,[10] Binder et al. assumed that their values should be adjusted upward from 7.5 g to 15.0 g. This adjustment in effect doubled their soil ingestion estimations.

The accuracy of the average daily freeze-dried fecal weight for a 1-2 year old child of 15 g based on the report of Lemoh and Brooke[10] is questionable. Their report included 17 subjects, aged 13-24 months, who exhibited a 42-fold range, 4-180 g, in individual stool weight. The mean stool weights calculated by Lemoh and Brooke were markedly influenced by a small number of extremely high value(s), as is apparent by the 4-180 range and the calculated mean of 15 g. In such a case, the median value would have been a more valid description of the central tendency. While it is not possible from their reported data to calculate the median, the highest value of their data set of 17 values actually contributed nearly 25% of the total sum of fecal weights for the 17 children. We believe that the present study with its N of 64 and nearly complete sample collection ($<0.5\%$ loss of food and excretory samples based on daily surveys of participating parents) will offer a more reliable estimate.

The present findings indicated that fecal sample mass increased directly with age and that stool mass in the present study was on average significantly greater than that reported by Binder et al. Collectively, children in the present study excreted fecal masses of approximately 15.0 g per day when calculations were made using only days feces produced and collected (Table 25), and 9-10 g per day when calculations were made based on the total fecal mass divided by the eight study days (Table 4a). However, the 1.0-1.5 year old children in the present study excreted fecal masses similar to the fecal masses reported by Binder et al. (Table 4b). It is important to note that 53% of the children in the Binder et al. study were 1.0-1.5 years of age. Thus, the subjects in that study were a much younger average age than the children in our study. This suggests that in the Binder et al. study a significant number of samples had not been lost and, due to the younger age of their subjects, had smaller fecal masses, and therefore there was no need to adjust their values upward.

Table 25.   Mean Daily Fecal Mass for Males and Females, from the UMass Study, Based on Average Daily Fecal Mass Reported (g freeze-dried weight/d).

Sex	n	Mean	Median	Geometric Mean	Std	Min–Max
Male	30	15.9	14.9	14.0	8.1	5.4–37.4
Female	34	16.0	15.5	14.0	7.9	3.5–34.1
Total	64	15.9	15.1	14.0	7.9	3.5–37.4

NOTE: Days with no fecal mass are excluded.

Taken collectively, if the fecal weights in the Binder et al. data had not been adjusted upward (which we feel, in retrospect, based on our data, was not justified) and the study had taken into account the presence of tracers in food and medicines, their soil ingestion estimates would have been reduced to a considerable extent and would have approached values reported in the present study.

In the present study, one aspect of the estimate of tracer ingestion was to attempt to determine whether the exposure to the tracers was originating from consumption of soil and/or dust, and the relative proportions of each. Since the dust samples did not significantly differ from the soil values with respect to the tracer element concentrations, it was not possible to adequately assess the relative contribution of each source. This remains an important research consideration since the children averaged only approximately 1.5 to 2 hours of outside time during the study (see Appendix).

In the designing and conduct of this study numerous methodological concerns and issues were raised that not only impact the present findings, but are of considerable use when contemplating future studies in this area. The study population of children aged 1–4 years was selected because of the widespread assumption that children of this age range have the highest hand-to-mouth activity and are likely to ingest more soil than older age groups. In the present study, children were not a randomly chosen population, but were selected from day-care centers and from nonrandomly identified volunteer families. While this presented some limitations, it also offered the advantage of involving families predisposed to being highly committed to the study. In fact, this may have contributed in large measure to the high compliance rates (>99%) in the study with respect to food, medicine, fecal and urine samples.

A two-week noninterrupted period was desired for collection of food and excretory samples to enhance sufficient completeness of the mass-balance methodology. However, this regimen was not attempted with the 1–4 year olds in our study because of the concern over the onerous demands that this would make on the children and their parents, which could result in breakdown in compliance and in dropping out of the study.

Instead, we collected data for food (Monday morning through Wednesday night) and excreta (Monday afternoon through Friday morning) over 2 weeks. In this manner the second week could confirm the findings in the first week and a stable and reasonably complete perspective would be provided without undue stress to the subjects or their parents.

Other issues of concern in the methodology included the identification of where and the extent to which the children played in the home and outside the home and the methods of soil sampling. In this study we identified where the children played during the period of the study in the home and yard by information supplied by parents. The parents identified the three most active outdoor play areas and they gave the relative percentage of the time spent by the child at each. The soil from these locations was mixed in the same time-weighted percentages to provide a composite average. Dust samples were vacuumed from the floor of the rooms in which the parent indicated the children frequently played.

While the approach taken for collection of soil samples seemed direct and rational, how to actually collect the soil was less clear. Should one take a surface sample of soil by vacuum, the upper 1 inch, the upper 2 inches, the upper 3 inches or something else? After much debate, the upper 3 inches was selected on the assumption that many children dig and play in soil at depths from 1 to 3 inches. Thus a collective sample would incorporate the range of behaviors. In retrospect, the ideal situation might be a personal tailoring of the collection protocol in accordance with each child's specific behavioral patterns, but this would be very resource-intensive.

It should be noted that the present study reported a child who ingested 5–8 g/day of soil. This study provides no data on the frequency of other such children in the general population or whether the observed soil ingestion by this child was an aberration in oral behavior or a consistent occurrence.

Finally, the community, the type of landscaping in the children's play areas, the degree of parental supervision, and the season of the year (late September through October) are all specific to this study. Nevertheless, if appropriate adjustments (e.g., food intake) are made in the Binder et al. and Clausing et al. studies, the results of their investigations and the present study would all be in remarkably close agreement.

CONCLUSIONS

In conclusion, the use of soil tracers to estimate soil ingestion has been validated using an adult volunteer study. The present findings indicate that the median soil intake by the children based on all eight tracers ranged from 9–96 mg/day with seven of the eight tracers having median values < 50 mg/day. The three most reliable tracers (i.e., Al, Si, Y) based on recovery studies revealed a range in median values of from 9 to 40 mg of soil ingestion/day. If the use of the mass-balance approach had not been employed, the present estimates would have been increased by factors of 2 to 6, depending on the tracer. The present findings suggest that the higher estimated values for soil ingestion by Binder et al. and Clausing et al. were due in large part to their not using a mass-balance approach. Also, the major discrepancy between Ti and other tracer elements of previous studies is now largely explained on the basis of high levels of Ti in food.

The results of Binder et al. and Clausing et al. are quite similar to the results found in this study when adjustments are made in the Binder et al. and Clausing et al. data.

Future studies to evaluate the interrelationships of dust on soil ingestion, seasonal influences on soil ingestion, social and cultural influences on soil ingestion, and the degree of exposed soil versus grass cover may be of significance in further defining soil ingestion in children.

ACKNOWLEDGMENTS

The investigators gratefully acknowledge the parents and children who participated in this study. Their efforts as co-investigators of this project resulted in a virtually complete, high quality data set. The research staff of the project, Linda Rosen, Pam Sleeper, Pat Rosigno, John Young, Jeff Smith, Carol Cady, Tom

Passa, Karen Hill, and the Day-Care staff are acknowledged for their efforts. We appreciate the useful suggestions from Ms. Dorrit Hoffer and the assistance of Dr. Pengyuan Yang in repairing the ICP-AES multichannel spectrometer and the corresponding software development. We acknowledge Ms. Cindy Anderau for the assistance with the Plasma 40 and Perkin-Elmer Corporation for the loan of Plasma 40. We appreciate the loan of the microwave digestion system (MDS) by the CEM Corporation. And, finally, we would like to thank Steve Bodine, Plant and Soil Science Department, Steve Feder, Kate Taylor, and Dina Yang for their help with sample handling. A special acknowledgment is appropriate for Judith Sanders and James Ryan, who coordinated field activities, often into the late evening and weekend, and to Eileen Keagan, who managed the data handling. Special thanks are in order for Dr. Robert Levin, Dr. Edward Sunderland, and Dr. Jesse Ortiz. Funding for this research was provided by Syntex Agribusiness, Incorporated. A shorter version of this study is published in *Regulatory Toxicology and Pharmacology*, Volume 10, Number 1, August 1989.

APPENDIX

Activity Times

The child's activities are recorded for Monday–Thursday of each week. This includes the number of hours and location where the child played. Both home and day-care locations are noted. Table 26 shows hours of outdoor activity in different locations, followed by the dataset description.

Table 26.   Hours of Childrens' Outdoor Activities by Location and Dataset Description.

N Obs	Variable	Minimum	Maximum	Mean	Std Dev
355	hours outside	0.00	8.00	1.56	1.63
		Skinner School			
137	hours at daycare	0.50	5.75	2.25	1.14
	hours out at daycare	0.00	1.50	0.63	0.45
	hours out at home	0.00	4.50	1.03	1.02
60	hours at daycare	5.25	6.75	6.00	0.50
	hours out at daycare	0.75	2.25	1.50	0.50
	hours out at home	0.00	3.00	0.40	0.76

REFERENCES

1. Spittler, T. Technical Resource Committee Meeting, Minnesota Pollution Control Agency, Roseville, MN, September 24, 1986.
2. Kimbrough, R. D., H. Falk, P. Stehr, and G. Fries. "Health Implications of 2,3,7,8-TCDD Contamination of Residual Soil," *J. Toxicol. Environ. Health* 14:47–93 (1984).

3. Lepow, M. L., L. Bruckman, R. A. Rubin, S. Markowitz, M. Gillette, and J. Kapish. "Role of Airborne Lead in Increased Body Burden of Lead in Hartford Children," *Environ. Health Perspect.* 7:99–102 (1974).
4. Lead in the Human Environment, National Research Council, Washington, DC, 1980.
5. Day, J. P., M. Hart, and M. S. Robinson. "Lead in Urban Street Dust," *Nature* 253:343–345 (1975).
6. Binder, S., D. Sokal, D. Maughan. "Estimating the Amount of Soil Ingested by Young Children Through Tracer Elements," *Arch. Environ. Health* 41:341–345 (1986).
7. Clausing, P., B. Brunekreff, and J. H. van Wijnen. "A Method for Estimating Soil Ingestion in Children," *Int. Arch. Occup. Environ. Med.* 59:73 (1987).
8. Schroeder, H. A., J. J. Balassa, and I. H. Tipton. "Abnormal Trace Elements in Man: Titanium," *J. Chron. Dis.* 16:55–69 (1963).
9. Bowen, H. J. M. *Environmental Chemistry of the Elements* (New York: Academic Press, 1979).
10. Lemoh, J. N., and O. G. Brooke. "Frequency and Weight of Normal Stools in Infancy," *Arch. Dis. Child.*, 54:719–20 (1979).

CHAPTER 31

Percutaneous Absorption of Benzo(a)pyrene from Soils With and Without Petroleum Crude Contamination

Joseph J. Yang, Timothy A. Roy, Andrew J. Krueger,
William Neil, and Carl R. Mackerer

INTRODUCTION

Concern regarding dermal exposure to soils contaminated with potentially toxic materials such as dioxins, pesticides, heavy metals, polynuclear aromatics (PNA), and petroleum products containing PNA, has prompted government and private sectors to examine and formulate dermal risk assessment methodologies for contaminated soils.[1-6] Generally, the methodologies that have been devised require information regarding the concentration of contaminant in the soil, dermal exposure to soil, the frequency of exposure, and the degree of absorption of soil-sorbed contaminant through the skin. Data for the last of these parameters, dermal absorption from soil, is rarely available from the literature and very difficult to estimate accurately.[7,8]

In the absence of experimentally determined percutaneous absorption values, many risk assessment schemes substitute available animal toxicology data on the pure contaminant, estimate dermal penetration of pure contaminant based on physicochemical models or, in the extreme, assume 100% bioavailability of the contaminant from soil.[9] Risk assessment based on these approaches is inherently inaccurate and likely to result in an overestimate of dermal bioavailability of contaminants from soil, particularly for the water insoluble lipophilic compounds found in petroleum products.

We, at the Mobil Environmental and Health Science Laboratory, have performed dermal bioavailability studies on a variety of lipophilic materials,[10,11] and special emphasis has been placed on the development and implementation of a modified in vitro technique[12] for evaluating the percutaneous absorption of PNA and PNA mixtures. The modified in vitro technique, designed specifically for lipophilic compounds, uses a 6% aqueous nonionic surfactant receptor solution and a 350 μm-thick skin preparation. Good correlation between in vitro and in vivo percutaneous absorption data for benzo(a)pyrene (BaP) and other PNA has been observed using this technique.[13-15] The study by Roy et al. also showed that the percutaneous absorption of many PNA is very similar to BaP and that the dermal penetration of PNA is largely correlated with the physicochemical properties that characterize lipophilic compounds.

Recently, we have measured the in vivo and in vitro percutaneous absorption of BaP in petroleum crude oil sorbed on soil in rats and have compared the results with those for the percutaneous absorption of BaP from crude oil alone. The air-dried loam soil used in these experiments was fortified with the ^{3}H-BaP-spiked crude at 1% (w/w). The BaP in the crude and crude-fortified soil was 100 and 1 ppm, respectively. In addition, we have also measured dermal bioavailability of soil-sorbed BaP alone (in vitro) at concentrations ranging from 0.1 to 1000 ppm. All experiments were carried out for 96 hr following a single dermal administration of the test materials.

MATERIALS AND METHODS

[1,3,6-^{3}H]-BaP and [G-^{3}H]-BaP, with specific activities of 30–70 mCi/mmol, were purchased from New England Nuclear (Wilmington, DE) and Amersham (Arlington Heights, IL), respectively. SRM 1582, a petroleum crude oil containing 1.1 ppm of native BaP, was obtained from the National Bureau of Standards (Gaithersburg, MD) and spiked with ^{3}H-BaP at 100 ppm. Reagent grade acetone and dichloromethane were purchased from Baker Chemical (Phillipsburg, NJ), carbon disulfide and thimerosal from Aldrich Chemical (Milwaukee, WI), and Volpo-20, a nonionic surfactant, from Croda (New York, NY). An air-dried loam soil was kindly provided by Dr. Harry L. Motto of the Soil Testing Laboratory, New Jersey Agricultural Experimental Station at Rutgers University, New Brunswick, NJ. The soil property data is presented in Table 1.

Table 1. Soil Properties.

Cation Ex. Capacity	9.97 mEq./100 g
Organic Content	1.64%
Sand	46%
Silt	36%
Clay	18%
Texture	Loam
pH	6.0

Soil particles of < 150 μm were used to prepare BaP and crude-fortified soils for percutaneous absorption experiments. Typically, the soil preparation involved adding BaP or the crude, in 10 mL dichloromethane, per gram of soil followed by removal of solvent on a rotary evaporator. Soil was fortified with the crude at 1% (w/w) or with BaP alone at 0.1, 10, and 1000 ppm. Soils containing 10 and 1000 ppm of BaP were prepared by adding unlabeled BaP to the soil previously fortified with ^{3}H-BaP at 0.1 ppm. Radioactivity in the crude was counted directly in a Beckman LS 9000 or 5801 liquid scintillation counter after addition of scintillation cocktail (Beckman HP/b). Radioactivity in BaP and crude-fortified soils was determined by a combustion method using a Harvey OX300 oxidizer (Hillsdale, NJ). Radioactivity in tritiated water, which was trapped in the cocktail after sample combustion, was counted directly. All prepared soils were used within 72 hr of the experiments.

Female Sprague-Dawley rats (Taconic Farms, Germantown, NY), 3–6 months old, were housed individually in wire mesh cages prior to treatment. Animal rooms were maintained at 70°F and 50% relative humidity with a 12-hr light/dark cycle. Food (Lab Chows, Ralston Purina, St. Louis, MO) and water were provided ad libitum.

In the in vitro experiments, the rats were sacrificed by exposure to 100% carbon dioxide. Dorsal skin was lightly shaved with an electric clipper before excision. Skin sections (~350 μm) were prepared from full-thickness skin with a Padgett dermatome (Kansas City, MO) and covered with saline-treated gauze before using within 2 hr. The thickness of the sections was measured with a pressure sensitive micrometer. An aqueous solution of 6% Volpo-20 and 0.01% thimersol antibacterial agent was used as the receptor fluid. Consoles containing 15 mm diameter Franz diffusion cells (Crown Glass, Somerville, NJ) were used in the experiments. The crude was administered in 70–145 μL acetone-carbon disulfide (1:1, v/v) to skin on the donor side of the diffusion cell, followed by evaporation of the vehicle with nitrogen or air. The BaP and 1% crude-fortified soils were weighed and applied evenly on the skin surface. In the experiment with a minimum amount ("monolayer") of soil coverage on the skin surface (~9 mg/cm^2), the excess soil was removed by gentle tapping of the inverted cell and weighed to obtain the exact amount applied. The temperature of the diffusion cells was maintained at 37°C by attaching the water-jacketed cells to a circulating water bath. After dosing, the receptor fluid in each cell was sampled (100 μL duplicates) once every 24 hr for four days. Radioactivity in the receptor fluid was counted directly after addition of cocktail.

In the in vivo experiments, the dorsal area of each rat was lightly shaved after mild anesthesia. Appropriate amounts of crude alone or 1% crude-fortified soil were applied over a 7 cm^2 skin surface area. The amounts administered per square centimeter of skin surface in the parallel in vivo and in vitro experiments were equivalent. The dosed area was covered with a nonocclusive glass cell specifically designed for this study (Krueger and Yang, filing for patent). The cell was attached to the skin with an epoxy adhesive and further secured with Elastoplast®

tape (Beiersdorf, Norwalk, CT). The dosed animals were individually housed in Nalgene metabolism cages and offered food and water ad libitum. Urine and feces were collected once daily for four days. At the termination of the experiment, samples of liver, kidney, small and large intestine, stomach, bladder, and blood were collected from each animal. Radioactivity in the urine, feces, blood, and tissue samples was determined by methods reported previously.[10,14]

A student's t-test ($p < 0.05$) was used to compare the various percutaneous absorption data. Linear regression analysis of the absorption data was carried out using SAS (Cary, NC).

RESULTS

A comparison of the two in vitro percutaneous absorption experiments with 1% crude-fortified soil is shown in Figure 1. The percentage of BaP absorbed (96 hr) was 8.4% and 1.3% at exposure levels of 9 ("monolayer") and 56 mg/cm² ($\sim 1:6$ ratio), respectively. The actual amount of BaP absorbed from 9 and 56

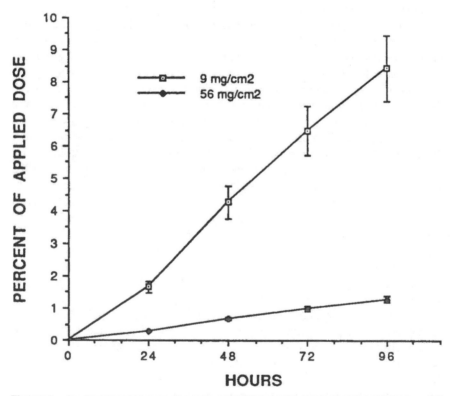

Figure 1.   In vitro percutaneous absorption of BaP from 1% crude-fortified soil (mean±SE for n=5 per time point).

mg/cm² of the fortified soil, however, was nearly identical (1.3 ng). A parallel in vitro experiment was conducted with 90 μg/cm² of crude alone, which is the same amount of crude contained in the 9 mg/cm² crude-fortified soil dose. In the absence of the soil matrix, approximately 38.1% of the BaP was dermally absorbed in 96 hr (Figure 2).

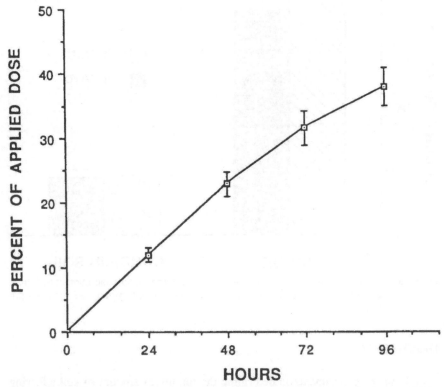

Figure 2. In vitro percutaneous absorption of BaP from petroleum crude (mean±SE for n = 5 per time point).

The parallel 96-hr in vivo percutaneous absorption experiments were conducted with 1% crude-fortified soil (9 mg/cm²) and crude alone (90 μg/cm²) containing the same amount of BaP. A total of 9.2% and 35.3% of the BaP was dermally absorbed from 1% crude-fortified soil and crude alone, respectively (Tables 2 and 3). A comparison of the results from corresponding in vitro and in vivo experiments with crude alone and 1% crude-fortified soil (Figure 3) show no statistical difference between the two results.

The 96-hr in vitro percutaneous absorption experiments with BaP-fortified soil (no crude oil) were carried out at 9 mg/cm². Three BaP concentrations in soils (0.1, 10, and 1000 ppm) were examined and preliminary results show that 1.7, 2.1, and 3.7% of the BaP in the applied doses were absorbed, respectively.

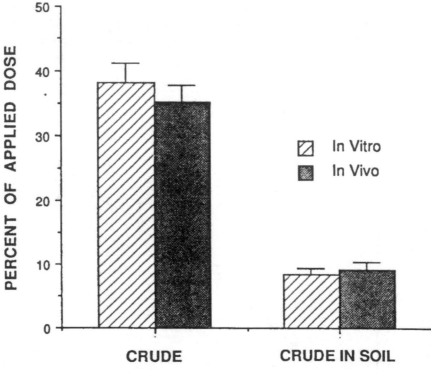

Figure 3. A comparison of 96-hr in vitro and in vivo percutaneous absorption of BaP from petroleum crude alone and 1% crude-fortified soil (mean±SE for n=5 per experiment).

DISCUSSION

Preliminary experiments to determine the minimum amount of soil adhering to the skin were conducted using the in vitro method. Close examination of the skin sections after removal of excess soil indicated that it was largely the silt and clay fractions (<50 μm particle size) that remained on the skin surface. It is in these fine fractions that the bulk of the soil organic carbon content resides.[16] Karickhoff et al.[17] have shown that the soil organic carbon is the dominant sorbent for lipophilic compounds, and that the silt and clay fractions are considerably more effective sorbents than the sand fraction (>50 μm particles). Indeed, we observed similar results when attempting to uniformly sorb crude on the native loam soil. Subsequent sieving of the fortified soil revealed that a majority of the BaP radioactivity resided in the soil fraction with particle size of <150 μm (data not shown).

Based on the results above, percutaneous absorption experiments with the crude sorbed soil were carried out exclusively with soil particles of <150 μm to represent the composition of the soil adhering to the skin surface. Approximately 9 mg/cm^2 of soil was found to be the minimum amount required for a "monolayer" coverage of the skin surface in both in vitro and in vivo experiments. This value is larger than the < 1 mg/cm² of soil (dust) reported for human skin.[18-20] The differences between the rat and human soil adhesion findings may result from differences in rat and human skin texture, the types of soils used, soil moisture content, or the methods of measuring soil adhesion.

The parallel in vitro percutaneous absorption experiments with 1% crude-fortified soil at exposure levels of 9 ("monolayer") and 56 mg/cm² were designed to evaluate the effect of excess soil (i.e., more than a "monolayer") on the dermal bioavailability of crude-sorbed soil. The results (see Figure 1) show the actual amount of BaP absorbed from the two doses (with 1 ppm BaP) was nearly identical even though a much higher percentage (8.4%) of BaP was absorbed from the smaller dose. The constant slope for both the 9 and 56 mg/cm² doses indicates that the small amount of BaP absorbed was entirely derived from the "monolayer" of soil in direct contact with the skin. Under the conditions of the present study, the degree of soil binding of BaP impedes the movement of BaP into skin to the extent that the migration of BaP from layers of soil above the "monolayer" is negligible.

Very good correlation was observed between the in vivo (Tables 2 and 3) and in vitro data (Figures 1 and 2) for dermal bioavailability of BaP. A comparison

Table 2. In Vivo Percutaneous Absorption of BaP from Petroleum Crude-Fortified Soil[a] in Female Sprague-Dawley Rats at a Dose of 9 μg/cm² of Skin Surface.

Total	Hours after Dosing	% Applied Dose Recovered[b]		
		Urine	Feces	Tissues
1.1 (0.3)	24	0.4 (0.1)	0.7 (0.2)	—
3.7 (0.8)	48	0.8 (0.2)	2.8 (0.6)	—
5.8 (1.0)	72	1.4 (0.2)	4.4 (0.9)	—
9.2 (1.2)	96	1.9 (0.2)	5.8 (1.0)	1.5 (0.2)

[a]The petroleum crude (SRM 1582 from the National Bureau of Standards) was fortified with ³H-BaP at 100 ppm. The final concentration of BaP in soil was 1 ppm.
[b]Values shown for 48, 72, and 96 hr are cumulative. Results are expressed as the mean (standard error) for five rats.

Table 3. In Vivo Percutaneous Absorption of BaP from Petroleum Crude[a] in Female Sprague-Dawley Rats at a Dose of 9 μg/cm² of Skin Surface.

Total	Hours after Dosing	% Applied Dose Recovered[b]		
		Urine	Feces	Tissues
5.5 (1.4)	24	1.6 (0.3)	3.9 (1.2)	—
20.1 (2.1)	48	3.2 (0.5)	16.9 (1.7)	—
27.6 (2.1)	72	4.3 (0.6)	23.3 (1.8)	—
35.3 (2.6)	96	5.3 (0.6)	27.5 (2.0)	2.5 (0.4)

[a]The petroleum crude (SRM 1582 from the National Bureau of Standards) was fortified with ³H-BaP at 100 ppm.
[b]Values shown for 48, 72, and 96 hr are cumulative. Results are expressed as the mean (standard error) for five rats.

of the results from percutaneous absorption experiments with crude alone (90 $\mu g/cm^2$) and 1% crude-fortified soil (9 mg/cm^2) is shown in Figure 3. The correlation supports the earlier results of Yang et al.,[14] which showed that in vitro techniques using nonviable skin sections closely mimic the in vivo percutaneous absorption of neat BaP. The present results indicate that the in vitro method will be useful for evaluating dermal bioavailability of PNA and other lipophilic compounds in a soil matrix.

The in vivo and in vitro experiments with the "monolayer" of 9 mg/cm^2 of 1% soil-sorbed crude and 90 $\mu g/cm^2$ of crude alone were intended to compare the absorption of equal amounts of BaP from either the oil matrix or the 1% oil-in-soil matrix. The results (see Figure 3) show that the absorption of BaP from the crude alone is 4–5 times higher than from the soil-sorbed crude.

Previously, Yang et al.[14] have reported that in vitro and in vivo percutaneous absorption of BaP in rats at 9–10 $\mu g/cm^2$ was approximately 45% of the applied dose after 96 hr. Results from the present in vitro experiments with soil-sorbed BaP show that BaP penetration through the skin when applied neat is approximately 10 times higher than from soil-sorbed BaP (2–4% of the applied dose). The percutaneous absorption of BaP from soil shows little change over a wide concentration range between 0.1 and 1000 ppm. Thus, the rate of BaP penetration at the three BaP-in-soil concentrations is effectively independent of concentration.

In conclusion, the results show a marked reduction of dermal bioavailability of BaP sorbed onto soil particles neat or as a component of petroleum. It is reasonable to expect a reduced risk to humans resulting from dermal exposure to soils contaminated with BaP and other PNA (with similar absorption properties) as compared to dermal exposure to the neat contaminant. The results suggest that only the "monolayer" of soil in intimate contact with the skin surface (rather than the total amount of soil on the skin) is important in the determination of risk associated with dermal exposure to soils contaminated with petroleum components.

REFERENCES

1. Kimbrough, K. D., H. Falk, P. Stehr, and G. Fries. "Health Implications of 2,3,7,8-Tetrachlorodibenzodioxin (TCDD) Contamination of Residential Soil," *J. Toxicol. Environ. Health* 14:47–93 (1984).
2. Schaum, J. "Risk Analysis of TCDD Contaminated Soil,"EPA Publication 600/8-84-031 (1984).
3. Hawley, J. K. "Assessment of Health Risk from Exposure to Contaminated Soil," *Risk Anal.* 5:289–302 (1985).
4. Calabrese, E. J., P. T. Kostecki, and D. A. Leonard. "Public Health Implications of Soils Contaminated with Petroleum Products," in D. D. Hemphill, Ed. *Proceedings of the University of Missouri's Annual Conference on Trace Substances in Environmental Health*, Vol 20, University of Missouri, Columbia, 1986, 261–297.

5. Paustenbach, D. J., H. P. Shu, and F. J. Murray. "A Critical Examination of Assumptions Used in Risk Assessments of Dioxin Contaminated Soil," *Fund. Appl. Toxicol.* 6:284–307 (1986).
6. Ryan, E. A., E. T. Hawkins, B. Magee, and S. L. Santos. "Assessing Risk from Dermal Exposure at Hazardous Waste Sites," in *Superfund '87. Proceedings of the 8th National Conference,* Hazardous Materials Control Research Institute, Silver Spring, MD, 1987, 166–168.
7. Poiger, H., and C. Schlatter. "Influence of Solvents and Absorbants on Dermal and Intestinal Absorption of TCDD," *Food Cosmet. Toxicol.* 18:477–481 (1980).
8. Shu, H., P. Teitelbaum, A. S. Webb, L. Marple, B. Brunk, D. Dei Rossi, F. J. Murray, and D. Paustenbach. "Bioavailability of Soil-bound TCDD: Dermal Bioavailability in the Rat," *Fund. Appl. Toxicol.* 10:335–343 (1988).
9. "Superfund Public Health Assessment Manual," U.S. EPA Publication 540/1-86/060 (1986).
10. Yang, J. J., T. A. Roy, W. Neil, A. J. Krueger, and C. R. Mackerer. "Percutaneous and Oral Absorption of Chlorinated Paraffins in the Rat," *Toxicol. Ind. Health* 3:405–412 (1987).
11. Yang, J. J., T. A. Roy , and C. R. Mackerer. "Evaluating Percutaneous Absorption Properties of a Light Middle Distillate in Mice," *Toxicologist* 8:124 (1988).
12. Bronaugh, R. L., and R. F. Stewart. "Methods for In Vitro Percutaneous Absorption Studies III: Hydrophobic Compounds," *J. Pharm. Sci.* 73:1255–1258 (1984).
13. Yang, J. J., T. A. Roy, and C. R. Mackerer. "Percutaneous Absorption of Anthracene in the Rat: Comparison of In Vivo and In Vitro Results," *Toxicol. Ind. Health* 2:79–84 (1986).
14. Yang, J. J., T. A. Roy, and C. R. Mackerer. "Percutaneous Absorption of Benzo[a]pyrene in the Rat: Comparison of In Vivo and In Vitro Results," *Toxicol. Ind. Health* 2:409–416 (1986).
15. Roy, T. A., J. J. Yang, and M. H. Czerwinski. "Evaluating Percutaneous Absorption of Polynuclear Aromatics Using In Vivo and In Vitro Techniques and Structure Activity Relationships," in A. M. Goldberg, Ed. *In Vitro Toxicology Approaches to Validation* (New York: Mary Ann Libert, 1987), pp. 471–482.
16. Brady, N. C. *The Nature and Properties of Soils,* 9th ed. (New York: Macmillan, 1984).
17. Karickhoff, S. W., D. S. Brown, and T. A. Scott. "Sorption of Hydrophobic Pollutants on Natural Sediments," *Water Res.* 13:241–248 (1979).
18. Lepow, M. L., L. Bruckman, M. Gillette, S. Markowitz, R. Robino, and J. Kapish. "Investigations into Sources of Lead in the Environment of Urban Children," *Environ Res.* 10:415–426 (1975).
19. Roels, H. A., J. P. Buchet, R. R. Lauwerys, P. Bruaux, F. Claeys-Thoreau, A. Lafontaine, and G. Verduyn. "Exposure to Lead by the Oral and the Pulmonary Routes of Children Living in the Vicinity of a Primary Lead Smelter," *Environ Res.* 22:8194 (1980).
20. Qee Hee, S. S., B. Peace, C. S. Clark, J. R. Boyle, R. L. Bornschein, and P. B. Hammond. "Evolution of Efficient Methods to Sample Lead Sources, Such as House Dust and Hand Dust, in the Homes of Children," *Environ Res.* 38:77–95 (1985).

CHAPTER 32

An Overview and Suggested Methodology to Determine the Adequacy of Cleanup of Contaminated Soils

Fred N. Rubel, Brian J. Burgher, and E. Corbin McGriff, Jr.

INTRODUCTION

Technical solutions to address environmental contamination events exit. New advances in this field occur daily, both as a result of actual experience gained, as well as from applied research being widely conducted regarding cleanup techniques and criteria. The greatest difficulty encountered in connection with environmental contamination events involve socioeconomic issues. The term "environmental gridlock" aptly describes this difficulty. A key issue in this regard is the decision about the adequacy of cleanup. Such decisions are generally subject to public and private sector scrutiny. Limited progress has been made in setting cleanup standards for soils (i.e., the polychlorinated biphenol [PCB] spill cleanup policy issued by the Environmental Protection Agency (EPA), under its Toxic Substances Control Act program, and the U.S Department of Energy guidelines for residual radioactivity at formerly utilized sites under its remedial action and remote surplus facilities management program), but these are limited. Preset cleanup standards for contaminated soil generally do not exist. There is a reliance on case-by-case determination, which may not always be consistent or systematic. It further appears that the U.S. EPA is no longer making a concerted effort (i.e., through dissemination of records of decisions for cleanup of Superfund sites), to record decisions regarding cleanup levels set at various sites.

Through experience spanning more than a decade of cleanup projects, and the parallel development of methods and criteria by others to address this topic, a logical sequence of questions and activities can be described to aid the practitioner in determining how to decide what level of contaminated soil cleanup is adequate.

An overview of approaches for decision making used under the federal Superfund, and the Resource Conservation and Recovery Act (RCRA) programs at EPA, and other approaches are reviewed. Perspective is provided through a review of data on the concentration of oil related compounds in the environment. Finally, an overall methodology providing a logic tree, incorporating key decision points, is consolidated.

SELECTING THE CLEANUP LEVEL

Figure 1 illustrates that the obvious "first cut" in the cleanup of oil contaminated soil is the removal of visually contaminated soil. The "second cut" could be the ignitability criteria that is used under RCRA to regulate the disposal of hazardous wastes.

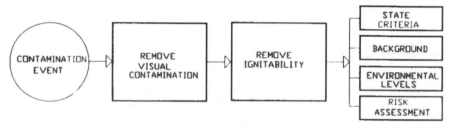

Figure 1. Logic sequence—oil contaminated soil cleanup.

At a spill site, oil moves downward in a spreading fashion, creating in some cases a pyramid of contamination (Figure 2). We also can portray a conceptual hierarchy of cleanup criteria (Figure 3) in the shape of a pyramid, based on the concentration of oil in the soil. The highest concentration (the base of the hierarchy of cleanup) will lend itself to visual criteria, while the lowest concentration (at the peak) will be subject to health risk criteria.

Having passed through the initial criteria for cleanup, one enters the amorphous world of risk assessment. Note that this may result in the most material removed (Figure 4), at the lowest concentrations, and can represent the highest cost. We are presented with essentially an inverted pyramid, and the most critical decision regarding cleanup level associated with relatively little guidance.

The EPA recently announced that regulations (under Subtitle D of RCRA) are being promulgated specifically to manage surface and underground injections of oil and gas drilling and production wastes. This may actually set future criteria for cleanup level consideration in the future.[1] But what of the present?

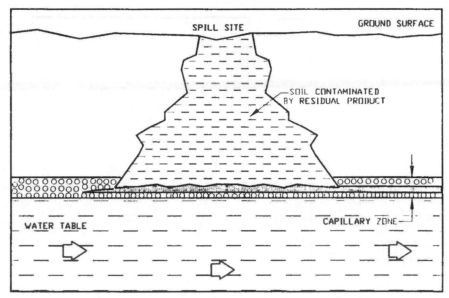

Figure 2. Behavior of product after spill has stabilized. Source: New York Department of Environmental Conservation.

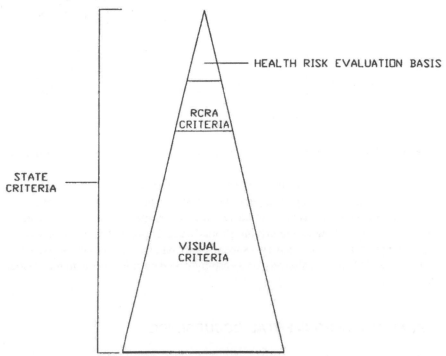

Figure 3. Hierarchy of determining adequacy of cleanup of contaminated soils (concentration).

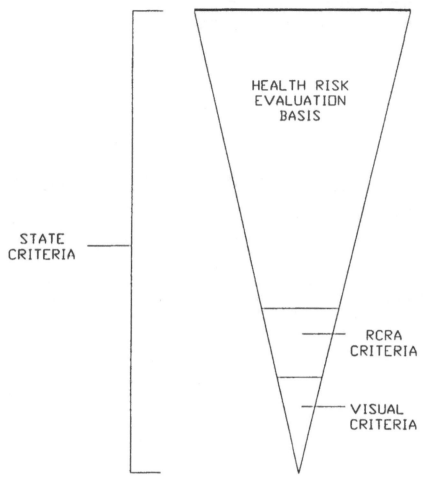

Figure 4. Relative proportions of contaminated soils removed (longer term incidents).

States are in various developmental stages of setting cleanup criteria. The state of Connecticut, for example, appears to utilize 50 ppm, total hydrocarbons, for defining a material as a hazardous waste. New Jersey may not be concerned about the ppm level, so long as underlying groundwater is protected, as evidenced by select empirical tests (soil slurry—sheen; paper bag test). Massachusetts, Pennsylvania, and California also rely on a case-by-case determination as to the cleanup level.

RELATIVE ENVIRONMENTAL OCCURRENCE

An environmentally important component of oil is the class of compounds identified as polynuclear aromatic hydrocarbons. They include carcinogens, and therefore represent a significant subset of data to focus on when setting a desired cleanup level for petroleum contaminated soil.

It can be helpful in this process to obtain a perspective on the levels of contamination one is dealing with by reviewing data collected. Others have reviewed the literature, and have reported the occurrence of from 2 ppb to 1.3 ppm of polynuclear aromatic hydrocarbons (PAH) in garden or forest soil.[2] Data was also cited indicating significantly higher levels of one PAH component, benzo(a)pyrene, in areas subject to intensive combustion exhaust. A correlation has been demonstrated in the vicinity of a Swiss mountain town between PAH content and proximity to a highway (4–8 ppm in the surrounding alps, versus up to 300 ppm in dry soil near a highway).[3]

Table 1 provides some relative concentrations of oil-related contaminants which have been recorded in the environment in a range of from zero to 3,498 ppm of PAH. Many of the data represent tainted conditions. The authors do not present such levels as acceptable or as a standard of cleanup. Nevertheless, these were levels which previously did not/have not resulted in cleanup or other regulatory action (i.e., banning of harvesting).

Table 1.   Relative Concentrations of Polynuclear Aromatic Hydrocarbons.

Concentration/Range (Min. Max.)	Media/Location/Reference
7 ppb	mussels, Charlotte Harbor, FL 1986 (4)
2 ppm	mussels, San Simeon Pt., CA 1986 (4)
2.6 ppm	mussels, N.Y. Bight 1986 (4)
5.3 ppm	mussels, L.I. Sound 1986 (4)
13 ppm	mussels, St. Andrew Bay, FL 1988 (4)
20 ppm	mussels, Hud./Raritan Estuary 1986 (4)
23 ppm	mussels, Elliot Bay, WA 1986 (4)
0–27.9 ppm	mud, St. Lawrence River, post oil spill (5)
2.3–29 ppm	fish, St. Lawrence River, post oil spill (5)
30–54 ppm	dehydrator pit soil, gas production New Mexico (6)
77 ppm	muskrat, St. Lawrence River, post oil spill (5)
45–349 ppm	produced water pit soil, gas production New Mexico (6)
557–683 ppm	pipeline drip pit, gas production New Mexico (6)
500–3,498 ppm	duck, St. Lawrence River, post oil spill (5)

No action was taken, presumably because of the difficulty involved in addressing massive areas/ecological components. The risk of ingestion by the human population as a whole, however, would appear to be greater for mussels, or duck, than it would be for soil. Bans on the taking of fish and shellfish have been issued on the basis of other parameters in the past (PCBs, fecal coliform). One could either conclude that cleanup standards which allow oil containing PAH in soil in the high ppm range are consistent with other regulatory decisions, particularly in view of data on its biodegradability, or that a new look should be taken at the regulation of shellfish and other foods, based on PAH content. Alternatively, one could take the view that such high levels of PAH are an undesirable consequence of bioaccumulation of environmental contamination that is in need of reduction through actions such as setting more strict cleanup standards.

It may also be helpful to compare any oil cleanup standard in soil with that for PCBs. For this especially notorious chemical, EPA's regulation requires cleanup to at least the 10 ppm level in a residential setting (provided that at least 10 inches of clean soil are placed on top), or as high as 50 ppm at a power substation, provided that notice of this contamination is posted. Variances to require lower levels in such cases can also be issued by the U.S. EPA.

The above adds to our perspective, but direct comparison remains difficult due to the many components of which oils are comprised, including benzene, benzo(a)pyrene, and other polynuclear aromatic hydrocarbons.

Perspective on a related subject was gained by Ames et al.,[7] who took a holistic look at the occurrence and relative potency of manmade and naturally occurring carcinogens. This opened for debate the decision making rationale regarding the relative importance of man-made chemicals contaminating our environment. Ames used the HERP index of possible human hazard. This expresses resulting human exposures as a percentage of the rodent TD_{50} for each carcinogen. [TD_{50} is the daily rate, in mg/kg, which results in tumors in half of the animal study population over their lifetime. HERP is the human exposure dose/rodent potency dose.] As an example, the HERP percent for benzene in indoor air, resulting in a dose of 155 ng/70 kg person daily for 14 hours per day, was 0.001. This compares to a 30-fold greater HERP percent of 0.03 for peanut butter (32 g; one sandwich per day), which contains aflatoxin at a dose of 64 ng/70 kg person daily. Ames has even gone so far as to be quoted as saying that exposure to manmade chemicals in water is "irrelevant to the causes of human cancer."[8] This leads one to question decisions regarding oil-contaminated soil cleanup levels set within the health risk assessment portion of the inverted criteria pyramid (Figure 4).

RISK ASSESSMENT

The U.S. EPA, through its guideline on corrective actions for leaking underground storage tanks[9] indicates that "the levels of acceptable risk should be established to assist in the determination of criteria for meeting cleanup objectives." It states that, "based on toxicity data and regulatory and environmental standards, the relative human and environmental hazard posed by exposure to various concentration levels can be assessed." No risk assessment methodology, however, is provided.

The EPA held a special hearing and received comments until October 3, 1988 on what constitutes acceptable risk levels for exposure to benzene in air. The alternative standards being considered comprised 90 pages of the *Federal Register*.[10] Four levels of risk are being considered in this proposal: one cancer case per 1,000 persons following lifetime exposure; one cancer case per year; one cancer case per 10,000; and one cancer case per 1,000,000.[11]

It is clear that this issue of acceptable risk is unresolved from a government regulatory standpoint. To further complicate any rigorous decision on cleanup levels, the selection of one (or more) model to produce the risk number also remains undecided.

In setting its policy for corrective action at solid waste management facilities, as well as for federal Superfund sites, the term "health standards" is used.[12] This relates primarily to groundwater, and the maximum contaminant levels (MCL) promulgated under the Safe Drinking Water Act. The provision of alternate concentration limits, or ACL, is allowed where less stringent cleanup is deemed to be safe. ACLs, again, remain a site-specific determination.

Under EPA's RCRA program, one determines the levels of contaminants in the plume and compares this to Appendices 8 and 9 of 40 CFR Part 261, and to MCLs, to determine if there is a basis for action. Where an MCL does not exist, removal to background is expected unless an ACL is requested and granted on the basis of risk assessment.

Ken Harsh of the Ohio EPA presented an excellent treatment of the entire subject of decisions regarding the level of cleanup in 1984.[13] Although he begins to speak of risk assessment, he highlights Best Professional Judgment (BPJ) as the practical approach, citing the costs of extensive sampling, and the complexities of risk assessment.

Edmond Frost of Kirkland & Ellis reviews criteria for determining the degree of cleanup required under remedial Superfund cleanups, and also notes that the state of the art continues to revert to case-by-case determinations.[14]

SUMMARY

Although a series of steps can be followed, and guideposts do exist for selecting the level of cleanup of oil-contaminated soils, with few exceptions this remains a case-by-case determination. The use of comparative data on the presence of oil and its constituents in the environment should be considered in setting any firm standard or site-specific cleanup level. Regulatory agencies should establish an agreed-upon risk assessment model for oil contaminated soils.

REFERENCES

1. "EPA Tightens Standards," *The Management of World Wastes*, Vol. 31, No. 9, September 1988, p. 20.
2. Zobel, C. E., "Sources and Biodegradation of Carcinogenic Hydrocarbons," *Joint Conference on the Prevention and Control of Oil Spills* (1972), pp. 441–451.
3. Blumer, M., et al., "Polycyclic Aromatic Hydrocarbons in Soils of a Mountain Valley: Correlation with Highway Traffic and Cancer Incidence," *Environ. Sci. Technol.* 11:(12), November 1977, pp. 1082–1084.
4. "National Status & Trends Program for Marine Environmental Quality—Progress Report, NOAA technical memorandum, NOS OMA 38, December 1987.
5. "Damage Assessment Studies Following the NEPCO 140 Oil Spill on the St. Lawrence River," EPA-600/7-79-256, December 1979.
6. Elceman, G. A., et al., "Depth Profiles for Hydrocarbons and Polycyclic Aromatic Hydrocarbons in Soil Beneath Waste Disposal Pits from Natural Gas Production," *Environ. Sci. Technol.* 20:5 (1986), pp. 508–514.
7. Ames, B., et al., "Ranking Possible Carcinogenic Hazards," *Science*, Vol. 236, pp. 271–280.
8. *Environment Reporter*. BNA, January 31, 1986, p. 1813.
9. *Underground Storage Tank Corrective Action Technologies*, U.S. EPA/625/6-87-015, January 1987, pp. 4–9.
10. *Federal Register*, July 28, 1988, pp. 28496–28586.
11. *Chem. Eng. News*, July 25, 1988, p. 6.
12. "Cleanup Criteria to be on Health Standards," World Wastes, April 1987, p. 18.
13. Harsh, K. M., "How Clean is Clean? How Toxic is Toxic?" *1984 Hazardous Material Soils Conference Proceedings*, pp. 205–213.
14. Frost, E. B., "Determining of Cleanup and Required Remedial Measures Under Superfund," *1984 Hazardous Material Soils Conference Proceedings*, pp. 386–392.

CHAPTER 33

Toward Economically Efficient Management of Underground Storage Tanks: A Risk-Based Approach

Bernhard H. Metzger

THE PROBLEM

Contamination of groundwater is being detected with increasing frequency and it is linked to adverse health, economic, environmental, and social impacts. But no source of groundwater contamination, save perhaps septic systems, is as pervasive and widely distributed as leaking underground petroleum storage tanks. Statistics published recently by the U.S. Environmental Protection Agency (EPA) illuminate the magnitude of the potential problem.[1,2]

The EPA estimates there are 1.4 million underground storage tanks (UST) owned and operated at over 500,000 facilities; over 95% of these are used to store petroleum. In addition, there are several million fuel and heating oil tanks at farms and residences. Eighty percent to 90% of UST are constructed of bare steel and thus are susceptible to corrosion. Most of them have been in the ground for more than 10 years; one third are over 20 years old. Some 10% to 30% of systems currently installed might be leaking already. Most releases are reported to be larger than 2,000 liters. Visual appearance and outdoor odor of product are the two most commonly cited means of leak detection. Impacts of leak incidents include, with high frequency, contamination of public and private wells, combustible fumes in confined areas, fire and explosion, human illness, and ecological damage.

In response, the EPA proposed for existing tanks a 10-year transition period to introduce corrosion protection and leak detection.[3] The proposed technical standards are uniform, i.e., they do not differentiate among sites of different vulnerabilities and different resource values and use. Given the scale of the problem, i.e., the sheer number, distribution, advanced age, propensity to failure, and high damage potential at vulnerable sites, the EPA's response appears inadequate. A response strategy which is technically more effective and more efficient economically is indicated—a strategy which incorporates the notion of differential protection, but which at the same time is politically and administratively feasible.

This chapter summarizes research done for the Illinois Department of Energy and Natural Resources (ENR).[4,5] The objective is to identify an economically efficient response strategy, to propose a more adequate approach, and to establish an analytical framework for its implementation. We then compare the estimated economic impact of the EPA's proposed policy to the expected impacts of a few alternative strategies. One of these alternatives is risk-based management. The latter requires criteria to classify, assess, and rank sites in accordance with the risk they pose to the public, the environment, and the economy. Next we present a system of such criteria and a conceptual framework for an algorithmic decision rule to identify the best preventive action at a particular UST site. We recommend areas of additional study, and, finally, offer conclusions.

THE ECONOMICS OF ALTERNATIVE RESPONSE STRATEGIES

The question before us is whether preventive management of UST should be based in part on site-specific factors or by ignoring differential site sensitivity and resource value. In the latter case, formulation, implementation, and enforcement of an equitable regulatory policy could be greatly simplified; uniform technical standards would suffice. Similarly, because financial risk to the carrier would be independent of site characteristics, an appropriate rate structure for liability insurance would be site-independent. If, on the other hand, risk-based preventive UST management improves economic efficiency, i.e., is more cost-effective, regulatory policy should be framed accordingly.

The appeal of uniform management lies in its administrative ease and simplicity. A risk-based policy requires criteria to assess site risk and a decision rule to translate risk into site-specific action. However, any acceptable mechanism for risk-based UST management must be administratively feasible, simple, reasonably accurate, and inexpensive enough to apply so as not to consume the margin of efficiency afforded by the approach.

Differential Site Sensitivity

A preventive strategy relying on only one technical standard for all tank sites may be sufficiently protective for some fraction of the site population, while perhaps being overprotective for low risk sites and underprotective for high risk sites. However, measures that are underprotective in some cases and overprotective in others can surely inflict a waste of resources. The magnitude of this potential waste depends on the spread of site risk, or damage potential (Figure 1), among the UST site population and on the range of costs for prevention. The cost of prevention tends to be positively correlated with the amount of risk reduction afforded by a particular technology. Site risk depends on hydrogeological site sensitivity, the value of threatened resources, land use, the probability of undetected UST failure, etc. The wider the spread of these contributing factors the greater the economic inefficiency of uniform strategies for preventive UST management.

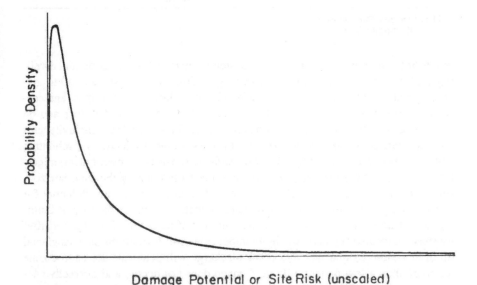

Damage Potential or Site Risk (unscaled)

Figure 1. Probability density of damage potential, or risk, for population of UST sites.

Damage potential of a UST is related to the amount of time it would take the leaking liquid, or components thereof, to migrate through the ambient geological formations to certain prescribed points at a distance from the tank. Travel times in the unsaturated (t_u) and saturated zones (t_s) can differ widely among different UST sites, their range extending easily over six or more orders of magnitude. Hydrogeological site sensitivity or vulnerability tends to increase with decreasing travel times. A site with saturated zone travel time of only a few years (for the leaked material to arrive at some sensitive point) can, if managed inadequately, lead to health, environmental, and economic disasters. However, sites with highly impermeable soils tend to have low damage potential.

Table 1 shows a simple but illustrative site classification scheme. It permits partitioning the population of UST sites into three very broad categories of hydrogeological site sensitivity based on the above characteristic times. High (H) sensitivity sites are settings with high potential for tank discharge to reach sensitive

Table 1.   Classification of Sites by Hydrogeological Sensitivity.

Site Category	Characteristics	Potential Impact	Criterion (time scales)
High sensitivity H	high hydraulic conductivity, no attenuation	high cost high EI high HI	$100 \text{ yr } \geq t_u + t_s$
Moderate sensitivity M	low hydraulic conductivity in saturated zone	some cost low EI low HI	$t_u + t_s \geq 100 \text{ yr } \geq t_u$
Low sensitivity L	very low hydraulic conductivity	low cost very low EI no HI	$t_u \geq 100 \text{ yr}$

NOTE: EI = environmental impact.
HI = health impact.

points or areas (e.g., wells, ecological systems, surface water, residential dwellings, etc.) within an economically meaningful time horizon of, say, 50 to 100 years. Contamination typically spreads quickly over large areas. Impact and corrective action costs depend crucially on the time of detection of failure and on use and value of the land and threatened resources. Moderate (M) sensitivity sites are characterized by geologic formations of low hydraulic conductivity, prohibiting tank discharge from advancing quickly through the subsurface environment; migration is so slow that sensitive points remain unaffected during the time horizon. But contamination of unsaturated and saturated zones does occur. Potential for offsite damage and cost of corrective action depends on the proximity of buildings and structures, and to a lesser extent on time of detection. Low (L) sensitivity sites are located typically in highly impervious soil formations such as glacial till or unweathered marine clay; a tank discharge will penetrate the vadose zone not more than a few meters. Offsite damage does not occur, and corrective action costs are very low and virtually insensitive to the time of detection; cleanup might be required to avoid vapor related onsite problems.

Figure 2 gives estimates of the distribution of subsurface sensitivities for the entire U.S. and for the state of Illinois. The analysis is based on the classification system for U.S. groundwater regions according to Heath[6] and on generic hydrogeological modeling assumptions used by the EPA to regulate landfills and hazardous waste tanks.[7] If we assume that the population distribution of UST sites is statistically independent of the distribution of groundwater regions and uniform across all regions, Figure 2 is a preliminary estimate of the distribution of UST site sensitivities. Because there probably is a causal relationship between UST site locations and hydrogeological regions (most UST sites are in urban or semiurban settings selected for their abundance of usable groundwater), one would expect an even higher fraction of tanks in high risk areas than indicated in Figure 2.

The fractions of H, M, and L regions, and by our assumption, UST sites, are practically the same for the U.S. and Illinois. More than 40% of sites appear

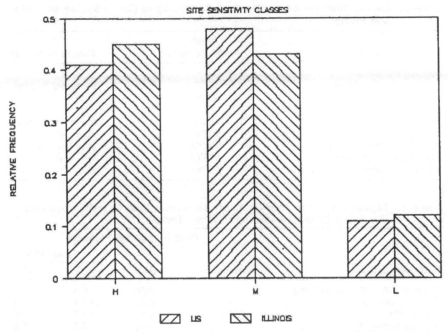

Figure 2. Relative frequency of three site sensitivity classes, for the U.S. and Illinois.

to be located at highly sensitive locations and a similar percentage at moderately sensitive sites, with the remainder (some 10%) in low sensitivity environments.

Cost of Prevention

The prevention and early detection (monitoring) options reviewed here assume a tank system at a generic gasoline retail station. Tank replacement involves overhaul of the entire facility, including new tanks, piping, pumps and backfill, and refinishing of the surface. Similarly, installation of a monitoring system is assumed to apply to the entire facility. Table 2 gives cost data for several preventive options.[5] Failure potential is assessed qualitatively. The EPA evaluated a number of detection methods for underground storage tanks, ranging from simple inventory monitoring to tank testing and groundwater monitoring.[8] Table 3 lists capital and operating costs for five different detection methods, assuming alternative discount rates of 3% and 10% and a 30-year time horizon.[5]

Generally, detection methods relying on noncontinuous monitoring or testing will permit releases of durations equaling the time between samples. According to the proposed tank rule, this can be 3 to 5 years for existing tanks. In Provincetown, Massachusetts, a 1-year leak at a high risk site resulted in cleanup costs of $5 million.[5] Had the leak been detected six months later, seven community

Table 2.   Capital Costs for Alternative Tank Technologies (Entire System of Tanks and Piping).

Tank Technology	Cost (1987 dollars)	Failure Potential
Retrofit old tanks with cathodic protection	15,000	moderate to high
New unprotected steel	40,000	high
New steel tanks with cathodic protection	45,000	low to moderate
New FRP, single-walled	50,000	moderate
New FRP, double-walled	95,000	very low
New steel, double-walled	85,000	very low
Closure of tanks	38,000	zero

Table 3.   Capital, O&M, and Total Costs for Alternative Leak Detection Systems (for Entire System Consisting of Three Tanks).

	Cost in '000 1987 Dollars			
			Total (PV)	
Detection System	Installation	O&M	3%	10%
Statistical inventory monitoring	—	0.45	8.8	4.2
Pressure testing	—	3.3	64.7	31.1
Inspections	—	2.4	47.0	22.6
Soil gas monitoring	7	—	7.0	7.0
Groundwater monitoring	3	1.6	34.4	18.1

Note: Semiannual sampling for discontinuous detection systems; 30 year time horizon.

wells would have been contaminated and corrective action would have been far more costly.

Cost of Correction

Existing data on corrective action cost for leaking UST are not very reliable, and the literature shows important discrepancies. Most recent reports agree, however, that the cost distribution is highly skewed toward high cost.

Rogers[9] asserts that 85% of all leaks at service stations remain on site due to early detection; these incidents incur cleanup costs of $20,000 to $30,000 (presumably in 1983 dollars). Another 10% of leaks cause some offsite damage costing $150,000 to repair. However 5% of tank failures lead to serious offsite damage, and the corresponding cleanup costs are $2.5 to $5 million, with some episodes running and as high as $11 million. Schwendeman cites a similar breakdown of costs but his estimates are slightly higher.[10] Both authors base their estimates on statistics provided by the American Petroleum Institute. However, these statistics are for past incidents when cleanup requirements were not based on real resource cost, and so they might underestimate real cost.

Hinchee et al., citing a study by the California Commission for Economic Development, report substantially higher figures for a similarly skewed distribution.[11]

The EPA, on the other hand, assumes very low cost estimates in its UST economic and regulatory impact analysis, neglecting completely the long tail of high risk sites with cleanup costs of $1 million and more.[3] This omission of the high risk tail might, incidentally, bias the proposed rule in favor of a uniform technical UST standard.

Figure 3 shows empirical frequency distributions of corrective action costs as reported in the literature. For subsequent analysis we assume they are equally likely and reliable because we know little about the data these distributions were derived from. Thus, the average of the means of the four distributions, $0.5 million per UST leak, gives an estimate of average corrective action costs for the UST population in the U.S. This average is sensitive to uncertainty in the tail of the distributions.

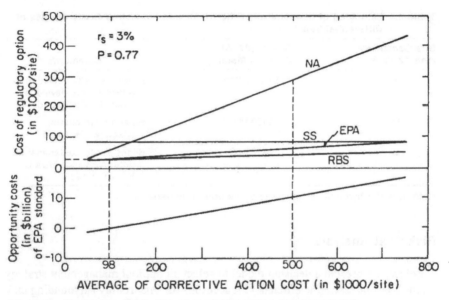

Figure 3. Frequency distribution of corrective action cost as reported.

Metzger estimates corrective action costs for UST leaks based on optimal remediation for sites of different sensitivity.[5] The resulting order of magnitude estimates are shown in Table 4 for the site sensitivity classes defined in Table 1. Each of these point estimates is surrounded by a band of uncertainty, accounting for ranges of hydrogeology, resource value, land use, detection time, and other factors. In logspace the band width might be ±0.5 with probability 1/3 (one standard deviation), which translates into a factor of 3 in dollar space. Remediation cost at high risk sites depends importantly on the duration of the contamination episode prior to detection.

The cost estimates in Table 4 compare favorably with the empirical data in Figure 3. The averages (using frequencies of site sensitivity classes shown in

Figure 2 as weights) are virtually identical. But the skews of the distributions differ because the high sensitivity sites (Table 1) contribute only fractionally a damage potential of more than $1 million. Suppose we divide the H class into two groups, one containing extremely high risk (HH) and the other moderately high risk sites, and lump the latter with the M and L classes into a combined MM class. Suppose this HH class comprises some 10% of sites (as indicated by Figure 3); given average corrective action costs of $0.5 million per incident, the respective means of corrective action cost for the HH and MM classes might range between $3 million (HH) and $250,000 (MM) for a small skew and between $4 million (HH) and $125,000 (MM) for a larger skew of the cost distribution. Sensitivity analysis suggests that principally the skew of the cost distribution determines the economic efficiency of alternative preventive management schemes.

Table 4. Estimates of Economically Efficient Corrective Action Costs at Sites of Differential Sensitivity.

Site Sensitivity (see Table 6)	Cost (1987, 3%) (Order of Magnit)	Comments
H	$1.0 million	Assumes early detection; offsite plume develops; typically: K > 10 m/d
M	$100,000	Small offsite plume; typically: K ≤ 1 m/d
L	$10,000	Slight onsite contamination; potential vapor hazard; K < 10⁻³ m/d

Note: K is the hydraulic conductivity of the aquifer material.

Risk-Cost Analysis

Below, we present a decision model to select an optimal management strategy on the basis of economic efficiency. It accounts for uncertainty surrounding tank failure and leak detection, and weights future costs and benefits using a social rate of discount to reflect society's time preference rather than the opportunity cost of capital. The optimal policy is defined as that conferring the largest social net benefit, or incurring the least real resource cost. An illustrative set of regulatory options is defined, followed by risk-cost analysis using the best available estimates of costs and effectiveness. Sensitivity analysis examines the robustness of the optimal policy with regard to uncertainty in key model parameters. Results, implications, and applications are discussed.

The Decision Model

The expected monetary present value, EMPV[TC(a)], of total cost, TC, associated with regulatory option, a, can be expressed for a given UST site as follows, assuming only one UST may fail at that facility and, at most, once during the economic time horizon of the analysis:

$$EMPV[TC(a)] = p(a)[q(a)C_r\alpha_{ar} + (1-q(a))(C_r\alpha_{pr} + C_{ca}\alpha_{ca})] + C_c(a)\alpha_c(a) \qquad (1)$$

where

$p(a)$ probability of tank failure $[0,1]$; depends on tank technology (regulatory policy, a), age and environment of tank;

$q(a)$ probability of detecting leak prior to occurrence of site damage $[0,1]$; depends on monitoring system (regulatory policy, a);

C_r cost of tank repair;

C_{ca} cost of economically most efficient corrective action; depends on site sensitivity, present and future resource use and value, discount rate, economic time horizon, etc;

$C_c(a)$ cost of complying with regulatory policy, a;

α_{ar} present value factor for repair of detected leak; depends on discount rate, time horizon, time of repair;

α_{pr} present value factor for repair of undetected leak; depends on discount rate, time horizon, time of repair;

α_{ca} present value factor for corrective action; depends on rate of discount, time horizon, time and duration of action;

$\alpha_c(a)$ present value factor for compliance; depends on policy, rate of discount, time horizon.

The cost relation in Equation 1 assumes further that leaks detected by the installed monitoring system are repaired immediately, and that no corrective action is needed. The value of EMPV[TC(a)] depends on site characteristics because probability of failure and cost of corrective action are functions of site-specific factors. A uniform regulatory policy has approximately the same attendant probability of detection and cost of compliance for all UST facilities. A risk-based policy implies differential compliance costs requiring risk-dependent tank technologies and monitoring systems. Because we are interested here in economic impact analysis rather than financial analysis from a UST owner/operator perspective, the cost terms in Equation 1 reflect real resource costs weighted at an appropriate social rate of discount.

Equation 1 applies for an individual site. However, for purposes of policy analysis the effects of policy on the entire population of sites must be evaluated. Therefore we lump the entire ensemble of sites into groups of similar sites exhibiting group-specific generic characteristics, for which aggregate cost and technical data are required. For simplicity we assume that Equation 1 remains valid at this aggregated level and that it can be used in a decision model to predict economic cost consequent to a particular policy choice.

Expected total economic cost, TEMPV[TC(a)], attendant on policy a is given by the product of two k-dimensional vectors:

$$TEMPV[TC(a)] = n\rho EMPV[TC(a)] \qquad (2)$$

where k is the number of risk groups, n is the number of UST sites in the population, ρ is the vector of fractions of sites in risk group i $(i=1..k)$; $\underline{EMPV[TC(a)]}$ is the cost vector whose components are evaluated from the aggregated variant of Equation 1 by substituting k-dimensional vectors for scalars p, q, C_{ca}, and C_c.

The objective is to find that policy, a°, of the set, A, of feasible policies which confers the largest social benefits or conversely, minimizes the value of TEMPV. The decision problem can thus be stated as the following nonlinear constrained optimization problem:

$$\text{Minimize TEMPV}[TC(a)], \text{ s.t. } a \in A \qquad (3)$$

There are four choice variables (assuming the interest rate is not negotiable), namely p, q, ρ, and k, of which only k is a scalar. Feasibility is defined by technological, economical, and political constraints. The term 'expected monetary value,' as defined in the theory of decision analysis, is used throughout to imply the premise of risk neutrality on the part of the decision making entity or authority.[12]

The Choices

The feasible region of the decision problem formulated in Equation 3 offers an infinite number of regulatory options. Systematic exploration of that space might be worthwhile to decide on the best UST policy. Here we intend merely to investigate whether or not risk-based management is superior to uniform regulation. We assume, for simplicity, a discrete set of feasible policies containing only four technical standards, including a base case referred to as 'no action':

NO ACTION (NA): This base case represents the underprotective status quo for UST management prior to the 1984 RCRA Reauthorization Amendments. It assumes there is no tank replacement for old tanks, and that new tanks are mostly bare steel, unprotected against corrosion. Leaking tanks are repaired when detected, and population averages apply for UST failure rate and expected cost of corrective action.

STRINGENT STANDARD (SS): This option is antipodal to the base case. It is uniformly stringent, and overprotective for some fraction of the UST population. It assumes the entire UST population is retired immediately and replaced with double-walled corrosion-protected steel tanks with interstitial monitoring. These systems are considered fail-safe, i.e., failure is detected immediately and damage is prevented; the rate of undetected failure and cost of corrective action are zero. This option is a useful limiting case for analysis, though practically infeasible due to the tank industry's inability to meet the demand.

EPA STANDARD (ES): This option approximates EPA's proposed rule[3] for existing UST, assuming all systems are retrofitted immediately with a continuous soil gas monitoring system, and that manual inventory control is performed. After a phasing-in period of 10 years, all UST systems are retrofitted with impressed current cathodic protection. Hence, this standard increases the probability of detection and lowers the probability of failure, thus reducing risk to some residual level presumably tolerable at all sites.

RISK-BASED STANDARD (RBS): A risk-based regulatory philosophy recognizes heterogeneity within the regulated ensemble, and tailors requirements accordingly. The standard defined here divides the UST population into $k=3$ risk classes. It requires the SS standard, i.e., rapid (immediate) replacement of tank systems, for the highest risk sites; associated expected damages are zero. Implementation of the ES standard is required for all other sites, except for those in low sensitivity areas. The probability of damage is the same as for the SS standard, but the expected costs of corrective action reflect the damage potential of this class of sites. Finally, UST at low sensitivity sites require no protective action because leaking product would never (within reasonable time scales) migrate beyond the site. Leaking tanks are repaired and minimal local corrective action is assumed necessary (Table 4).

Results

In order to simplify evaluation of the regulatory options we assume that leaks occur on average at an age of 25 years for the UST population.[13] Instantaneous damage detection and implementation of corrective action is assumed. Costs are expressed in real 1987 dollars.

Table 5 contains a sample set of site parameter estimates to solve the decision problem in Equation 3. The risk-based policy assumes the HH fraction of the high risk group to comprise 10% of sites. The low risk group is estimated to contain 10% of UST sites. The remaining 80% of sites constitute the MM group. A conservative skew is chosen for the distribution of remedial cost, assuming $3 million per incident at HH sites and $250,000 at MM sites. Derivation of the other parameter values is discussed in Reference No. 4.

Figure 4 is a graphic representation of the decision problem stated in Equation 3 with parameter values from Table 5. The column of numbers to the right of the decision tree lists total costs. The numbers within the tree are expected costs for the branches they are associated with; probabilities (failure or detection) are shown at chance nodes for each branch. All costs are in $1000 per UST site.

Clearly, the risk-based standard is economically most efficient, i.e., $a^\circ=RBS$, and the associated value of the objective function is TEMPV/$n=$ $44,000/site. The cost of the next best option, the EPA standard, is 50% higher at $65,000/site. The highest social costs are associated with the 'No Action' policy. Consequently, nationwide implementation of the RBS, as opposed to the ES, would

Table 5.   Assumed and Estimated Values of Decision Variables and Parameters (Equation 3; Reference No. 7).

Parameter	Symbol	Value
Social interest rate	r_s	3%
Tank repair cost	C_r	$6,000
Compliance cost	C_c	
Detection system		$7,000
New tanks (corrective protection)		$45,000
Retrofitted tanks		$15,000
Double-walled tanks		$85,000
Corrective action cost	C_{ca}	
Population average		$500,000
High risk class		$3 million
Medium risk class		$249,000
Low risk class		$10,000
Risk class frequency		
High risk class	ρ_{HH}	0.1
Medium risk class	ρ_{MM}	0.8
Low risk class	ρ_L	0.1
Tank system failure rates	p	
Unprotected system		0.77
New/retrofitted system		0.77/2
Double-walled system		0.77/10
Detection rates	q	
No detection system		0
Detection system		0.7
Double-walled tanks		1.0
Times of occurrence	t	
Installation of detection system		0 years
Retrofitting of tanks		5 years
Installation of double-walled tanks		0 years
Leak		8–10 years
Leak detection		10 years
Repair/corrective action		10 years
Present value factors		
Retrofitting	α_c	0.863
Repair	α_{sr}, α_{pr}	0.744
Corrective action	α_{ca}	0.744

produce savings of societal resources worth approximately $10 billion (multiply $EMPV_{ES} - EMPV_{RBS}$ by the size of the UST site population in the U.S., n=500,000 sites). Savings for Illinois would amount to some $460 million (n=23,000 sites). This point estimate is surrounded by uncertainty due to model simplicity and imprecision of parameter estimates.

Sensitivity analysis reveals the most influential parameters of the decision problem: (1) the skew of the distribution of corrective action costs as expressed by the size of the highest risk UST group, ρ_{HH}, and the attendant expected corrective action cost; (2) the average expected corrective action cost across the entire UST population, i.e., $C_{ca} = \Sigma_i \rho_i C_{ca,i}$; (3) the rate of discount, r_s; and (4) the

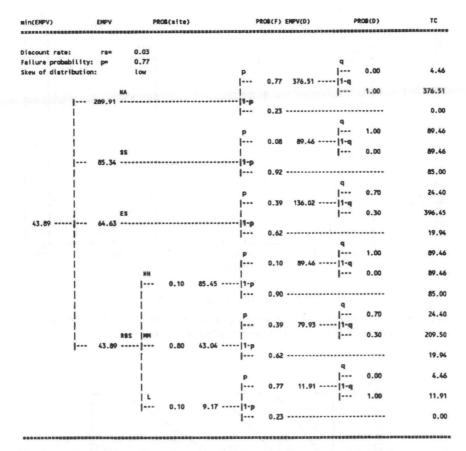

Figure 4. Decision problem (Equation 3) with parameter values of Table 5.

failure rates of the various tank technologies. The value of the objective function in Equation 3 is sensitive to all of these parameters.

If the skew of the distribution of actual cost of corrective action is larger than that assumed in Table 5, the efficiency advantage of the risk-based standard increases, and conversely. The empirical data in Figure 3 suggest a real skew at least as large as that assigned in Table 5. Figure 5 illustrates the sensitivity of the cost of the several policies with respect to average corrective action cost, C_{ca}. The risk-based standard is optimal for values of $C_{ca} \geq \$100,000$. Below this value the EPA standard is more efficient. The lower portion of Figure 5 shows the savings realized by implementing the RBS rather than the EPA standard.

Figure 6 demonstrates that RBS is a robust optimal policy for a wide range of values of key model parameters. The contours are the iso-cost loci of the lowest average corrective action cost, C_{ca}, for which RBS is still the optimal regulatory

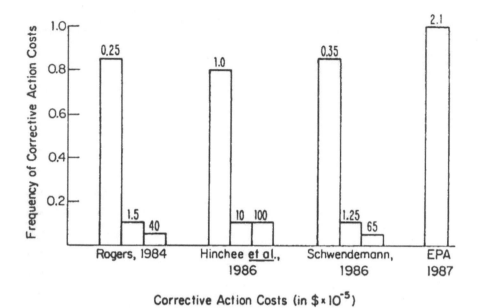

Figure 5. Total cost of regulatory options (top) and opportunity cost of EPA standard (bottom) as a function of average corrective action cost.

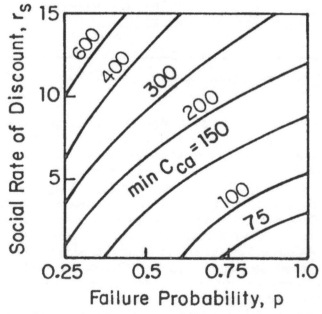

Figure 6. Sensitivity of optimality of risk-based UST management associated with uncertainty in average corrective action cost (in $1000/site), social rate of discount and UST failure probability.

policy, given a tank failure rate, p, and a social discount rate, r_s. For example, at $p=0.77$ and $r_s=3\%$, RBS is optimal so long as true average corrective action cost is $C_{ca} > \$98,000/\text{site}$.

Implications and Applications

Risk-based management of UST appears to be more cost-effective than any approach neglecting differential site sensitivity and resource use/value. The major prerequisite for implementation of a risk-based policy is a set of criteria to facilitate site classification by site risk. The cost (CI) of obtaining the requisite site information must not exceed the value of perfect information (VPI), which is defined as the opportunity loss incurred by implementing the second best policy. Assuming an estimated VPI of $10,000 to $20,000 per site, almost any classification scheme short of requiring physical inspection of individual sites would appear to be able to meet this requirement, i.e, yield VPI > CI. Both VPI and CI depend on the complexity of the classification requirements. Also, because any practical classification system will be imperfect there will be economic inefficiencies associated with misclassification. The best classification scheme is that which produces the largest net savings relative to the case of using no classification, i.e., that scheme which at the margin balances efficiency gains and efficiency losses associated with site classification.

The occurrence of differential compliance costs resulting from risk-based UST management might raise delicate political problems. Questions of equity and distribution of costs and benefits of environmental regulation arise; the financial survival of marginal UST owners and operators at high risk sites could be threatened. But different risks and the notion of differential cost sharing are well known in other spheres of society (car insurance, life insurance, etc.) and there, ways were found to deal with the attendant problems.

Administratively, a risk-based policy could be implemented either by means of technical standards imposed on the regulated community or through a system of incentives and penalties; the insurance industry could, for instance, act as enforcement agent, levying risk-based insurance premiums on all UST sites. Premiums commensurate with site risk would induce owners and operators to reduce risk toward a balance between regulatory compliance cost and expected corrective action cost. Both risk-based technical standards and an incentive system would drive the system to a higher level of economic efficiency.

The finding of optimality of risk-based UST management has implications beyond the realm of regulation. For if it is more economical for society to regulate UST on the basis of risk, then it is also desirable for the insurance industry to establish liability premiums on a risk basis, and for large UST owner/operators to schedule and organize UST monitoring, inspection, and replacement programs on the basis of risk. Risk-based management of UST could also be beneficial to local governments. Zoning, permitting, and siting/removal ordinances for new and existing UST could be rendered more protective of public health and the environment and economically more efficient, if based on the notion of site risk.

A METHODOLOGY FOR RISK-BASED UST MANAGEMENT

In this section we present a methodology for assessing and ranking UST sites according to their propensity to inflict damage, and for identifying the optimal action for a given site. First, we introduce a multiattribute site classification system based on the notions of resource value and land use and on the concept of differential hydrogeological site sensitivity. A mathematical model developed by Metzger,[5] and refined by Metzger and Fiering,[4] is used to assess a UST site's potential human health or environmental impact on the basis of site sensitivity and cancer risk. A multivariate decision scheme is outlined to map the site score into economically efficient action at a given site. A procedure is laid out for UST site assessment and management. A hypothetical example illustrates data requirements and site parameterization.

Site Assessment

The following three scales are convenient measures for a site's damage potential from UST failure:

- Resource Value (RV): this indicator is used as a surrogate measure for health, environmental, and economic damage potential associated with long-term degradation of groundwater quality;
- Land Use (LU): this indicator is a surrogate measure for acute damage potential associated with fire and explosion hazards; and
- Hydrogeological Sensitivity (HS): this indicator is a surrogate measure for the magnitude of long-term degradation of groundwater quality and the likelihood of acute fire and explosion hazard, and thus is a complement to the other two indicators.

The population of all UST sites can be unambiguously and completely partitioned by each of these classification metrics. All three scales are discrete. Damage potential of each and every site can be expressed by a three-dimensional vector $\underline{D} = \underline{D}(RV, LU, HS)$. Minimal data requirements permit rapid classification of large numbers of sites. Comparison and ordering of sites based on $\underline{D}$ require a value function specifying trade-offs among the RV, LU, and HS scales.

Resource Value (RV)

The present or future economic or intangible value of affected resources (groundwater, land, environment) at a UST site is a determinant of the type of damage caused by groundwater contamination. For ease of site classification we define three distinct types of resources, each susceptible to degradation of groundwater quality depending on hydrogeological site sensitivity:

MW   groundwater resources used primarily for municipal water supply;
PW   groundwater resources used primarily for private water supply;
      and
EE   other economic and environmental resources including land, and
      aquatic, plant and wildlife habitats.

Thus, the RV scale of damage potential, $\underline{D}$, consists of three discrete values, whose order depends to some extent on subjective valuation of the three types of resources and so cannot be established a priori.

Land Use (LU)

Land use determines the extent of a site's potential for acute safety, fire, and explosion hazard. We define three types of land use:

U   urban settings with built-up residential or commercial and indus-
      trial areas around the UST site and utility trenches criss-crossing
      the subsurface;
SU   semiurban settings with moderate development around the UST
      site; and
R   rural areas with low density development.

The LU scale of damage potential, $\underline{D}$, consists also of three discrete values. But unlike the RV scale, objective ordering of $\underline{D}$(RV, LU, HS), marginal with respect to the LU scale, is possible:

$$\underline{D}(.,U,.) > \underline{D}(.,SU,.) > \underline{D}(.,R,.)$$

i.e., urban settings (U) have the highest hazard potential and rural settings (R) have the lowest.

The 3×3 matrix in Figure 7 illustrates the partitions of the UST site universe based on resource value and land use. Each cell represents a class of sites with a particular type of groundwater-related resource value and a particular safety, fire, and explosion hazard (high, medium, low).

Hydrogeological Sensitivity

Hydrogeological descriptors of the subsurface are used as metrics for the rate of spread of contaminants, and thus as indicators for the magnitude of potential resource damage and for the extent of potential safety, fire, and explosion hazard. Thus, hydrogeological site sensitivity (HS) adds a third dimension to the classification system shown in Figure 7, and so the population of all UST sites can be classified by means of a three-dimensional array. The HS scale of damage

RESOURCE VALUE (RV)

		MW	PW	EE
L A N D U S E (LU)	U			
	SU	A5	ij	
	R			

U, SU, R: urban, semi-urban, rural setting;
MW: municipal groundwater supply;
PW: private groundwater supply;
EE: economic or environmental impact;
ij: cell address; i = RV and j = LU;
A5: example of Section 3.3

Figure 7. Resource value and land use classification matrix for UST sites.

potential, $\underline{D}$, developed below, consists of four discrete values, marking the following hydrogeologically distinct classes of UST sites:

I	Exposure and Offsite Damage	(highly sensitive)
II	Offsite Damage	(moderately sensitive)
III	Onsite Damage, Chance of Exposure and Offsite Damage	(moderately sensitive)
IV	Onsite Damage	(slightly sensitive)

The order of $\underline{D}$(RV, LU, HS) marginal with respect to the HS is:

$$\underline{D}(.,.,I) > \underline{D}(.,.,II) \sim \underline{D}(.,.,III) > \underline{D}(.,.,IV)$$

i.e., marginal damage potential is highest at category I sites, lowest at category IV sites and approximately equal among category II and III sites.

The metric for hydrogeological site sensitivity, HS, is based on Darcy's law for flow in porous media. The rate of travel of water in the subsurface is used as a surrogate for the time it takes contaminants to reach sensitive areas, and so estimates of travel times can be used to measure a site's hydrogeological sensitivity. Two composite hydrogeological parameters are employed, namely travel times of water in the unsaturated and saturated zones. Use of a suitable economic time horizon to be chosen by the decision maker renders the travel times into dimensionless numbers, permitting simple distinction of various hydrogeological classes. The underlying conceptual model the subsurface presupposes a homogeneous, isotropic, unconfined and unconsolidated or unfractured aquifer; these assumptions can be relaxed by invoking the notion of equivalence when specifying parameter values. (Symbols are defined at first appearance).

For estimating travel time, t_u, in the unsaturated zone we assume a vertical hydraulic gradient, i.e., $J = 1$, saturated flow behind the wetting front and zero initial soil moisture. Normalized by the time horizon, T, we obtain:

$$t_u/T = z_u n_u / K_u T \tag{4}$$

where z_u is depth of the groundwater table below the tank bottom, n_u is porosity, and K_u is hydraulic conductivity of the geological material in the vadose zone. For the saturated zone we have:

$$t_s/T = x n_s / K_s J T \tag{5}$$

where x is the horizontal distance between the UST site and the receptor, n_s is porosity, K_s is hydraulic conductivity, and J is the hydraulic gradient in the saturated zone.

Figure 8 illustrates the grouping of UST sites into the above four classes of hydrogeological site sensitivity. The hydrogeology of site category I is such that within the economic time horizon there is important potential for human exposure to contaminants through drinking water and/or for substantial environmental damage. These sites have earlier (Table 1) been referred to as highly sensitive sites. Sites falling in quadrant II are characterized by very slow groundwater flow in the saturated zone amounting to, say, a few meters during time T; but the unsaturated zone is sufficiently permeable to allow contaminants to reach the groundwater table. Category III sites have geological formations in the unsaturated zone which normally do not permit tank release to reach the groundwater table in time less than T. But the underlying saturated zone is highly permeable (comparable to category I), so that human exposure or considerable environmental damage could occur if the contaminants did for some reason (parameter estimation error, nearby borehole, soil-contaminant interaction) reach the groundwater table. Finally, sites in quadrant IV share hydrogeological features which make it highly unlikely, if not impossible, for contaminants ever to leave (within T) the site. Very minor onsite damage, or perhaps soil gas formation, might occur.

Site Risk

Sites falling in category I (Figure 8) all have potential for human exposure and offsite damage, but their hydrogeological sensitivities range over more than three orders of magnitude. A fraction of these sites, those with very short travel times, high resource value, and intensive land use, pose the greatest threat; efficient management of the entire UST population requires that this subgroup of sites be

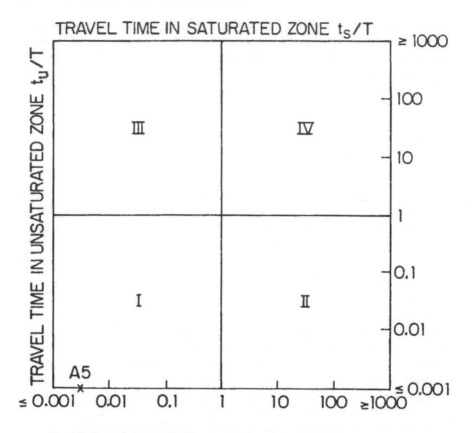

I   EXPOSURE AND OFF-SITE DAMAGE
II OFF-SITE DAMAGE
III ON-SITE DAMAGE (Chance of Exposure and
                                Off-Site Damage)
IV ON-SITE DAMAGE

Figure 8. Hydrogeological classification of UST sites.

clearly identified and appropriately handled. A risk assessment model developed by Metzger[5] and refined by Metzger and Fiering[4] is used to further evaluate category I sites on a continuous scale using cancer risk as a surrogate measure for site comparison and ranking. It bases assessment of a site on a set of characteristics including tank type, hydrogeology, geochemistry, toxicology, and exposure; output is a composite score for the site's potential for health or environmental risk.

The conceptual premise of the model is that there is a UST leaking into groundwater presently used in one of the three ways defined above (MW, PW, EE).

For a given set of site characteristics, using a linear dose-response model, population excess cancer risk, R_p, is computed from the population size, N, the chemicals' (soluble components of the leaking liquid) carcinogenic potencies, B_i ($[mg/kg-day]^{-1}$), and the lifetime average daily dose, d_i (mg/kg-day), for each chemical i:

$$R_p = N \ \Sigma_i \beta_i d_i \tag{6}$$

The BTX compounds (benzene, toluene, xylenes, ethyl benzene; and EDB for leaded gasoline) are used as model substances for hydrocarbon contamination of groundwater. Other compounds or other floating immiscible NAPLs (nonaqueous phase liquids) could be included by augmenting the computer code of the model.

The estimate of risk in Equation 6 is a scalar, but uncertainty in any of the independent variables in Equation 6 renders it stochastic. This resulting uncertainty must be specified to allow a decision maker adequately to exercise her/his risk preference when using this metric. Therefore, the risk assessment model employs Monte Carlo analysis to estimate the risk density, from which the decision maker's risk preference can be translated into percentiles for site ranking.

The performance of this risk assessment model is compared elsewhere[4] with several existing ranking schemes, including EPA's Hazard Ranking System[14] and a system of LeGrand's,[15] and it is demonstrated that for ex ante assessment and ranking of UST sites, this model is (by design) best suited. It permits explicit accounting for data uncertainty and is responsive to a decision maker's risk preference. The scheme is based on defensible science and allows clear distinction between objective assessment and subjective management. More so than other systems it is capable of evaluating the important physical, chemical, and exposure-related aspects of UST failure and groundwater contamination.

Site Management

Consider a menu of potential actions including the immediate or subsequent installation of a double-walled tank, the initiation of a rigorous program of monitoring, no action whatever, etc. If each of these actions reduces site risk and is associated with a cost, it is a standard problem in decision analysis to array and compare the several options by noting the reduction in risk with increasing expenditure. Actions which are more costly but less effective than any other are said to be dominated, and are of no further interest. At the margin, when the expenditure required for the next unit of remediation has more utility than the associated marginal reduction in risk, the rational decision-maker will no longer invest in risk reduction and the problem is solved.

However, it is inconvenient or perhaps even impossible to perform such an analysis for each of the thousands of UST which are potential sources of environmental damage and threat to human health. The economic costs for the several options are highly site-specific, and it is unreasonable to expect every tank owner or operator, or every jurisdiction, to undertake the required statistical analysis.

What is needed instead is an approximation to this detailed analysis which can be used, e.g., by a regulatory authority in a given jurisdiction, by a large tank owner/operator, or by an insurance company, to make a rapid assessment of a given tank site on the basis of a few readily determined parameters which define the tank and its environment, and to prescribe the best action. The purpose of this portion of the analysis is, thus, to collapse the dimensionality of the assessment problem into a single index or scalar whose value can unambiguously point to a ranking or, ultimately, to an action.

Univariate Case

Suppose a particular state or other jurisdiction has within its boundaries a large number N, say 10,000, of underground storage tanks. From these we draw a random sample of size n, say 100 or 200, for which there are available all the data required to classify the tank, its physical and demographic environment, and the discounted costs of each of several alternative preventive and remedial actions. We define a policy as a combination of actions to be taken under particular conditions; it is in effect an algorithm for specifying a choice.

Suppose some univariate risk metric, R (e.g., the expected number of cancers in the absence of any preventive action), is calculated for each of the n tank sites. If the sample is a representative random sample of the entire population of N tanks, the average risk for the sample, when scaled by the size of the population, represents the expected risk for the entire jurisdiction of N. Suppose we classify sites according to the estimated risk level at each. We divide the entire range of risks into several intervals or groups and then simply distribute the sample sites according to the range in which its risk falls. Suppose we calculate for each group of sites the average cost associated with each of the several alternative policies or action combinations under consideration. The result is an envelope suggested in Figure 9 which identifies the optimal policy for each segment of the risk axis.

That is, for each risk category or sampling bin we calculate the *average* performance of each policy by calculating the numerical performance index for all the sites characterized by that risk level, and compute the mean. Any policy above (below) the minimization (maximization) envelope is dominated by that envelope and should not be implemented. If the objective is to minimize the average cost (or more generally, the system response), then policy C is preferred in the range of R_1 and R_2; policies A and B are equally acceptable in the range of R_3 and policy A is preferred in the range of R_4.

The assumption on which our methodology rests is that the sample of n sites is representative enough of the population of N sites so that identification of the dominant or optimal policy for the sites in the ith risk range will hold as well for all the population sites in that same range. If so, the time-consuming and

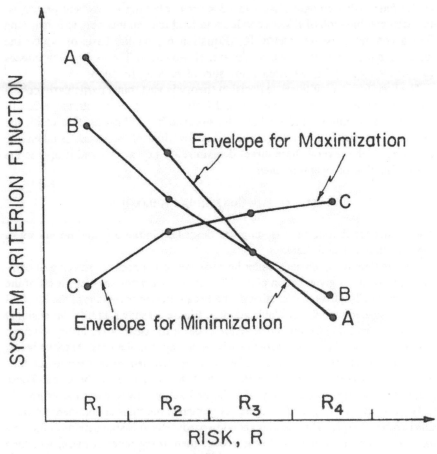

Figure 9. Performance for various policies.

difficult economic analysis need not be done for all new sites in the population of N-n unsampled locations, or in any sites which subsequently come to our attention. All that is necessary is calculation of the risk as a function of the few easily measured variables which describe the site, from which a reasonable estimate of the optimal policy can be made. The problem of misclassification and efficiency losses associated with uncertainty in the ranking function is addressed in Reference No. 4.

Multivariate Case

In fact, our classification scheme is based on three arguments: hydrogeology, resource value, and land use. We allow four hydrogeologic classes and three each for resource value and land use, giving $4 \times 3 \times 3 = 36$ classes or cells. For one of the four hydrogeologic classes ($1 \times 3 \times 3 = 9$ cells) the assessment performs, for each combination of resource value and land use, an unambiguous mapping into a continuous scalar metric R_p (Equation 6), on the basis of which the material in the previous section is relevant. However, for three of the four classes (or $3 \times 3 \times 3 = 27$ cells) we need some sort of ranking function.

The literature on multivariate decision theory is vast, but a particularly coherent overview is contained in Keeney and Raiffa.[16] They demonstrate the conditions and assumptions under which the multivariate utility function of all the decision variables is a function of the utility functions of the variables taken one at a time. If the several (here three) decisions are (x_1, x_2, x_3) and if $u(\cdot)$ is the utility function or operator, then

$$u(x_1, x_2, x_3) = f[u_1(x_1), u_2(x_2), u_3(x_3)].$$

The details for finding the mapping are described in the text and do not merit reproduction at length here.

Our problem is somewhat simpler because we do not need an ordering of combinations within a given action class. That is, we wish only to place each of the 27 discrete cells into one or another of the broad groups or regions of the $3 \times 3 \times 3$ space associated with the available actions. Figure 10 shows a three-dimensional rectilinear space divided into three action classes; the clear boxes justify the most benign action a_1, the dotted boxes justify action a_2, and the cross-hatched boxes justify the most severe action a_3. The axes are oriented so that the lower left-hand corner is best in terms of the three attributes which define the space. These increase in severity (or urgency of indicated action) away from the origin.

There are $3^{27} = 7.63 \times 10^{12}$ possible assignments, but rules of dominance reduce this number sharply. The recommended procedure is to assign actions at the best and worst corners of the rectilinear space, then at the other corners, and then along the edges of the $3 \times 3 \times 3$ box. Interior spaces will largely be determined by dominance, with perhaps a few uncertain borderline decisions to be made.

Consider the following simplified explanation of the multi-attribute decision process required in this analysis. Let there be three attributes (x_1, x_2, x_3) which together define or describe an outcome, and suppose that more of any one of these attributes is better than less. On the basis of these attributes, or this system classification, a decision must be made to take one of (say) three available actions a_1, a_2, or a_3. Let all the attributes be non-negative and let them be bounded by (b_1, b_2, b_3), respectively, so the origin (0, 0, 0) is the worst allowable system and the point (b_1, b_2, b_3) is the best. Starting at the origin we increase x_i alone until that action appropriate for (0, 0, 0) is displaced by another candidate; action a_3, optimal at the origin, gives way to a_2. Now we return to the origin and increase x_2 alone until a_2 is once again invoked. Finally, we return once more and increase x_3 alone until a_2 displaces a_3. These three transition points define a plane which contains a convex wedge within which action a_3 is best.

Worst Corner

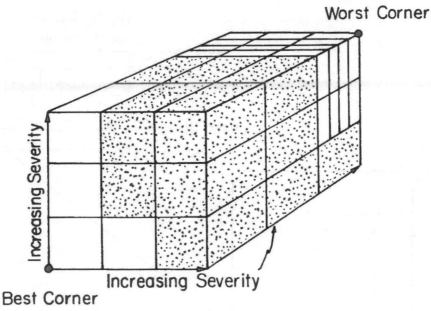

Increasing Severity

Best Corner

Figure 10. Multidimensional rating space.

Similarly, starting at (b_1, b_2, b_3) and decreasing x_1, x_2, and x_3 in turn gives a wedge within which action a_1 is best. The remaining portion of the rectilinear space is associated with action a_2.

This is shown in Figure 11, which incorporates several assumptions. First, the surfaces separating the actions are taken to be planes, implying that the optimal strategy in each convex segment is derived from a scalar index which is a linear combination of the attributes x_i. In fact, the separation surface might be sharply warped to reflect nonconstant trade-offs among the attributes. Second, the planar or warped separation surfaces suggest continuous variation in any and all attributes x_i, when in practice one or more attribute might be known to fall in an aggregated interval as shown in Figure 10. The surfaces of separation must now accommodate the 3-dimensional platform arrangement in the figure.

Our recommended scheme is somewhat simplistic in that we assume an unambiguous decision to switch from one action to another can be taken as we increase or decrease one attribute alone, leaving the others fixed. It might be difficult or impossible to separate the attributes and consider them independently because of complex correlation structures among them; these and related problems are considered in the standard literature on the subject.[16]

If it happens that the cost of the replacement and rehabilitation program which evolves from any particular assignment of actions a_i to the various spaces is unmanageable, marginal adjustments to the assignments might be made to reduce

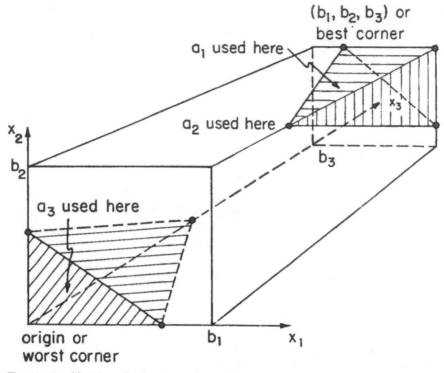

Figure 11. Mapping of rating space into action space.

the economic burden. In other words, assignment of a policy by specification of the action required in each of the 27 cells can be viewed as a trial solution in multidimensional space, with the explicit understanding that this trial is iterated, modified, improved, and generally brought into conformity with social and political objectives, economic constraints and technical reality.

The total decision space now consists of 10 segments. Nine of these are similar in that they are characterized by continuous risk functions, one for each combination of resource value and land use; the tenth segment contains the $3 \times 3 \times 3 = 27$ discrete combinations of resource value, land use, and hydrogeologic classification. If, on the basis of hydrogeologic risk, the project under study falls into the high risk category, the optimal action is based on analysis of one of the nine continuous risk formulations as described for the univariate case. On the other hand, if the hydrogeologic risk places the system in one of the three smaller risk categories, the coarse-grid discrete analysis encompassing the 27 cells is utilized.

Procedure and an Example

Figure 12 and an example carried forward step-by-step illustrate the procedure developed in the sections on Site Assessment and Site Management to classify,

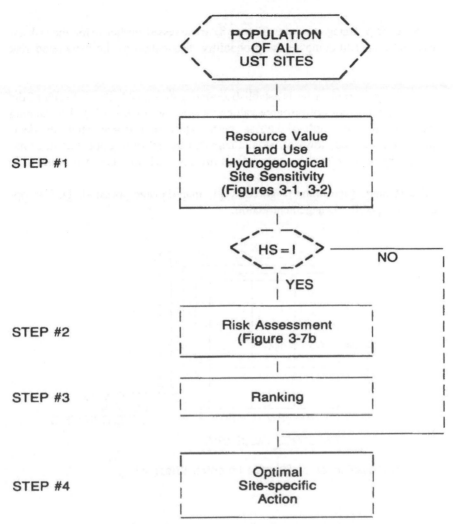

Figure 12. Procedure for UST site assessment and management.

assess, and manage a UST population. The first two steps constitute the assessment stage. The third (ranking of $3 \times 3 = 9$ types of category I sites) and fourth steps (prescription of action) belong in the realm of risk management, requiring specification of the decision maker's risk preference (step #3) and value structure (step #4).

First Step. Resource value (RV), land use (LU), and hydrogeological site sensitivity are determined in accordance with the section on Site Assessment. Each site is assigned to one of $3 \times 3 \times 4 = 36$ categories defined by the array for damage potential, $\underline{D}$, in Figure 13a.

Second Step. Category I sites (Figure 8) are assessed further using the risk assessment model to compute the probability distribution of the associated risk metric.

Third Step. A rank order is established among category I sites for each of the $3 \times 3 = 9$ combinations of resource value and land use (Figure 13b). The ranking criterion is that percentile of the risk metric which reflects adequately the decision maker's attitude toward risk and captures the effect of imperfect information on the rank order. The risk indices are rescaled for ease of presentation.

Fourth Step. Algorithms (Figures 9, 11) map damage potential, $\underline{D}$, into optimal site specific management action.

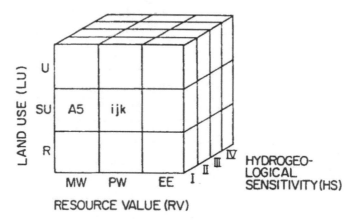

a) Classification by Damage Potential (Steps #1,2)

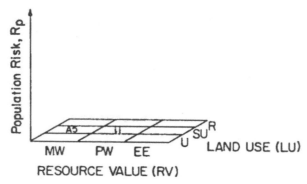

b) Risk Assessment of Category I Sites (Step #3)

Figure 13. Multivariate assessment of UST sites.

Example

First Step. Water use information such as available from a state water agency is sufficient to classify the site according to resource type (MW, PW, EE). Population maps can be used to determine the land use (U, SU, R) at the site. Geographic information systems (GIS) can be helpful in specifying the site parameters.

Suppose we have a hypothetical retail gasoline station (Site A5) which is found to be near a municipal well field (MW) and census data for that area indicate that land use is semiurban (SU). Thus resource value is RV=MW and land use is LU=SU; consequently, the site falls in cell #1,2 of the classification matrix in Figure 7.

Classification according to hydrogeological sensitivity (Figure 8) requires specification of seven parameters (Table 6) to describe the physics and one (time horizon) to express political and economical priorities. The requisite parameter values can be determined from USGS maps and other readily accessible databases. The data for MW sites should be as site-specific as possible because of potentially high regret of misclassification (high corrective action cost of UST failure). Data requirements for PW and EE sites are less stringent.

Table 6.   Data Requirements For Hydrogeological Site Sensitivity Assessment (Step #1).

	Parameter	Symbol	Unit
Unsaturated Zone	Hydraulic conductivity	K_u	M/D
	Porosity	n_u	—
	Depth to water table below tank	z_u	M
Saturated Zone	Hydraulic conductivity	K_s	M/D
	Porosity	n_s	—
	Hydraulic gradient	J	—
	Distance to well/CS[a]	x	M
Economics	Time horizon	T	YR

[a]CS = Some compliance surface for 'EE' sites.

Dimensionless travel times:

$$\frac{t_u}{T} = \frac{z_u n_u}{365 K_u T} \quad \text{and} \quad \frac{t_s}{T} = \frac{x n_s}{365 K_s J T}$$

Suppose medium sand is found to dominate the geological material at the hypothetical site in both the unsaturated and saturated zones of the subsurface. Values of K = 100 m/d for saturated hydraulic conductivity and n=0.3 for porosity are assigned. Suppose, further, that the groundwater table is found at a depth of approximately z_u =5 m below the bottom of the tanks and the nearest well is located at a distance of x=300 m from the retail station. From the static and dynamic water level readings of the well an approximate hydraulic gradient in the saturated zone of, say J=0.01, is inferred. If the decision making agency assumes a T=50 year economic time horizon we obtain the following:

$$\frac{t_u}{T}=\frac{z_u n_a}{365 K_u T}=\frac{5 \cdot 0.3}{365 \cdot 100 \cdot 50}=2.7.10^{-7}$$

$$\frac{t_s}{T}=\frac{x n_s}{365 K_s JT}=\frac{300 \cdot 0.3}{365 \cdot 100 \cdot 0.01 \cdot 50}=4.9.10^{-3}$$

The site plots in quadrant I in Figure 8, and is thus identified as a highly sensitive site. Damage potential is $\underline{D}$=(MW, SU, I) and the site is assigned to cell #1,2,1 of the array in Figure 13a.

Second Step. Risk assessment for category I sites requires additional information to estimate time-to-failure of the tank, quantify contaminant mass fluxes, and characterize the extent of exposure. Data sources and site parameterization are discussed in detail in Reference No. 4. The computer model can handle a variety of petroleum substances if the chemical composition is specified. The model uses built-in values for carcinogenic potency and other chemical-specific parameters of the BTX compounds and ethylene dibromide. It calls for measures of uncertainty for key parameters such as hydraulic conductivities, hydraulic gradient and organic carbon in the saturated zone.

Suppose the analyst derives the following parameter values for the UST site (Table 7; see Reference No. 4 for details of site parameterization): the aggressiveness of the tank backfill material at Site A5 is estimated to be high (e.g., pH=5.0). The backfill material consists of fine sand; hydraulic conductivity is estimated at K_b= 10 m/d and the actual (or effective) value is believed with 95% confidence to fall within the range 2 m/d $\leq K_b \leq$ 50 m/d, and so uncertainty is $\rho_{LOG(Kb)}$=0.7. Suppose there are 10,000 gallon tanks which are 10 years old and bare steel. Because the ground surface of the hypothetical retail station is sealed, residual soil moisture in the unsaturated zone is assumed to be zero. Uncertainty of the hydraulic conductivity (K=100 m/d) in both zones is estimated at $\rho_{LOG(K)}$=0.7. The saturated zone is assumed free of organic carbon, and uncertainty of the estimate of the hydraulic gradient is taken to be $\rho_{LOG(J)}$=0.3. The groundwater table is estimated to fluctuate D=$\pm$1 m over the course of a year. The well closest to the UST is threatened; well discharge is known to be Q_w=1220 m³/d for each of five wells and so the mixing factor (mixing of contaminated with uncontaminated well water) is ζ=0.2. The stored product is gasoline.

The computer model performs a Monte Carlo analysis and computes a probability density (pdf) for the risk metric R_p in Equation 6. In the hypothetical example the 50th and 98th percentiles, for instance, scaled by a factor of 1000, are, respectively:

$$R_{p,50}=83$$

$$R_{p,98}=333$$

Table 7.   Data Requirements for Site Risk Assessment (Step #2).

	#	Parameter	Unit
Tank Environment	T1	Soil aggressiveness	H,M,L[a]
	T2	Tank capacity	US Gal
	T3	Tank age	Yr
	T4	Hydraulic conductivity backfill	M/D
	T41	Uncertainty in T4	—
Unsaturated Zone	U1	Depth to water table below tank	M
	U2	Soil moisture	—
	U3	Hydraulic conductivity	M/D
	U31	Uncertainty in U3	—
	U4	Porosity	—
Saturated Zone	S1	Organic carbon content	—
	S11	Uncertainty in S1	—
	S2	Hydraulic conductivity	M/D
	S21	Uncertainty in S2	—
	S3	Porosity	—
	S4	Water table fluctuation	M
	S5	Hydraulic gradient	—
	S51	Uncertainty in S5	—
	S6	Well discharge	M³/D
	S7	Distance to well/CS[b]	M
Exposure	E1	Population	People
	E2	Mixing	—
	E3	Substance	—
	E4	Economic time horizon	Yr

[a]H = high; M = medium; L = low.
[b]CS = Compliance surface for 'EE' sites.

Third Step. The ranking criterion chosen by the decision maker may be any percentile of the pdf of R_p.

Fourth Step. The algorithm shown graphically in Figure 9 is applied to identify the economically efficient action for Site A5.

FUTURE WORK

This study suggests that the EPA's proposed UST program[3] may be considerably less economically efficient, or cost-effective, than an alternative approach which we call risk-based management. The EPA's economic analysis rests on the key assumption that the distribution of damage potential for the population of UST sites in the U.S. is only slightly skewed, spanning just one order of magnitude (in U.S. dollars). Our analysis reflecting empirical evidence (Figure 3) suggests that the distribution is highly skewed and that damage potential spans three orders of magnitude.

These different assumptions lead to policy recommendations with substantially different economic consequences. Assuming the true distribution of damage potential is skewed (to the extent shown in Table 5), we compute an opportunity loss associated with implementing the EPA's policy (ES) of as much as $10 billion (present value) for the U.S., or $460 million for Illinois with its 23,000 UST sites. If the true distribution of damage potential is as assumed by the EPA, the loss of economic efficiency, or regret, associated with implementing a risk-based policy (RBS) is negligible because slightly higher compliance costs are offset by lower corrective action costs. Hence, by most decision criteria and virtually regardless of the likelihood of the occurrence of a true skewed distribution of damage potential, risk-based UST management is the action of choice. This is illustrated by the regret matrix in Figure 14.

	State of Nature		
Policy	**Skew**	**No Skew**	**Regret**
EPA Standard	21†	0	21
Risk-based Standard	0	1‡	1 ←minmax value

† 65–44 (Figure 2-4)
‡ assumption based on Figure 2-4

Figure 14.   Regret matrix (in $1000 per UST Site) associated with EPA policy and risk-based policy.

The findings and methodologies presented in this study can be translated directly into a UST program for the U.S. as a whole, for individual states or jurisdictions, for large UST owners/operators, or for insurance companies, after verifying key assumptions, collecting cost and impact data, testing and calibrating procedures, and formulating rules and guidelines. This might include the following:

• survey and analysis of UST failure incidents
• pilot study on UST site classification and risk assessment
• identification of a menu of control options
• decision rule for site-specific UST management

Assuming potential savings from a risk-based policy of, say, $5 billion to $10 billion for the U.S. (or between $200 million and $400 million for the state of Illinois) it would not be unreasonable to commit the required additional funding for further study. Implementation of a practical risk-based policy might cost 0.5% or 1% of potential savings (on the order of $100 per site).

CONCLUSIONS

There may be 500,000 UST sites in the U.S. (23,000 in Illinois) and the average age of the facilities is over 14 years. Most of the tanks are of unprotected bare steel construction and, according to national statistics, 10% or more may be leaking at this time. A large fraction of them are installed in urban areas, posing safety threats and fire and explosion hazards, and affecting environment and land values. Many of the UST are located near private wells or near shallow municipal wells that draw water from surficial aquifers.

Economic analysis of alternative regulatory policies suggests that EPA's proposed UST rule is ineffective and economically inefficient. Its principal shortcoming is that it is underprotective for highly vulnerable and valuable sites where damage potential is high and corrective action can cost $1 million or more. Risk-based management is found to be more efficient because it identifies high risk sites at which to provide adequate protection. Expected efficiency gains associated with risk-based prevention may amount to $5 billion to $10 billion for the U.S. ($200 million to $400 million for Illinois; or $10,000 to $20,000 per site), depending on the true value of some of the determinants.

The major reason that the EPA favors a uniform technical standard for all UST sites, regardless of site risk, appears to be that the agency in its economic model assumes corrective action costs never to exceed $388,000 at any site; this contradicts empirical evidence (see, e.g., the case of Provincetown, Massachusetts[5]), which suggests that in 5% to 10% of sites corrective action can cost $1 million, and in some cases as much as $10 million. Had these high risk sites been entered, the agency's recommendation would likely be the same as ours.

Nondifferential regulation as proposed by the EPA is desirable also for political reasons because the same compliance costs are imposed on all members of the regulated community. Differential regulation penalizes owners/operators at high risk sites, and thus could force some marginal businesses out of existence unless society uses public funds to aid them. But efficiency losses associated with subsidizing prevention pale in comparison with efficiency gains achieved by avoiding subsequent corrective action.

Another reason for EPA to be apprehensive about risk-based UST regulation turns on the greater effort required to implement and enforce such a policy. Site classification adds a level of complexity to regulation of USTs, perhaps causing some efficiency loss. But the potential benefits appear far to outweigh administrative costs so long as the instruments required for site classification remain simple. Classification costs on the order of $100 per site are believed reasonable and attainable.

The material presented earlier in this chapter lays the groundwork for implementation of risk-based UST management. Algorithms are developed for rapid, sufficiently accurate, and inexpensive classification, assessment, and ranking of UST sites, and a conceptual framework is presented for identifying the best protective action for any given site. These algorithms have not been tested yet under field conditions, and the mechanism for prescribing action has to be fully developed and calibrated using field data. Therefore, a number of tasks have been identified and proposed which warrant further work before a risk-based policy can be formulated.

REFERENCES

1. "Underground Motor Fuel Storage Tanks: A National Survey," Office of Pesticide and Toxic Substances, U.S. Environmental Protection Agency, EPA 560/5-86-013, 1986.
2. "Summary of State Reports on Releases from Underground Storage Tanks," Office of Underground Storage Tanks, U.S. Environmental Protection Agency, EPA 600/M-86/020, 1986.
3. "Underground Storage Tanks; Proposed Rule," 40 CFR Parts 280 and 281, U.S. Environmental Protection Agency, Vol. 52(74) *Federal Register*, Friday, April 17, 1987.
4. Metzger, B. H., and M. B. Fiering, "Management of Health and Environmental Risks from Underground Storage Tanks in Illinois," Environmental Systems Program, Harvard University, prepared for Illinois Department of Energy and Natural Resources, ILENR/RE-EH-87, 1988.
5. Metzger, B. H.,"Management of Health Risk from Groundwater Contamination," PhD dissertation, Harvard University, Cambridge, MA, 1987.
6. Heath, R. C.,"Groundwater Regions of the United States," U.S. Geological Survey Water Supply Paper 2242, 1984.
7. ICF, Inc., "Hazardous Waste Tanks Risk Analysis," Draft Report, U.S. Environmental Protection Agency, Office of Solid Waste, 1986.
8. "Hazardous Waste Management System; Standards for Hazardous Waste Storage and Treatment Systems; Proposed Rule," 40 CFR Part 260 et al., *Federal Register*, U.S. Environmental Protection Agency, Vol. 51(123) Wednesday, June 26, 1985.
9. Rogers, W., Personal Communication, cited in *Protecting the Nation's Groundwater from Contamination*, Vol. I, II, Office of Technology Assessment, OTA-O-233 (Washington, D.C: U.S. Government Printing Office, October 1984).
10. Schwendeman, T., "Implementation of Tank Management Programs," the Fourth Conference on Geotechnical Engineering, Annual ASCE Convention / Exhibition, Boston, MA, October, 1986.
11. Hinchee, R. E., H. J. Reisinger, D. Burris, B. Marks, and J. Stepek, "Underground Fuel Contamination, Investigation, and Remediation: A Risk Assessment Approach to How Clean Is Clean," in Proceedings of the NWWA/API Conference, Houston, TX, 1986, pp. 539-564.
12. Raiffa, H., *Decision Analysis*, (New York: Random House, 1968).
13. Rogers, W., "Report on the Statistical Analysis of Corrosion Failures; Unprotected Underground Steel Tanks," Warren Rogers Associates, Inc., Newport, RI, undated.
14. "National Oil and Hazardous Substances Contingency Plan," 40 CFR Part 300, *Federal Register*, U. S. Environmental Protection Agency, Vol. 47(137) Friday, July 16, 1982.
15. LeGrand, H. E., "A Standardized System for Evaluating Waste-Disposal Sites," National Water Well Association, 1980.
16. Keeney, R. L., and H. Raiffa, *Decision with Multiple Objectives: Preferences and Value Tradeoffs*, (New York: John Wiley & Sons, 1976).

PART VII

Regulatory Considerations

The California Leaking Underground Fuel Tank Field Manual: A Guidance Document for Assessment of Underground Fuel Leaks

Seth J. Daugherty

INTRODUCTION

An effort to develop more consistent, systematic, and defensible procedures to assess the significance of soil contamination from underground fuel tanks, and its relationship to potential groundwater pollution, was initiated in California in October 1985. A result of this effort is the "California Leaking Underground Fuel Tank (LUFT) Field Manual,"[1] a 121-page document that was issued in December of 1987 and updated in May of 1988. The manual was developed by a 40-member task force coordinated by the State Department of Health Services (DHS) and the State Water Resources Control Board (SWRCB). State, regional, and local environmental regulatory agencies were represented. The intent was to apply the California Site Mitigation Decision Tree[2] process to the case of subsurface contamination associated with underground fuel tanks. Over 20 meetings of the task force or its committees were held. In addition, 320 comments were received and considered as a result of two public workshops. The LUFT Field Manual is thus a compromise, and, insofar as possible, a consensus document.

The LUFT Field Manual procedures are directed toward site-specific risk assessment and risk management. The manual is a nonmandated, nonexclusive technical guidance tool that uses a stepwise screening procedure to assess sites for regulatory purposes. LUFT procedures are applicable only to contamination from underground fuel tanks and focus on the ingestion (via groundwater) exposure pathway. The methodology is limited to gasoline and diesel fuel, and develops procedures that result in numerical cleanup levels.

Underground Storage Tank and Groundwater Regulation in California

In California, local agencies are responsible for enforcing the underground storage tank law. The California Health and Safety Code requires that no person shall close an underground storage tank unless the person "demonstrates to the local agency that there has been no significant soil contamination resulting from a discharge in the area surrounding the underground storage tank or facility." There are 100 local agencies in California that implement the underground storage tank law. These agencies vary from health departments familiar with the state of the art and science of environmental fate modeling and public health evaluation, to fire agencies and planning departments having less detailed expertise in the cleanup of toxic substances. From a local agency viewpoint, the LUFT document can be considered an attempt to better define "significant soil contamination."

On a regional level, there are nine California Regional Water Quality Control Boards that designate beneficial uses of water within their jurisdictions. Groundwater basins are often designated for the highest use, that of municipal and domestic drinking water supply. Although regional boards may set specific site groundwater cleanup levels, the most important guidance for setting cleanup objectives are specified water quality objectives that are similar throughout the state, and the state nondegradation policy. With respect to petroleum products, the most important water quality objective is that of toxicity, stated as "all water shall be maintained free of toxic substances in concentrations that produce detrimental physiological responses." Numerical criteria often used to specify the toxicity water quality objective are the Department of Health Services Drinking Water Action Levels. As related to petroleum, these action levels are based on chronic toxicity and established as 0.7 parts per billion (ppb) of benzene, 100 ppb toluene, 620 ppb xylene, and 640 ppb ethylbenzene. The level for benzene has been established by the state as the concentration that would result in a lifetime cancer risk of one in one million.

The generalized conservative response to present regulatory perspectives and hydrogeologic uncertainties is to place the point of application of the water quality objectives at or near the first groundwater encountered below ground surface. Many entire groundwater basins are designated as drinking water supply sources, without vertical or horizontal differentiation. This designation may include perched or semiperched zones (groundwater in a less permeable zone overlying an aquifer, but with a saturated condition throughout, i.e., no intervening unsaturated zone) where the hydraulic continuity, or lack thereof, is unknown.

The great importance of California's groundwater basins indicates the appropriateness of the general strategy of strict aquifer protection. This perspective has been incorporated into the environmental fate modeling for the standard LUFT risk appraisal.

DESCRIPTION OF THE CALIFORNIA LUFT MANUAL

Purpose, Assumptions, and Status

The stated purpose of the California LUFT task force was to establish procedures for determining whether an underground storage fuel tank site is clean enough to protect public health and the environment. The procedures provide a means of determining whether fuel contaminated soil may pose a threat to groundwater, or if groundwater has already been affected. The California LUFT Field Manual presents procedures that are intended to give practical guidance to regulatory agencies with respect to investigating leaks, assessing risks to human health, determining soil cleanup levels, and need for remedial actions.

A basic assumption of the California LUFT evaluation is that, although desirable, the complete cleanup of all traces of contaminants at many sites cannot be done, due to both economic and technological limitations. A further assumption is that cleanup of all contaminated soil and dissolved hydrocarbons is not always necessary to protect public health and the environment. The task force decided to develop a general, broadly applicable procedure to establish numerical site-specific cleanup levels that take basic site features into account, rather than set a single numerical level or qualitative criteria for the entire state. The field manual is a guidance document that presents one approach to the problem of setting cleanup levels at fuel contaminated sites. This approach is a conservative compromise that represents the best collective efforts of the task force. Use of the manual is not mandated and explicitly does not preclude the use of other approaches. A summary of the overall focus and perspective of the manual is given in Table 1.

Table 1.   California LUFT Manual Focus and Perspective.

• Not Mandated	• Groundwater Impacts
• Not Exclusive	• Public Health
• Technical Guidance	• Risk Assessment
• Categorical Screening	• Risk Management
• For Underground Fuel Tanks	• Regulatory Purposes

Only gasoline and diesel fuels are considered. Identification of fuel components is limited to the analysis of total petroleum hydrocarbons (TPH) according to a gas chromatograph-flame ionization detector (GC-FID) method given in the manual; and benzene, toluene, xylene, and ethylbenzene (BTXE) by EPA Method

8020. Tests for organic lead were recommended only for significant leaks of leaded gasoline or where there may be danger of direct exposure. Experience also indicated that ethylene dibromide was usually found in only trace amounts in soil. Experience with ethylene dichloride, methyl tertiary butyl ether, and other additives in soil was too limited to make firm recommendations.

Selection of Site Category

The initial step in the LUFT procedure is the designation as a Category 1 (no evidence of contamination), Category 2 (known or suspected soil contamination), or Category 3 (known groundwater pollution) site (Figure 1). Different bases for site investigations and the need for a review of site history are recognized.

Category 1: No Evidence of Soil Contamination

Category 1 was originally established to allow sites to be screened, based on the response of hydrocarbon vapor monitoring instruments. However, a lack of confidence that field tests for head space vapor would effectively detect contamination, and a perceived need for more definitive indication of contamination, resulted in a consensus that vapor levels should be used only to help decide where to take soil samples for laboratory analysis. The Category 1 procedure then became identical with the initial portion of the Category 2 investigation.

Category 2: Suspected or Known Soil Contamination

An overview of the Category 2 investigation is given in Figure 2. The procedures include two distinct phases: (1) the Leaching Potential Analysis (LPA), which includes sampling of the bottom or sides of the excavation (both tank and piping excavations), including the collection and analysis of samples, application of TPH and BTXE limits, and determination of whether the soil contamination is minor; and (2) the General Risk Appraisal (GRA), which includes determining the vertical profile of soil contamination by borings and soil core sampling, determining if groundwater is at risk due to soil contamination, and, if necessary, determining groundwater gradient, collecting groundwater samples, and evaluating the need for remedial action. The excavation sampling and the LPA determines if the TPH and BTXE limits are exceeded. If the limits are exceeded, then the results of core samples obtained throughout the zone of contamination are used to determine if groundwater is at risk using the GRA. The "standard" GRA is based on the SESOIL and AT123D models, although alternative state-of-the-art appraisals (which could include different models or more site-specific inputs into those same models) are explicitly allowed.

The LPA for gasoline is shown in Table 2. A table for diesel fuel is identical, except that the maximum allowable TPH levels are increased by an order of magnitude. This analysis is based on "worst case" soil samples collected one to two

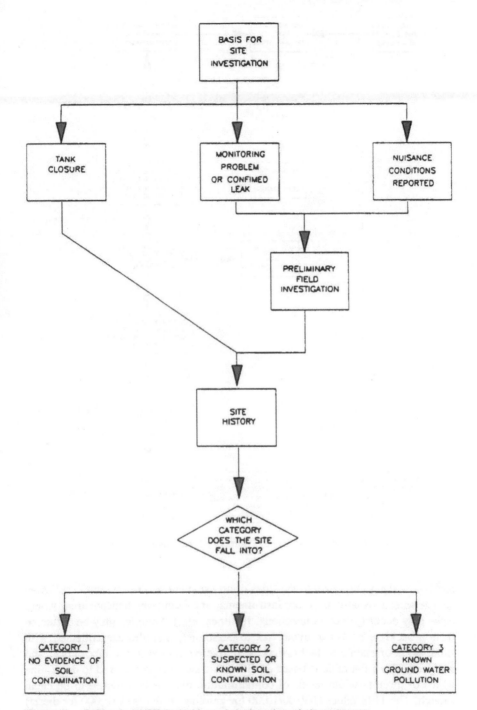

Figure 1. California LUFT Field Manual designation of site category.

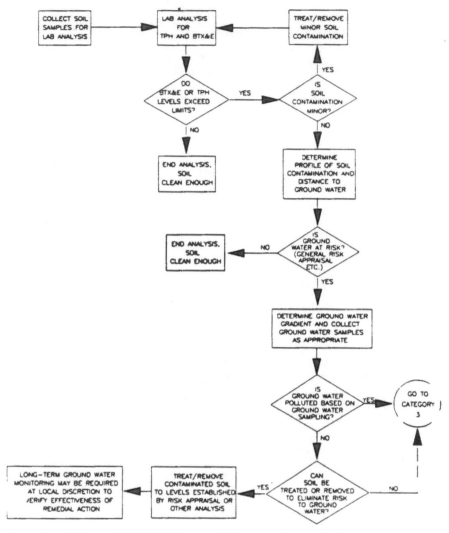

Figure 2. California LUFT Field Manual Category 2 site investigation.

feet below the bottom or into the sides of an excavation. The "worst case" may be determined visually, by vapor instruments, or by sampling beneath areas where leaks may occur (joints, connections, fill lines, etc.). Samples may be collected from a backhoe bucket to avoid risk to personnel. Samples are collected with small brass or stainless steel cylinders, delivered chilled to the laboratory, and analyzed. The LPA table is based on five basic site features scored to determine which of three possible levels of contamination may be considered to be clean enough. The TPH values (10/100/1,000 for gasoline, 100/1,000/10,000 for diesel) follow historical standards used in California. The 0.3 mg/kg benzene and toluene

Table 2. California LUFT Field Manual Leaching Potential Analysis for Gasoline.

Leaching Potential Analysis for Gasoline Using Total Petroleum Hydrocarbons (TPH) and Benzene, Toluene, Xylene, and Ethylbenzene (BTX&E)

The following table was designed to permit estimating the concentrations of TPH and BTX&E that can be left in place without threatening groundwater. Three levels of TPH and BTX&E concentrations were derived (from modeling) for sites which fall into categories of low, medium or high leaching potential. To use the table, find the appropriate description for each of the features. Score each feature using the weighting system shown at the top of each column. Sum the points for each column and total them. Match the total points to the allowable BTX&E and TPH levels.

Site Feature	Score	Score 10 Pts. If Condition Is Met	Score	Score 9 Pts. If Condition Is Met	Score	Score 5 Pts. If Condition Is Met
Minimum depth to groundwater from the soil sample (feet)	____	>100	____	51–100	____	25–50[1]
Fractures in subsurface (applies to foothills or mountain areas)	____	None	____	Unknown	____	Present
Average annual precipitation (inches)	____	<10	____	10–25	____	26–40[2]
Manmade conduits which increase vertical migration of leachate	____	None	____	Unknown	____	Present
Unique site features: recharge area, coarse soil, nearby wells, etc.	____	None	____	At least one	____	More than one
Column Totals **Total Pts.**	____	+	____	+	____	= ____
Range of total pts.	49 pts. or more		41–48 pts.		40 pts. or less	
Maximum allowable B/T/X/E levels (ppm)	1/50/50/50		.3/.3/1/1		NA[3]	
Maximum allowable TPH levels (ppm)	1000		100		10	

[1]If depth is greater than 5 ft. and less than 25 ft., score 0 points. If depth is 5 ft. or less, this table should not be used.
[2]If precipitation is over 40 inches, score 0 points.
[3]Levels for BTX&E are not applicable at a TPH concentration of 10 ppm.

levels were determined to be a routinely achievable detection level by the DHS Hazardous Materials Laboratory. This value was controversial, as many laboratories reported much lower detection limits. However, the 0.3 value is very similar to the practical quantitation limit for benzene and toluene in a high level soil matrix given in Method 8020 (method detection limit of 0.2 μg/1, matrix factor of 1250). Other aromatics values are derived from the GRA. No values for aromatics are associated with the lowest TPH level to allow use of a lower limit of detection (the matrix factor for low level soils is 10) and reflects the expectation of essentially no detectable aromatics at the low TPH levels.

It is important to understand that at this stage, the procedure is a screening tool to determine if the site should go on to the next phase of analysis, and not primarily a process to determine a cleanup level (although it may in fact do so). Risk management is implied by taking the position that borings do not have to be done at every site.

The approach is nonetheless conservative. Note that the screening level is only 10 TPH and essentially no BTXE at gasoline contaminated sites where groundwater is less than 25 feet below the samples (often 35 to 40 feet below ground surface).

If no contamination is found in the bottom or sides of tank and piping excavations, the site may be closed. However, in order to minimize the number of sites where excavation sampling may fail to indicate contamination at a deeper level, the LPA must strictly be applied. Indeed, several cases have been encountered in Orange County where sites were elevated to the next phase (i.e., GRA) due to TPH or xylene being a few mg/kg in excess of the LPA values (no benzene or toluene detected), and very extensive contamination was subsequently found (up to several thousand mg/kg TPH and several hundred mg/kg of the aromatics through a 40-foot zone). Another consideration is that of sites where organic vapor instrument readings are quite high, while laboratory results are very low. The concern is not only that of public safety due to fire or explosion, but also of redissolution into groundwater, possibly in an upgradient direction.[3] The response to this type of situation, unusual in Orange County, has been to use a vapor level essentially the same as background to determine soil cleanup levels.

An important risk management decision was to disregard any effects of standard asphalt and concrete paving. Only properly designed and maintained barriers may modify the rainfall infiltration components of either the LPA or GRA. The rationale was that rainwater may collect from a considerable area and pond at low points over breaks or cracks in the paving. Thus, portions of standard paved surfaces could act more as a funnel than as a barrier and result in infiltration far in excess of average rainfall at some areas.

If the site does not meet the limits specified in the Leaching Potential Table, then LUFT GRA must be completed. This appraisal is based on a profile of soil contamination and depth to groundwater. Two mathematical models, SESOIL and AT123D, were adapted and interfaced to generate tables of acceptable cumulative soil contamination levels for BTXE. Acceptable levels for benzene and xylene are shown in Tables 3 and 4. The values for toluene fall between those of benzene and xylene, while those for ethylbenzene are similar to xylene. Data requirements are average annual precipitation, distance from surface to groundwater, depth of soil sample, and concentration of BTXE in samples collected at five-foot vertical intervals throughout the zone of contamination. A work sheet is included (Table 5). A minimum of three borings is required, at or next to the

Table 3.   California LUFT Field Manual Acceptable Cumulative Soil Contamination Levels for Benzene.

Benzene Acceptable Cumulative Soil Contamination Levels for Protection of Ground Water Qualified California Basin Sites

Stop: Do not use this table unless the site in question has been screened using the applicability checklist (LUFT manual) for general risk appraisal to protect ground water.

Distance to Highest Ground Water from Soil Sample in Feet

	Mean Annual Inches Precipitation																					
	0 to 5	5.1 to 6	6.1 to 7	7.1 to 8	8.1 to 9	9.1 to 10	10.1 to 11	11.1 to 12	12.1 to 14	14.1 to 16	16.1 to 18	18.1 to 20	20.1 to 22	22.1 to 24	24.1 to 26	26.1 to 28	28.1 to 30	30.1 to 32	32.1 to 34	34.1 to 36	36.1 to 38	38.1 to 40
5-9.9	1	1	0	0	0	0	0	0	0	0	0	0	0	0	0	0	0	0	0	0	0	0
10-14.9	5	3	2	1	0	0	0	0	0	0	0	0	0	0	0	0	0	0	0	0	0	0
15-19.9	10	10	6	3	1	0	0	0	0	0	0	0	0	0	0	0	0	0	0	0	0	0
20-24.9	60	40	20	10	5	2	1	0	0	0	0	0	0	0	0	0	0	0	0	0	0	0
25-29.9	200	100	60	30	10	7	3	1	0	0	0	0	0	0	0	0	0	0	0	0	0	0
30-34.9	800	400	200	100	40	20	10	4	1	0	0	0	0	0	0	0	0	0	0	0	0	0
35-39.9	1000	1000	700	300	100	60	20	10	3	1	0	0	0	0	0	0	0	0	0	0	0	0
40-44.9	1000	1000	1000	1000	400	100	80	30	9	2	1	0	0	0	0	0	0	0	0	0	0	0
45-49.9	1000	1000	1000	1000	1000	500	200	100	20	4	2	1	0	0	0	0	0	0	0	0	0	0
50-54.9	1000	1000	1000	1000	1000	1000	600	200	50	9	5	2	1	0	0	0	0	0	0	0	0	0
55-59.9	1000	1000	1000	1000	1000	1000	1000	700	100	20	10	5	2	1	0	0	0	0	0	0	0	0
60-64.9	1000	1000	1000	1000	1000	1000	1000	1000	300	40	30	10	4	3	1	0	0	0	0	0	0	0
65-69.9	1000	1000	1000	1000	1000	1000	1000	1000	700	80	60	20	8	5	3	1	0	0	0	0	0	0
70-74.9	1000	1000	1000	1000	1000	1000	1000	1000	1000	100	100	40	10	9	6	3	1	0	0	0	0	0
75-79.9	1000	1000	1000	1000	1000	1000	1000	1000	1000	300	200	100	40	20	10	6	3	1	0	0	0	0
80-84.9	1000	1000	1000	1000	1000	1000	1000	1000	1000	700	400	200	80	40	20	10	6	3	1	0	0	0
85-89.9	1000	1000	1000	1000	1000	1000	1000	1000	1000	1000	700	400	200	100	50	40	10	10	2	1	0	0
90-94.9	1000	1000	1000	1000	1000	1000	1000	1000	1000	1000	1000	800	400	200	100	90	40	20	4	2	1	0
95-99.9	1000	1000	1000	1000	1000	1000	1000	1000	1000	1000	1000	1000	800	400	200	100	90	30	10	5	3	1
100-104.9	1000	1000	1000	1000	1000	1000	1000	1000	1000	1000	1000	1000	1000	800	400	200	100	40	20	8	4	2
105-109.9	1000	1000	1000	1000	1000	1000	1000	1000	1000	1000	1000	1000	1000	1000	800	400	200	100	30	10	7	3
110-114.9	1000	1000	1000	1000	1000	1000	1000	1000	1000	1000	1000	1000	1000	1000	1000	800	400	200	50	10	10	5
115-119.9	1000	1000	1000	1000	1000	1000	1000	1000	1000	1000	1000	1000	1000	1000	1000	1000	800	400	90	20	10	7
120-124.9	1000	1000	1000	1000	1000	1000	1000	1000	1000	1000	1000	1000	1000	1000	1000	1000	1000	800	100	40	20	10
125-129.9	1000	1000	1000	1000	1000	1000	1000	1000	1000	1000	1000	1000	1000	1000	1000	1000	1000	1000	200	60	30	10
130-134.9	1000	1000	1000	1000	1000	1000	1000	1000	1000	1000	1000	1000	1000	1000	1000	1000	1000	1000	300	100	40	20
135-139.9	1000	1000	1000	1000	1000	1000	1000	1000	1000	1000	1000	1000	1000	1000	1000	1000	1000	1000	400	100	60	30
140-144.9	1000	1000	1000	1000	1000	1000	1000	1000	1000	1000	1000	1000	1000	1000	1000	1000	1000	1000	500	200	100	40
145-149.9	1000	1000	1000	1000	1000	1000	1000	1000	1000	1000	1000	1000	1000	1000	1000	1000	1000	1000	700	300	100	60
150+	1000	1000	1000	1000	1000	1000	1000	1000	1000	1000	1000	1000	1000	1000	1000	1000	1000	1000	1000	500	200	90

Note: Individual concentrations for any soil sample cannot exceed 100 ppm. The numbers in this table do not represent soil concentrations; they reflect the accumulation of pollutant mass in contaminated soil. These numbers can be derived from the LUFT manual worksheet.

Table 4.   California LUFT Field Manual Acceptable Cumulative Soil Contamination Levels for Xylene.

Xylene Acceptable Cumulative Soil Contamination Levels for Protection of Ground Water Qualified California Basin Sites

Stop: Do not use this table unless the site in question has been screened using the applicability checklist (LUFT manual) for general risk appraisal to protect ground water.

Mean Annual Inches Precipitation

Distance to Highest Ground Water from Soil Sample in Feet	0 to 5	5.1 to 6	6.1 to 7	7.1 to 8	8.1 to 9	9.1 to 10	10.1 to 11	11.1 to 12	12.1 to 14	14.1 to 16	16.1 to 18	18.1 to 20	20.1 to 22	22.1 to 24	24.1 to 26	26.1 to 28	28.1 to 30	30.1 to 32	32.1 to 34	34.1 to 36	36.1 to 38	38.1 to 40
5-9.9	100	100	90	60	40	30	20	10	10	9	7	7	6	6	5	5	5	4	4	4	3	3
10-14.9	600	400	300	200	100	80	50	30	20	10	10	10	10	9	8	8	7	6	5	5	4	4
15-19.9	1000	1000	1000	600	400	200	100	70	40	20	20	10	10	10	10	10	10	9	8	7	6	5
20-24.9	1000	1000	1000	1000	1000	600	300	100	80	50	30	30	20	20	20	20	10	10	10	9	7	6
25-29.9	1000	1000	1000	1000	1000	1000	700	300	100	90	50	40	40	30	30	20	20	20	10	10	9	8
30-34.9	1000	1000	1000	1000	1000	1000	1000	700	300	100	90	70	60	50	40	40	30	30	20	20	10	9
35-39.9	1000	1000	1000	1000	1000	1000	1000	1000	600	200	100	100	100	60	60	50	50	40	30	30	10	10
40-44.9	1000	1000	1000	1000	1000	1000	1000	1000	1000	500	200	200	100	100	90	70	70	50	40	20	10	10
45-49.9	1000	1000	1000	1000	1000	1000	1000	1000	1000	800	400	300	200	100	100	100	100	80	60	30	10	10
50-54.9	1000	1000	1000	1000	1000	1000	1000	1000	1000	1000	700	500	300	200	200	200	100	100	80	40	20	20
55-59.9	1000	1000	1000	1000	1000	1000	1000	1000	1000	1000	1000	600	600	400	300	300	200	100	100	50	30	20
60-64.9	1000	1000	1000	1000	1000	1000	1000	1000	1000	1000	1000	1000	900	600	400	400	300	200	100	60	30	20
65-69.9	1000	1000	1000	1000	1000	1000	1000	1000	1000	1000	1000	1000	1000	1000	700	600	400	300	100	70	40	30
70-74.9	1000	1000	1000	1000	1000	1000	1000	1000	1000	1000	1000	1000	1000	1000	1000	1000	600	400	200	90	60	40
75-79.9	1000	1000	1000	1000	1000	1000	1000	1000	1000	1000	1000	1000	1000	1000	1000	1000	900	600	300	100	70	50
80-84.9	1000	1000	1000	1000	1000	1000	1000	1000	1000	1000	1000	1000	1000	1000	1000	1000	1000	700	400	100	90	60
85-89.9	1000	1000	1000	1000	1000	1000	1000	1000	1000	1000	1000	1000	1000	1000	1000	1000	1000	1000	600	200	100	70
90-94.9	1000	1000	1000	1000	1000	1000	1000	1000	1000	1000	1000	1000	1000	1000	1000	1000	1000	1000	800	200	100	90
95-99.9	1000	1000	1000	1000	1000	1000	1000	1000	1000	1000	1000	1000	1000	1000	1000	1000	1000	1000	1000	300	100	100
100-104.9	1000	1000	1000	1000	1000	1000	1000	1000	1000	1000	1000	1000	1000	1000	1000	1000	1000	1000	1000	400	200	100
105-109.9	1000	1000	1000	1000	1000	1000	1000	1000	1000	1000	1000	1000	1000	1000	1000	1000	1000	1000	1000	500	300	100
110-114.9	1000	1000	1000	1000	1000	1000	1000	1000	1000	1000	1000	1000	1000	1000	1000	1000	1000	1000	1000	700	400	200
115-119.9	1000	1000	1000	1000	1000	1000	1000	1000	1000	1000	1000	1000	1000	1000	1000	1000	1000	1000	1000	900	500	300
120-124.9	1000	1000	1000	1000	1000	1000	1000	1000	1000	1000	1000	1000	1000	1000	1000	1000	1000	1000	1000	1000	600	400
125-129.9	1000	1000	1000	1000	1000	1000	1000	1000	1000	1000	1000	1000	1000	1000	1000	1000	1000	1000	1000	1000	800	500
130-134.9	1000	1000	1000	1000	1000	1000	1000	1000	1000	1000	1000	1000	1000	1000	1000	1000	1000	1000	1000	1000	1000	600
135-139.9	1000	1000	1000	1000	1000	1000	1000	1000	1000	1000	1000	1000	1000	1000	1000	1000	1000	1000	1000	1000	1000	800
140-144.9	1000	1000	1000	1000	1000	1000	1000	1000	1000	1000	1000	1000	1000	1000	1000	1000	1000	1000	1000	1000	1000	1000
145-149.9	1000	1000	1000	1000	1000	1000	1000	1000	1000	1000	1000	1000	1000	1000	1000	1000	1000	1000	1000	1000	1000	1000
150+	1000	1000	1000	1000	1000	1000	1000	1000	1000	1000	1000	1000	1000	1000	1000	1000	1000	1000	1000	1000	1000	1000

Note: Individual concentrations for any soil sample cannot exceed 40 ppm. The numbers in this table do not represent soil concentrations; they reflect the accumulation of pollutant mass in contaminated soil. These numbers can be derived from the LUFT manual worksheet.

Table 5.   California LUFT Field Manual Environmental Fate Worksheet.

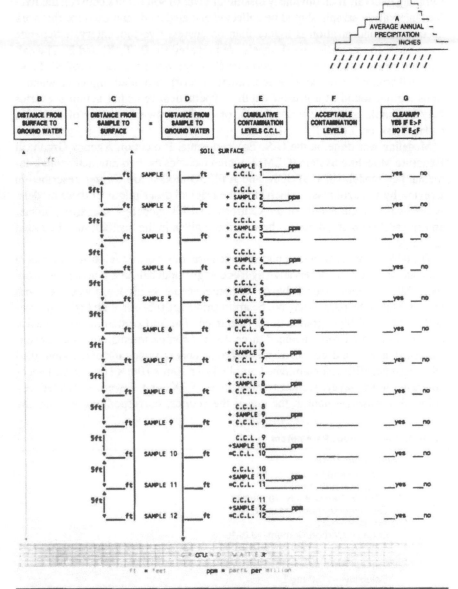

ft = feet
ppm = parts per million

*Note: Concentrations for any single soil sample cannot exceed 100 ppm for Benzene, 80 ppm for Toluene, 40 ppm for Xylene and 40 ppm for Ethylbenzene in order to be used with the general risk appraisal. The last sample to be included in the calculations for cumulative contamination must be at or above the detection limit; do not include bottom samples which have concentrations less than the detection limit.

tank site. Laboratory analyses for BTXE are required for all samples. The value used for the work sheet should be the highest value obtained from each vertical sampling interval. If an obviously dissimilar layer of soil occurs between the five-foot intervals, a sample should be collected and analyzed, and used for the work sheet if its concentration is higher than its nearest five-foot interval. Separate cumulative concentration work sheets are prepared for B, T, X, and E. The calculation of cumulative concentration levels stops when the values of B, T, X, or E fall below 0.3 mg/kg (the equivalent of zero), provided higher concentrations are not found at a greater depth. Hypothetical remedial action scenarios may be simulated on the work sheets, or on the software version of the GRA called "risk on a disc."

Modeling was done on the U.S. Environmental Protection Agency Graphical Exposure Modeling System (GEMS), which includes the two interactive environmental fate models. The SESOIL and AT123D models are further described in a review by Bonazountas,[4] which cites the original descriptions of these models. A discussion of the details of the California LUFT modeling scenarios, inputs, and spreadsheet manipulations, beyond those given in the manual, may be found in Hubbard.[5]

The tables of acceptable cumulative concentrations were developed from model simulations based on conservative hydrogeologic input data and climate data from four California cities (see Table 6). A homogeneous sandy loam was used with the caveat that the model may not be appropriate for subsoils with higher permeability. Minimal biodegradation was assumed. The soil contamination scenario for the unsaturated zone included three distinct compartments: an upper uncontaminated layer 5 m deep, corresponding to an underground tank excavation filled with clean backfill; a contaminated middle layer $10m \times 10m \times 1m$; and an uncontaminated lower layer. Concentrations of B, T, X, and E were simulated at a point 10 m downgradient at the top of the aquifer, corresponding to a nearby

Table 6.   Model Input Parameters.

Soil Parameters

Type—Sandy Loam
Density—1.35 g/cm^3
Intrinsic Permeability—0.2 cm^2
Disconnectedness Index—6.3
Porosity—0.25
Organic Carbon Content—0.02%
Clay Content—10%

Chemical Parameters for Benzene

Solubility—1780 mg/L
Organic Carbon Adsorption Coefficient—69
Diffusion Coefficient in Air—0.089 cm^2/sec
Biodegradation Rate—0.02%/day
Henry's Law Constant—0.0055 m^3 atm/deg K mol
Adsorption Coefficient—0.00

shallow well. One hundred grams of each contaminant was added to the 100 m³ middle layer. Attenuation factors were derived from two simulations at each location by placing the contaminated layer at 1 m and at 3 m above groundwater. Concentrations of the contaminants in soil that would result in the DHS action levels in groundwater were calculated by multiplying the ratio of the initial soil contamination level to the simulated groundwater contamination level by the appropriate action level. Attenuation factors were assumed to be constant. Simulations were carried out for a 10-year period. Finally, other cumulative concentration values were interpolated or extrapolated, rounded downward, and an arbitrary cumulative concentration of 1000 chosen as an upper limit. The use of cumulative concentration levels effectively moves all the contamination down to the next lower level and is, again, a conservative approach to account for increased contaminant mass.

Table 7 shows the applicability checklist used to judge whether the standard GRA is appropriate for a specific site. Note that a single or a few "no" answers do not suggest that the model is entirely inappropriate, but rather that the appraisal may be less valid. The single sample concentration limits of 100 ppm benzene, 80 ppm toluene, and 40 ppm xylene and ethylbenzene were derived from a conservative application of the CEPAC Steady State Model at the University of California at Davis, and estimate the boundary between aqueous and nonaqueous phases.

The final portion of the Category 2 analysis is determining the groundwater gradient, and collecting groundwater samples if the GRA indicates that groundwater is at risk. In general, installation of four short-screened wells or piezometers, allowing the independent solution of four three-point problems, are suggested. Piezometers should be located in a configuration that allows proper determination of gradient and detection of free product.

The LUFT GRA explicitly provides for use of an alternative environmental risk analysis approved by the appropriate regulatory agency. Alternative analyses could include more site-specific inputs to the same models, or the use of different models. An alternative analysis that appears to be consistent with LUFT methodology was recently presented by Crell and Garner.[6] The manual also provides for long-term groundwater monitoring at the discretion of the regulatory agency to verify the effectiveness of remedial actions.

Category 3: Known Groundwater Pollution

This category involves a determination of whether free product is present, whether a fire safety hazard exists, and usually the subsequent removal of recoverable undissolved product. The manual then directs consultation with the Regional Water Quality Control Board, and reproduces policy statements from California law. No specific guidance is given regarding the cleanup of dissolved product other than to "consult with the affected Regional Board to ensure that anticipated

Table 7. California LUFT Field Manual General Risk Appraisal Applicability Checklist.

GENERAL RISK APPRAISAL FOR PROTECTION OF WATER QUALITY: APPLICABILITY CHECKLIST	YES	NO
1. Is the site in a mountainous area? (shaded moist areas &/or areas with rocky subsurface conditions)		
2. Is the site in an area that could collect surface runoff or intercept water from a source other than the natural precipitation?		
3. Does the areal extent of soil contamination exceed 100 meters?		
4. Do the concentrations of fuel constituents in any soil samples exceed the following amounts: benzene—100 ppm, toluene—80 ppm, xylene—40 ppm, ethylbenzene—40 ppm.		
5. Are there any records or evidence of man-made or natural objects which could provide a conduit for vertical migration of leachate?		
6. Do any boring or excavation logs show the presence of fractures, joints or faults that could act as a conduit for vertical migration of leachate?		
7. Do any boring logs show that contaminated soil could be within 5 ft. of highest groundwater?		
8. Do any boring logs show the presence of a layer of material, 5 ft. thick or more, which is more than 75% sand and/or gravel?		

Directions:
1. Boring logs taken during the general risk appraisal can be used to answer questions 5–8. In addition, analytical results of the soil samples taken during the general risk appraisal can be used to answer questions 3 and 4.
2. Lateral migration of constituents to problem areas should also be considered in questions 5–8.
3. The above checklist contains questions which are designed to identify sites with environmental conditions which could produce a greater risk to ground water than was modeled. The results of the general risk appraisal are most applicable if all of the questions on the checklist can be answered "no" with reasonable certainty. If any of the questions on the checklist cannot be answered "no", then the results of the general risk appraisal may be less valid.

remedial action is consistent with the applicable water quality control plan(s) and policies." With regard to groundwater pollution, the LUFT document further states that "Due to the many factors influencing a particular beneficial use assessment, it is impossible to generalize about the process in a meaningful way."

DISCUSSION

The California LUFT Field Manual methodology is an attempt to apply state-of-the-art environmental fate modeling to the practice of evaluating sites with soils contaminated by petroleum fuels. The methodology analyzes a general case that can be used to establish site-specific cleanup or screening levels. The procedures were strongly influenced by decisions to limit the scope to gasoline and diesel, to develop a procedure that results in numerical cleanup levels, and to focus on the protection of groundwater resources. The decision to develop numerical standards was a response to the need to give meaning to quantitative analyses of petroleum hydrocarbon constituents reported for contaminated soils. The document has been, and still is, somewhat controversial, but properly used, can give regulatory agencies and responsible parties a basis to conclude that contamination at a site is within an acceptable risk level; or that contamination at a site could cause groundwater to exceed a regulatory level of concern and that further investigations should establish a more site-specific basis for closure, or that remedial action should be initiated. The response to the uncertainties inherent in both the science of environmental fate and risk analysis, as well as the difficulties in characterization of hydrogeological sensitivity and contaminant mass at individual sites, was to perform a very conservative evaluation of the general case. The relative sophistication of the methodology, in that it is based on quantitative procedures and environmental models rather than tenuous qualitative statements as the basis for exposure assessments, sets a standard of quality for further evaluation.

Release of the California LUFT document has been followed by at least two evaluations of the methodology. Marks et al.[7] questioned the use of drinking water standards if groundwater is not actually being used as a drinking water supply. They also concluded that the LUFT procedures underestimated risk in a case where no soil contamination was detected from soil borings "made in the area of the tank field," but where groundwater was found to be contaminated, possibly due to a product line leak. This conclusion is probably in error, as the LUFT procedure does require systematic survey and sampling of pipelines. They further stated that the California LUFT set two precedents that may bring the leaking tank problem into manageable proportions: the use of risk assessment procedures for site evaluation and the position that detectable residual levels of contamination above drinking water action levels may be acceptable. August and Mercer[8] concluded that the LUFT approach was too conservative and that portions (e.g., SESOIL output) could not be validated. However, using the newer 1987 version of SESOIL, they obtained larger groundwater concentrations and lower attenuation factors than obtained by the California LUFT model, which used the 1984 SESOIL. They also recommended against developing new cumulative concentration levels, due to the significant effort required.

In conclusion, the California Leaking Underground Fuel Tank Field Manual was developed by a multiagency task force and reflects the environmental conditions, aspects of perceived public protection expectations, and present environmental regulatory perspectives in California. The document must be considered as one evaluation tool that may give some foundation to regulatory agency requests for further investigations, or conversely, some consistent defensible bases for site closure. The interest in the California LUFT for those outside California is perhaps not in the numbers generated, or in specific assumptions or input parameters, but in the overall approach of attempting to develop a rational process for site analysis by applying a combination of practical procedures and quantitative environmental fate models to the evaluation of soil contamination. The more obvious modifications for those who may attempt to develop similar methodology includes the use of other environmental fate models, assumptions, and input parameters more applicable to the area of concern, and placement of biological receptors in other appropriate locations.

ACKNOWLEDGMENTS

The California LUFT Field Manual was developed by a 40-member task force sponsored by the California Department of Health Services (DHS) and the State Water Resources Control Board (SWRCB). The following individuals must be acknowledged for their very special efforts: David J. Leu, Ph.D., Chief, Alternative Technology Section, DHS, whose gifts of time, energy, and intellect, as well as fiscal resources, were absolutely essential; Terry Brazel, the primary representative of the SWRCB, editor, organizer, and principal wordsmith; Wayne Hubbard (SWRCB), who courageously developed the GRA environmental fate model; and Brian Zamora, Director of Environmental Health, County of Solano, whose special leadership guided the development of the best possible document. Invaluable contributions of logic and expertise were given by Cheri Eir, M.P.H. (County of San Diego), Charles Fischer (SWRCB), Paul Hadley, P.E. (DHS), and Stephen Reynolds, C.E.G. (DHS). Other task force members who played especially important roles were Gordon Lee Boggs (Central Valley RWQCB), Michael Falkenstein (SWRCB), Peter Johnson (San Francisco Bay RWQCB), Robert Holub, P.E. (Santa Ana RWQCB), George Rose (DHS), Bart Simmons (DHS), Dan Tempelis (San Francisco Bay RWQCB), and Jeff Wong, Ph.D. (DHS).

REFERENCES

1. "Leaking Underground Fuel Tank Field Manual: Guidelines for Site Assessment, Cleanup and Underground Storage Tank Closure," State of California, Leaking Underground Fuel Tank Task Force, Sacramento, CA, May 1988.
2. "The California Site Mitigation Decision Tree Manual," Toxic Substances Control Division, Department of Health Services, Sacramento CA (1986).
3. Kreamer, D. K. "Vapor Transport and Its Implications to Underground Tanks," in *Processes Affecting Subsurface Transport of Leaking Underground Tank Fluids*, USEPA Report CR810052, USEPA, Las Vegas, NV, June 1987.
4. Bonazountas, M. "Mathematical Pollutant Fate Modeling of Petroleum Products in Soil Systems," in *Soils Contaminated by Petroleum: Environmental and Public Health*

Effects, E. J. Calabrese, P. T. Kostecki, and E. J. Fleischer, Eds. (New York: John Wiley & Sons, 1988), pp. 31–97.

5. Hubbard, W. "Methodology for Developing Acceptable Cumulative Concentration Levels for Protection of Water Quality (Groundwater)," California State Water Resources Control Board Report, Sacramento, CA (1987).

6. Crell, W. M., and E. Garner. "Evaluation of Benzene Migration and Biodegradation to Assess the Environmental Risks Posed by Petroleum Hydrocarbons in Groundwater," in *Proceedings, Second National Outdoor Action Conference on Aquifer Restoration, Ground Water Monitoring and Geophysical Methods*, NWWA, OH (1988).

7. Marks, B. J., R. M. Gray, R. W. Greensfelder, R. E. Hinchee, and C. A. Presley. "The California LUFT Manual Compared to Risk Assessments for 16 Underground Gasoline Storage Tank Sites," paper presented at Hazmacon 88 Conference, EA Engineering, Science and Technology, Inc., Lafayette, CA (1988).

8. August, L. L., and J. W. Mercer. "Reproduction of Acceptable Concentration Levels in Soil." Geotrans Report to Mobil Oil Corporation (1988).

CHAPTER 35

Letting the Sleeping Dog Lie: A Case Study in the No-Action Remediation Alternative for Petroleum Contaminated Soils

Evan C. Henry and Michael E. F. Hansen

INTRODUCTION

In selected circumstances where geological and hydrological conditions are favorable, a "no-action remediation" may be considered the appropriate alternative to the cleanup of petroleum contaminated soils. This case study presents the rationale for leaving some petroleum contaminated soil in place and untreated, based on the understanding of site conditions and potential impacts as drawn from environmental investigations performed at the site.

The subject site is located in the greater Los Angeles Basin within the flood plain of the Santa Ana River. As such, the site is underlain primarily by coarser-grained river-deposited alluvium. Despite the proximity to the river at a distance of approximately one-half mile, depth-to-groundwater is greater than 100 feet due to the combination of natural conditions and long-term groundwater withdrawal from the area.

The site is located within a municipality which adopted its own regulations for underground storage tanks prior to the passage of the legislation by the state of California. Therefore, the local agency has primacy to regulate underground storage tanks as long as the intent of the state legislation is served by those local regulations. Within the local municipality, as an adjunct to existing authority over the fire safety aspects of underground storage tank operation, the municipal fire prevention department was given the task of administering the underground tank regulation. The local fire department, although not governed by the state or county health departments, has relied heavily on technical support from state and county agencies for assistance in evaluation of the environmental considerations.

DISCOVERY

Three 8000-gallon steel tanks were excavated and removed from the location shown in Figure 1, with the intention that they would be replaced with three dual wall tanks within the same excavation. During removal it was observed that the

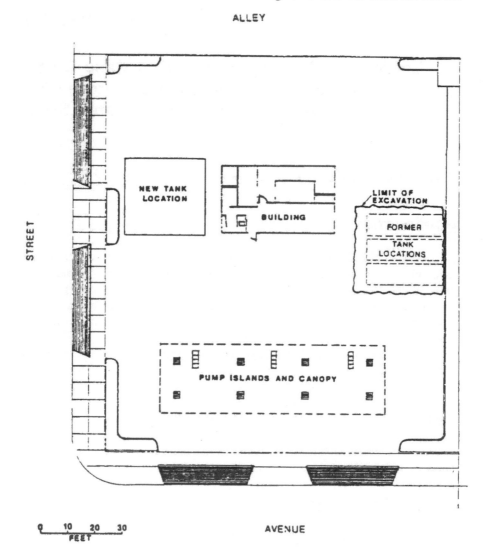

Figure 1. Site plot plan.

tank backfill soils exhibited a strong gasoline odor, particularly in the vicinity of the tank turbine and fill ports at one end of the tanks. Soils were removed from around the tanks and stockpiled adjacent to the excavation on a paved part of the station site. Based on visual observations the fuel hydrocarbons in the soils appeared to be from over-spillage during filling at unsealed fill boxes or minor piping leakage at the western end of the tanks (see Figure 1). Six soil samples were obtained according to local regulatory requirements from beneath the former location of the three tanks. The locations of the soil samples are illustrated in Figure 2. Analysis of these soil samples indicated that hydrocarbons were present below the western ends of the tanks at a sampling depth of 14 feet. Hydrocarbon concentrations ranged from nondetected at a detection limit of 1 to 7200 mg/kg, and were identified as gasoline in character. Based on a regulatory action limit of 100 mg/kg for fuel hydrocarbons in the soils, it was evident that the presence of hydrocarbons below the tanks had to be addressed and potentially remediated.

INITIAL REMEDIATION ACTIONS

From field observations during the tank removal process the extent of hydrocarbons in the soils appeared to be limited. Although the depth to which hydrocarbons extended could not be determined at the time, the lateral extent appeared to be confined to an area of approximately 15 by 30 feet. In keeping with a policy of active remediation at the time of tank removal, the decision was made to excavate all affected soils to the maximum practical depth. The excavation was extended to a depth of 20 feet in the western end of the excavation. No additional excavation was performed in the eastern end, where hydrocarbons had not been detected in the soils. The depth of 20 feet was approximately 8 feet below the bottom of the original excavation left from the tank removal. Approximately 500 cubic yards of soil were removed from the excavation and stockpiled on the site. These soils were subsequently aerated in batches to reduce the concentrations of volatile organics to acceptable concentrations for disposal at a local landfill as a nonhazardous material.

Additional soil samples were obtained at the practical limit of excavation, which upon analysis indicated that fuel hydrocarbons were still present at concentrations of 1000 to 2000 mg/kg. As these concentrations exceeded the agency standard of 100 mg/kg, the local regulatory agency required further characterization. With the understanding that it would be practically infeasible to remove additional soils without endangering the adjacent building and canopy structure, the active remediation by excavation of soils was halted. As this particular station is one of the highest sales volume facilities in the company retail network, there was a strong emphasis on returning the station to active service as soon as possible. To expedite the process, the new tanks were installed in an excavation in an unaffected area on the opposite end of the site (Figure 1). Because the former tank excavation presented a potential safety hazard and to assist in access for subsequent investigation, the former tank excavation was backfilled with crushed

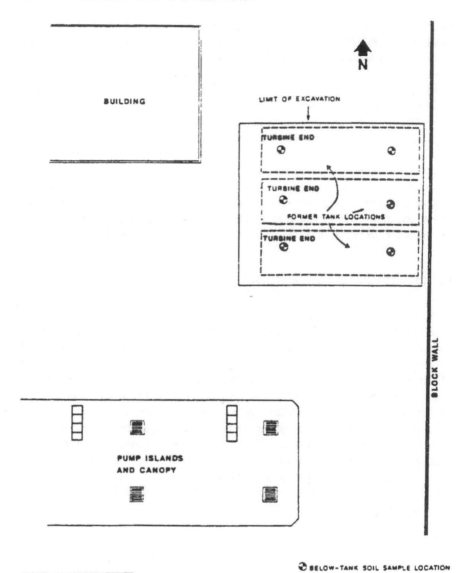

Figure 2. Excavation soil sample locations.

rock. At the time of backfilling, four vertical standpipes of 4-inch diameter poly-vinylchloride (PVC) well screen were installed within the crushed rock backfill to serve as vapor sampling ports and potential vapor extraction wells for use in the future, if necessary.

SUBSURFACE CHARACTERIZATION

To characterize the subsurface geologic conditions and the distribution of fuel hydrocarbons in the subsurface, four vertical borings were drilled in the area of the former tank excavation. One boring was located within the excavation and was drilled through the crushed rock backfill. Three other borings were drilled immediately adjacent to the tank excavations through natural materials. The locations of the four borings are shown in Figure 3. The borings were drilled with hollow stem auger drilling equipment, with soil samples obtained with a 2-inch inner diameter split-tube drive sampler outfitted with brass soil sample retainers.

Based upon onsite drilling observations, natural soils consisted of a variety of materials typical of river-deposited alluvium including silty sand, sandy silt, sand and gravel. In general, silty sand or sandy silt was encountered from the ground surface to a depth of 25 feet; coarse sand and gravel were encountered to an approximate depth of 65 feet. The sand and gravel were underlain by sand at an approximate depth of 65 feet. A schematic subsurface cross-section is shown in Figure 4. No groundwater was encountered at the maximum drilled depth of 65 feet.

Analysis for hydrocarbon presence for the three borings drilled just outside the excavation indicated that the hydrocarbon concentrations for the 20 and 50 foot samples were below the detectable limits. This indicated that there had been no significant lateral spreading of hydrocarbons with depth to areas adjacent to the excavated area. Hydrocarbons at a concentration of 3000 mg/kg were detected in the soil sample retrieved at the 20-foot level (just below the level of crushed rock in the excavation) from the one boring drilled within the tank excavation. In contrast, the samples obtained at 55 and 65 feet showed concentrations to be below detectable limits. The presence of gravel with cobbles prohibited acquisition of soil samples suitable for laboratory analysis between depths of 20 and 50 feet in all of the borings.

Based on the results of the soil sampling, it appeared that the presence of hydrocarbons was confined to a relatively shallow zone of limited areal extent, as shown in the cross-section in Figure 4. Although sampling limitations prohibited the determination of the exact bottom of the affected soils, the sampling indicated that the bottom of the hydrocarbons did not extend below a depth of 55 feet. Based on field screening of soil samples with a portable photoionization detector at the time of drilling, it appeared that the base of the hydrocarbons was located at a depth of approximately 25 feet. At this level there was a stratigraphic contact between overlying sandy silt and underlying sand and gravel.

A report of investigation was prepared based on these findings, which acknowledged that hydrocarbon-contaminated soils were present in the subsurface but that the extent of these hydrocarbons was limited. The conclusions presented based on the results of the investigations included:

1. Hydrocarbons have apparently not spread laterally beyond the

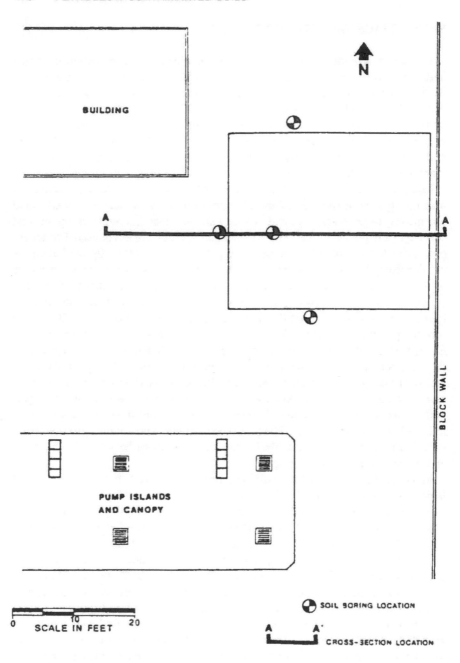

Figure 3. Soil boring locations.

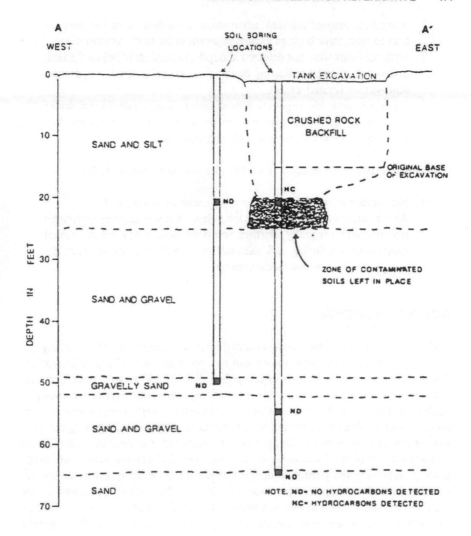

Figure 4. Schematic cross-section (A-A') of tank excavation.

immediate area of the tank excavation, as indicated in the data retrieved from three borings located adjacent to the tank excavation area.

2. Hydrocarbons were not detected at depths immediately below the tank excavation, thereby indicating downward vertical migration hydrocarbon was of limited extent,

3. Based on available regional groundwater data, groundwater under water table conditions appeared to be at least 100 feet deeper than the deepest detected presence of hydrocarbons.

Recommendations developed from the site investigation included:

1. No further onsite investigation was recommended, and

2. As the extent of hydrocarbon in the subsurface was apparently limited laterally and vertically, further remedial action beyond the initial removal of soil during tank excavation, (which had already been accomplished), was not recommended.

AGENCY RESPONSE

Following the lead of the county health department standards, the local regulatory agency rejected the conclusions and recommendations of the investigation. The agency responded with its hard-line policy that any and all soils with hydrocarbon concentrations above the set limits of 100 mg/kg would require remediation. Such remediation would have to be undertaken (either started or potentially completed) prior to obtaining a permit to operate the station. From a timing perspective, when this decision by the agency was received, the new tank installation was almost complete. Therefore, waiting for remedial actions to be undertaken prior to receiving the permit to operate would have severely impacted the reopening date of the station. The reasoning for the "hard-line" promoted by the regulatory agency was that this was their only tool for ensuring that remediation took place. Once a permit to operate was approved, the agency had to undertake legal action to close the station if adequate remediation of the soil contamination was not subsequently undertaken.

Despite the agency attitude toward requiring remediation, based on the practicalities of the site, it was felt than a no-action alternative for remediation was the most prudent. In discussion with the agency personnel, it was negotiated that the station could be reopened with a written commitment from the company to address the fate and transport of the remaining hydrocarbons in the soils in a timely manner. Remediation needs would be determined based on the results on the fate and transport study. The intent of the fate and transport study was to estimate the disposition of the hydrocarbons over time if left in place.

FATE AND TRANSPORT ASSESSMENT

The purpose of the fate and transport investigation was to:

1. To acquire research and technical data to address the environmental fate and potential for migration of residual hydrocarbons within shallow soils at the tank excavation site; and
2. To determine the potential for adverse impact on groundwater quality in the area.

The scope of work for the investigation involved two aspects: research of the technical aspects of hydrocarbon presence and migration in unsaturated soils, and assessment of site-specific characteristics relative to the current technical understanding of hydrocarbons fate and transport in the subsurface. Research included a literature search to develop an understanding of hydrocarbon characteristics and potential for migration in soils.

DISCUSSION OF FATE AND TRANSPORT OF HYDROCARBON IN THE SUBSURFACE

It is inferred that the hydrocarbon presence detected in the subsurface during the tank removal process was primarily the result of spillage at the unsealed tank fills during transfer of product from tank trucks to the underground storage tanks. This inference is the result of observation of hydrocarbon staining of tank backfill materials at the fill end of the excavation and the lack of hydrocarbon in the subsurface at the other end of the tanks.

From spillage near the ground surface, hydrocarbon product is inferred to have moved vertically downward within the tank backfill around and ultimately to beneath the underground storage tanks. Sufficient hydrocarbon was present to have moved from the backfill materials into the underlying natural soils.

The downward migration of liquid hydrocarbon product typically takes place where sufficient hydrocarbon product is present to allow for liquid flow. As product moves downward through the soil, a small amount attaches itself to the soil particles contacted and remains behind the main body of product. Where the spill is small relative to the surface area available for contact in the zone of migration, the body of product is exhausted on the way down until the degree to which it saturates the soil reaches a relatively low point called the "immobile" or "residual" saturation. At this point the product essentially stops moving.[1]

The volume of soil required to immobilize a given amount of product depends on two primary factors: (1) the porosity of the soil; and (2) the nature of the hydrocarbon as reflected in its characteristic "maximum residual saturation."

At or below its maximum residual saturation, the product will not move as a liquid in the soil. The residual saturation for various hydrocarbon products have been empirically derived and are estimated as follows: gasoline—10%; diesel and light fuel oil—15%; and lube and heavy fuel oil—20%.[1]

With the above residual saturation percentages, theoretical hydrocarbon concentrations (in units of milligrams per kilogram, mg/kg) were calculated for various soil porosities. The hydrocarbon concentrations were calculated using the following formulae:

(1) Hydrocarbon Concentration = $\dfrac{\text{Weight of Hydrocarbon}}{\text{Weight of Hydrocarbon plus Weight of Soil}}$

(2) Weight of Hydrocarbon = Unit Weight of Water (62.4 lb/ft)
  × Specific Gravity of Gasoline (0.80)
  × Residual Saturation (0.10)
  × Porosity (varies from 0.20 to 0.60)

(3) Weight of Soil = 146 lb/ft

The theoretical concentrations are listed in Table 1. A unit weight of 146 pounds per cubic foot for sandy and silty soils[2] and a specific gravity of 0.80 for gasoline were used in the calculations.

From Table 1, the possible range of residual hydrocarbon concentration in the soils is from approximately 3,400 to 60,864 mg/kg. The concentration of hydrocarbons in the soil at a minimum would have to exceed the low value of this range for movement of the hydrocarbon as a liquid. The maximum concentration of hydrocarbons detected in the soils still remaining in place was 3,000 mg/kg. Based on comparison of the concentration of product in the soil to the possible range of residual saturation, the hydrocarbon in the soil was not anticipated to be mobile as a liquid.

From the initial investigation, no hydrocarbons were detected in samples outside the tank excavation area nor at depth. Samples were not obtained directly beneath the excavation between 20 and 55 feet, due to the presence of coarse gravel in the formation which precluded the acquisition of samples with the drive sampler. However, no hydrocarbons were evident in the relatively finer grained

Table 1.   Theoretical Residual Hydrocarbon Concentration.

Soil Porosity (percent)	RESIDUAL SATURATION (percent of total)					
	5	10	15	20	25	30
20	3,400	6,800	10,200	13,600	17,000	20,400
30	5,100	10,200	15,300	20,400	25,400	30,500
40	6,800	13,500	20,300	27,000	33,800	40,500
50	8,512	17,024	25,536	34,304	42,560	51,072
60	10,144	20,288	30,432	40,576	50,720	60,864

(a) All results are reported in milligrams per kilogram.
(b) For estimated specific gravity of gasoline equal to 0.80 and for estimated unit weight of silty sand of 146 lbs/cubic foot (Lambe and Whitman[2]).

soils at depths of 55 and 65 feet. The data from these samples indicated that hydrocarbons had not migrated to this depth. Field observations during drilling also supported the conclusion that hydrocarbons had not migrated below a depth of 25 feet. Therefore, the bottom of the immobile residual hydrocarbons was projected to be greater than 75 feet above the projected depth to groundwater of 100 feet.

The locations of the borings and the sample depths are considered sufficient to have detected whether any lateral spreading had occurred, or if alternate pathways for migration were present. There were no soil layers of lower permeability which could possibly have altered the potential for downward migration of hydrocarbons. In one documented case, the mass transfer of hydrocarbons through sand/clay interfaces was not found to produce appreciable lateral spreading.[3] This further supports the conclusion that the extent of the presence of residual hydrocarbons in the subsurface at the site was limited and localized.

Although in the liquid phase hydrocarbons are immiscible with water, some hydrocarbons and other associated organic compounds are slightly soluble in water. As a result, it is possible for some hydrocarbon constituents to be carried downward if percolating waters come into contact with the hydrocarbon. At the station, the soils beneath the tank excavation were described variably as dry, slightly moist, or moist. No samples were described as wet, which would be taken as indicative of being saturated enough to allow for water flow to the maximum drilled depth of 65 feet. Since the tank excavation was paved, infiltration of surface water into the affected subsurface area has been eliminated. Therefore, it is not expected that any water will enter the soils to serve as a vehicle for further downward migration of dissolved organic compounds. Hydrocarbons also have the potential for migration in the soils in the vapor state. The amount of vapor generation from liquid hydrocarbon is a function of the vapor pressures of the individual organic compounds, and the total pressure of the atmosphere (in this case the soil atmosphere). The "lighter" organics will volatilize faster, thereby decreasing the overall concentration but increasing the relative proportion of "heavier" compounds in the residual product. This process is frequently referred to as "weathering."

The relative percent of certain organic compounds found in gasoline are listed in Table 2. Table 2 also lists the percentage of selected compounds relative to the total hydrocarbon concentration detected in the soil vapor from the crushed rock excavation backfill. The percentage of certain organic compounds was significantly less in the soil vapor from the crushed rock excavation fill than in the product itself. In addition, the percentage of benzene was significantly less than toluene and xylene in the vapor samples. The vapor pressure of benzene is much higher than that of toluene and xylene, and therefore it will volatilize more rapidly, resulting in a lower residual percentage of this compound being present over time. In addition, the three compounds as a group are generally lighter than many of the polyaromatic hydrocarbons found in gasoline. As a result, the percentage of these compounds relative to the total amount of residual hydrocarbon was expected to diminish over time. The hydrocarbon product appears to have been in

Table 2.   Relative Concentrations of Hazardous Compounds in Gasoline and
Percentages Found in Soil Vapors.

Compound	Relative Concentration in Product (% by Weight)	Relative Concentration in the Soil Vapor from Crushed Rock Backfill[a] (% by Weight)
Benzene	0.81 to 1.35	0.013
Toluene	5.92 to 12.3	0.11
o-xylene	1.94 to 2.05	NA[b]
m-xylene	3.83 to 3.87	NA
p-xylene	1.54 to 1.57	NA
Total Xylene	7.31 to 7.49	0.36
Cyclohexane	0.17 to 0.36	NA

Source: Reference No. 4
[a]Data reported is from vapor wells installed in the tank excavation area.
[b]NA = Not Available.

the soil for a relatively long time, at least long enough to allow the escape of
a significant percentage of the volatile aromatic compounds present in the soils
in the vapor state.

From field observations using a portable photoionization detector, no significant hydrocarbon vapors were detected in soil samples taken from borings drilled
adjacent to the tank excavation. However, by qualitative analogy, a reading of
"no detection" on the photoionization detector was taken to indicate a relatively
low, if not undetectable, level of hydrocarbons in the soil vapor of the natural
soils outside the excavation. In contrast, the organic content of the soil vapor
in the newly emplaced crushed rock backfill was relatively high.

Transport of volatile organic compounds through the soil is a function of the
concentration of the gas at the source, and a diffusion coefficient. The diffusion
coefficient in soil is a function of the soil characteristics which affect the tortuosity of the path which the organic molecules must follow. The soil characteristics affecting the tortuosity include the air volume of the soil and the porosity
of the soil. The contrast in the inferred soil vapor concentration of the surrounding natural soils and the crushed rock backfill was probably the result of differences in the diffusion coefficient of the two soil materials. The natural materials
at the 20-foot level were characteristically a silty sand with up to 50% silt. The
tank backfill was crushed rock with particle size characteristic of fine gravel.
No specific estimates were developed, but it was inferred that the crushed rock
would be more transmissive of vapors than the finer-grained natural soils as indicated by the levels of vapors seen in the samples from the crushed rock backfill
area.

The hydrocarbons which remain in the soils are expected to biodegrade over
time. It is well known that many organic compounds, including hydrocarbons
found in gasoline, are broken down by naturally occurring bacteria in the soils.
The hazardous compounds such as benzene and toluene are also broken down
aerobically.[5] Under aerobic biodegradation, the nonhazardous end products of
organic breakdown are water and carbon dioxide. Since hydrocarbons occur within

the unsaturated zone well above the water table, they will be subject to long-term aerobic conditions and are expected to diminish in concentration over time, due to biodegradation.

CURRENT STATUS

The preceding discussion was presented in a report of the fate and transport investigation. The report was submitted to the local regulatory agency for their review, comment, and action.

A period of one year transpired between the submittal of the report of the environmental fate and transport to the regulatory agency and the preparation of the present case study. During that time, the retail service station has been in full operation with the new tank and piping system in compliance with the current local and state regulations. No specific further actions have been requested by the local regulatory agency in response to the report submittal.

It is our opinion that the above-outlined actions represent satisfactory environmental compliance and are an appropriate remedial alternative. The environmental conditions were quantified through scientific investigation. Based on the understanding of the potential fate and transport of the residual subsurface hydrocarbons, it was concluded that no detrimental impact would result from leaving the affected soils in place. As such, in this case the "no-action alternative" was the appropriate response.

REFERENCES

1. Mull, R., "Migration of Oil Products in the Subsoil with Regard to Groundwater Pollution by Oil," in Jenkins, Ed. *Advan. Water Poll Res., Proc. Int. Cong.*, (Oxford: Pergamon Press, Inc., 1971).
2. Lambe, T. W., and R. V. Whitman, *Soil Mechanics*, (New York: John Wiley & Sons, Inc. New York, 1969).
3. Kuhlmeier, P.D.D., and G.L. Sunderland, "Distribution of Volatile Aromatics in Deep Unsaturated Sediments," in *Proceedings of Conference on Characterization and Monitoring of the Vadose (Unsaturated) Zone*, National Water Well Association, November, 1985 pp. 198–214.
4. Marley, M. C., and G. E. Hoag, "Induced Soil Venting for Recovery/Restoration of Gasoline Hydrocarbons in the Vadose Zone," in *Proceedings Petroleum, Detection and Restoration, National Water Well Association*, November, 1984.
5. Patterson, J. W., and P. F. Kodukala, "Biodegradation of Hazardous Organic Pollutants," *Chem. Eng. Progress*, 77:4 April 1981, pp. 48–55.

Council for Health and Environmental Safety of Soils (CHESS): A Coalition to Standardize Approaches to Soil Contamination Problems

Paul T. Kostecki and Edward J. Calabrese

Soil contamination has become recognized as a significant environmental and public health concern over the past decade. The range of contaminants in soil is now known to be very broad, including petroleum, heavy metals, dioxins, pesticides, organic solvents, and others. The presence of such contaminants in soil is affecting the use of land throughout the United States, including types of development and insurance policies. The costs associated with cleanup activities have escalated enormously, often to the point of having major impact on business and residential development. Along with these increased costs of cleanup is the emerging concern that soil contamination may present significant health concerns as a result of groundwater contamination, soil ingestion, crop contamination, and localized air pollution.

Recent surveys have indicated important differences at the state and federal level concerning how to deal with soil contamination. Widespread inconsistencies between and within states have led to confusion in the private sector concerning what to expect from regulatory agencies and what the scientific/technical basis is for their actions.

On November 9th and 10th, 1987, the International Society of Regulatory Toxicology and Pharmacology convened a meeting at the University of Massachusetts, Amherst MA, to assess the need to create an expert committee or council to

Lheaddevelop a consensus risk assessment methodology with respect to soil contamination. The attendees at the meeting included representatives from the federal and state public health and environmental agencies, the private sector, and Society representatives. More specifically, these representatives were from the University of Massachusetts, United States Environmental Protection Agency (USEPA), Agency for Toxic Substances and Disease Registry (ATSDR), New Hampshire Department of Public Health, the Association of State and Territorial Health Officers (ASTHO)—Environmental Subcommittee, New Jersey Department of Environmental Protection (NJDEP), Electric Power Research Institute (EPRI), American Petroleum Institute (API), McLaren-Chemrisk, Inc., and the Society's vice president, secretary, and legal counsel.

The attendees at the meeting agreed that there was a crucial and immediate need for a consensus risk assessment methodology for assessing soil contamination. Furthermore, they concluded that the Society should play a significant leadership role in this area by creating a council comprised of recognized experts in the area of soil contamination and relevant disciplines to develop a consensus methodology which would then be made available to federal and state agencies, the private sector, and the scientific community.

Thus, on November 24, 1987, the International Society of Regulatory Toxicology and Pharmacology created the Council for Health and Environmental Safety of Soils—CHESS. The Council's goal would be to provide leadership in soil contamination issues by:

- providing consensus guidelines on analytical techniques, risk assessment methodologies, and remediation of contaminated soils
- conducting scientific evaluations and analyses and providing recommendations
- providing technical information transfer
- providing education and training functions
- enhancing dialogue among affected groups

ORGANIZATION

The Society's first action was to appoint as Council Chairman, Edward J. Calabrese, Ph.D., Professor, University of Massachusetts, Amherst MA, and as Managing Director, Paul T. Kostecki, Ph.D., University of Massachusetts, Amherst MA. It is the responsibility of the Chairman to oversee the overall conduct of CHESS, provide continuity between the Governing Board, Council, and Expert Committees, and ensure that the Governing Board's directives are implemented. The Managing Director is responsible for day-to-day activities and working with the expert committees in coordinating technical as well as administrative information.

The Council for Health and Environmental Safety of Soils (CHESS) oversees the actions of the Expert Committees and is directed by a Governing Board through the Chairman and Managing Director. The Governing Board is appointed by the International Society of Regulatory Toxicology and Pharmacology.

The organizational relationship of the Society, Governing Board, Council, and Expert Committees is shown in Figure 1.

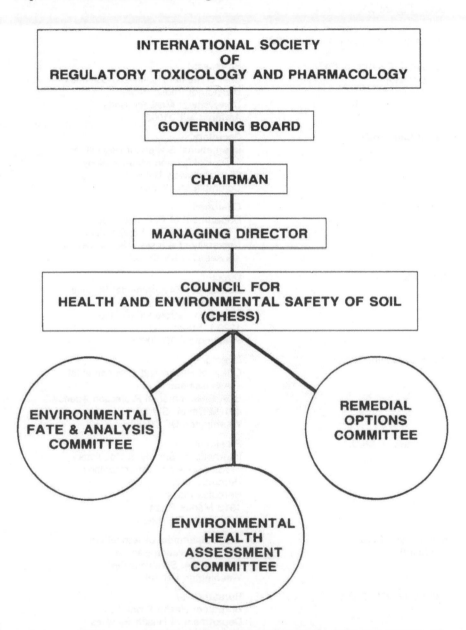

Figure 1. Relationship of the International Society of Regulatory Toxicology and Pharmacology, Governing Board, Council, and Expert Committees.

Governing Board

To ensure that the goals and objectives of CHESS are met, the Society, in conjunction with the Chairman, established an overall Governing Board of 17 members (Figure 2).

Figure 2.   Governing Board—CHESS.

Edward J. Calabrese, Ph.D. (*ex officio*)	Professor Environmental Health and Sciences Division of Public Health University of Massachusetts Amherst MA 01003
Jelleff Carr, Ph.D.	Secretary International Society of Regulatory Toxicology and Pharmacology 6546 Belleview Drive Columbia MD 21046
John Doull, Ph.D.	Chairman Department of Pharmacology, Toxicology and Therapeutics University of Kansas Medical Center Kansas City KS 66103
Robert Drew, Ph.D.	Director Health and Environmental Science Department American Petroleum Institute 1220 L Street, NW Washington DC 20005
William Farland, Ph.D.*	Director Office of Health and Environmental Assessment U.S. Environmental Protection Agency 401 M Street, SW Washington DC 20460
John Frawley, Ph.D.*	President International Society of Regulatory Toxicology and Pharmacology Hercules, Inc. Hercules Plaza 1313 Market Plaza Wilmington DE 19894
Klaus Flach, Ph.D. (retired)	U.S. Department of Agriculture Soil Conservation Service Room 5204, South Building Washington DC 20013
Peter Galbraith, D.D.S.	Bureau Chief Bureau of Health Promotion Department of Health Services Washington Street Hartford CT 06106

Barry Johnson, Ph.D.*	Assistant Director Agency of Toxic Substances and Disease Registry Centers for Disease Control Atlanta GA 30333
Donald R. Lesh, Ph.D.	Executive Director Global Tomorrow Coalition 1325 G Street, NW Washington DC 20005-3104
David Loveland	Director Natural Resources League of Women Voters 1730 M Street, NW Washington DC 20036
Donald Lyman, Ph.D.	Director Protection Division Department of Public Health Preventive Medical Services Division 714 P Street, Room 450 Sacramento CA 95814
Anthony G. Marcil*	Director Technical Assistance Program World Environment Center 419 Park Avenue South Suite 1403 New York NY 10016

Paul Means
  (retired)
	Director
	Department of Pollution Control and
	  Ecology
	8001 National Drive
	P.O. Box 9583
	Little Rock AK 72209

Gordon Newell, Ph.D.*	Senior Program Manager Environmental Risk and Health Science Department Electric Power Research Institute 3412 Hillview Avenue Palo Alto CA 94303
James Solyst	Senior Policy Analyst National Resources Policy Studies National Governor's Association 4444 North Capitol Street Washington DC 20001
Robert Tucker, Ph.D.	Director Office of Science and Research Department of Environmental Protection Trenton NJ 08625

*Members of Executive Committee

Governing Board members were selected from relevant federal, state, and private sector organizations to provide oversight and guidance of all aspects of Council activities, including approval of operational procedures and ensuring that procedures are followed.

In addition, the Governing Board assures that committees are balanced and unbiased, that the Council's activities are mission oriented, that scientific reviews are done properly, and that adequate funds are available. Decision-making by the Governing Board is by a simple majority with a quorum present.

Council

The Governing Board selected Council membership (Figure 3) by voting from a list of nominees developed by the Governing Board and the Chairman. The Governing Board determined that Council membership should reflect the areas of interest: analysis, environmental fate, public health, remediation, and decision theory, as well as balancing federal, state, and private sector perspectives.

Figure 3.   Council for Health and Environmental Safety of Soils.

James Dragun, Ph.D.	The Dragun Corporation 3240 Coolidge Berkley MI 48072-1634
Allan Hatheway, Ph.D.	University of Missouri School of Engineering RFD 4, Box 66 Rolla MO 65401
Dorothy Keech, Ph.D.	Chevron Oil Field Research Co. P.O. Box 446 3282 Beach Blvd. La Habra CA 90631
Renate Kimbrough, M.D.	U.S. Environmental Protection Agency, A-101 401 M Street, SW Washington DC 20460
Robert Menzer, Ph.D.	Director Mees Program Room 0313 Symons Hall University of Maryland College Park MD 20742
Warner North, Ph.D.	Decision Focus 4984 El Camino Real Los Altos CA 94022
Dennis Paustenbach, Ph.D.	McLaren-ChemRisk 980 Atlantic Avenue Suite 100 Alameda CA 94501
Brian Strohm, Ph.D.	Assistant Director of Public Health Service Department of Health and Welfare Division of Public Services Concord NH 03301

OPERATION

CHESS policy is set by the Governing Board, which provides oversight and guidance to all aspects of Council activities, including approval of operational procedures, and assurances that procedures are followed. The Governing Board meets three to four times a year to review and modify, if necessary, operational procedures.

Technical supervision is provided by the Council, which assists in the development of the Expert Committees, provides oversight and guidance to the expert committees, reviews committee reports, and serves as a quality assurance component during product development. The Council meets several times a year, or as often as necessary, depending on funding, in the conduct of their duties. The Expert Committees are charged with the development of CHESS products by working with the Chairman and Managing Director.

The products will be developed by initial identification and review of information presently available in each area. This may include federal and state reports and in-house documents, as well as private sector material. This information will be identified through literature searches and through the knowledge each Council and Committee member will bring to his/her Committee. Each Committee will then thoroughly review the information and:

- recommend an existing document (methodology)
- modify and/or consolidate existing documents
- develop an entirely new document

Since a great deal of fragmented work has been done in a number of areas, it is anticipated that the second course of action will be the most common. CHESS' goal is to develop peer-reviewed, consensus procedures; therefore, acceptance of a given procedure is as important as its technical merit. This may necessitate working with sections of existing documents that meet both criteria.

Expert committees will meet as often as necessary, depending on funding, to ensure successful completion of the tasks. Committee responsibilities in relation to the anticipated reports are outlined in Figure 4.

Upon completion of the products by the committees, the methodology will be subjected to a rigorous peer-review process. The Council Chairman will oversee the review process, which will be coordinated by the Managing Director.

PRODUCT

The ultimate product of CHESS will be a comprehensive decision-making methodology (tree) which will be applicable to all types of soil contamination sites, including those related to Superfund sites, leaking underground storage tanks, surface spills, mining operations, and many others. The methodology will incorporate a decision-making framework for site-specific considerations from initial

COMMITTEE   DELIVERABLE

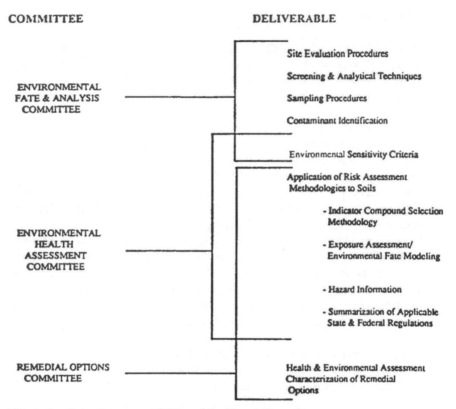

Figure 4. Reporting responsibilities of the Expert Committees.

analyses to final corrective action. The methodology will include concern for sensitive environmental receptors such as groundwater, public health risks, land use options, as well as other factors.

The decision tree will be designed to be readily adopted for use by multiple organizations, with particular emphasis on state agencies which are responsible for site evaluation, contamination determination, environmental and public health risk evaluation, and remediation and cleanup.

Development of the decision methodology will generate independent reports in a number of areas:

Site Evaluation Procedures

This report will describe the necessary steps for a proper site evaluation for contaminated soils. Ideally, the approach should be applicable to a wide range of environmental and contaminant situations, as well as levels of effort (i.e., screening, in-depth evaluation). In the decision-tree type of format a first step screening may lead directly to a risk assessment or even a cleanup action or to a more intensive site evaluation, depending on the circumstances. The site evaluation will contain, at a minimum, such site characteristics as: historical use; hydrogeologic profiles; sociodemographic information; land use practices; and soil and climatic information.

Screening and Analytical Techniques

This report will describe appropriate applications of various field screening (olfaction, portable organic vapor analyzers, portable GC/MS, etc.) and laboratory analytical techniques. It will identify and discuss the strengths and weaknesses of each methodology, and recommend ways to standardize their application.

Sampling Procedures

This report will describe how sample collection should be performed in order to ensure the collection of representative and useful information. The report will cover the best soil sampling methodologies for a particular contaminant or groups of contaminants, their transport and storage, as well as developing sampling patterns to ensure statistical strength and the determination of contaminant extent.

Contaminant Identification

This report will describe the procedures necessary to identify unknown soil contamination, the approximate age of that contamination, as well as the degradation products.

Environmental Sensitivity Criteria

This report will establish environmental sensitivity criteria, which would consist of: prioritizing site characteristics from the site evaluation deliverable, e.g., hydrogeologic, climate, soil parameters, biotic factors such as vegetation and wildlife presence; synthesis of bioconcentration and environmental fate information; identification of current and future land use practices (residential, agricultural, recreational, commercial, industrial, etc.); incorporation of sociodemographic parameters.

Risk Assessment

This report will provide a standardized risk assessment methodology for soil contamination. It will identify and recommend and/or direct users to: data bases for hazard assessment values for various contaminants, indicating compound selection methodologies for streamlining the procedure; exposure values (environmental fate data—field or modeling); and a methodology for the synthesis of this information into a decision-tree format.

Health and Environmental Assessment—Remedial Options

This report will identify the types of remedial options available for the cleanup of contaminated soils, as well as their relative effectiveness and cost. The report will consider the occupational and community health risks associated with each option. This information will be an important factor in the risk management process.

The reports, as well as the final decision-tree, may be produced in several formats including:

- hardcopy
- interactive software
- video workbooks

REVIEW

The organizational structure of CHESS is designed to permit peer review of relevant documents at two sequential levels. The first level of review for technical committee reports will be by the Council. A second level of review, which will incorporate not only scientific evaluation but also relevance to federal and state agencies, will be conducted by the Governing Board. In addition, both the Governing Board and Council will have at their disposal the option to obtain external review from independent experts.

SUPPORT

Funding for CHESS will be sought from a broad base of organizations, including state departments of public health and environmental protection, federal agencies, foundations, and the private sector. Ideally, funds would be distributed equally among these groups.

All members of the Governing Board and Council are encouraged to solicit support. An ad hoc Fund-Raising Committee consisting of some Governing Board members has been established to solicit support from various industries and foundations: Dr. Robert Drew has been contacting petroleum companies; Dr. Gordon Newell, electric utilities; Dr. John Frawley, chemical and manufacturing; Dr. Anthony Marcil, foundations. Ongoing fund-raising activities are expected to be part of the Council Chairman's and Managing Director's responsibilities.

CHESS contributions are made directly to the Society, which provides the University of Massachusetts with a grant to support the Chairman and Managing Director's activities.

A first year target budget of $300,000 has been developed; however, activities will be scaled to support. As of the beginning of 1989, the Society has received

CHESS support from a number of groups including: federal—Agency for Toxic Substances and Disease Registry, USEPA/Office of Health & Environmental Assessment (forthcoming); private—Chevron, Eastman Kodak, Ford Motor Co., Gillette, Goodyear, Hercules, Hoechst Celanese, Morton-Thiokol, Shell, and Union Carbide.

TECHNICAL INFORMATION EXCHANGE

Following successful completion of the CHESS products, the Council will conduct a comprehensive program of technical information transfer to federal, state, and private sectors, as well as other organizations for which the product is intended.

Information transfer activities will include a combination of the following:

- journal articles through the Society's journal, *Regulatory Toxicology and Pharmacology*
- presentations and special sessions at major conferences such as the annual *Petroleum Contaminated Soils Conference* at the University of Massachusetts, Amherst MA
- regional workshops devoted to bringing the CHESS concept and deliverables to targeted groups
- a book comprised of the CHESS deliverables, published and distributed by a major publishing firm.

Glossary of Acronyms

AAR	Association of American Railroads
AIC	Acceptable intake, chronic
ADEQ	Arizona Department of Environmental Quality
ADI	Acceptable daily intake
API	American Petroleum Institute
ATSDR	Agency for Toxic Substances and Disease Registry
AWQC	Ambient water quality criterion
BaP	Benzo(a)pyrene
BDAT	Best Demonstrated Alternative Technology
BPJ	Best professional judgment
BTC	Breakthrough curve
BTX	Benzene-toluene-xylene
CDD	Chlorinated dibenzo-dioxins
CDF	Chlorinated dibenzofurans
CERCLA	Comprehensive Environmental Response, Compensation, and Liability Act of 1980
CHESS	Council for the Health and Environmental Safety of Soils
COD	Chemical Oxygen Demand
CTM	Carcinogen testing matrix
DEHP	Bis(2-ethylhexyl)phthalate
DEQE	Department of Environmental Quality Engineering (Massachusetts)
DER	Department of Environmental Regulation
DOC	Dissolved organic compound
DOE	Department of Energy
EBOS	Environmental Behavior of Organic Substances
ECRA	Environmental Cleanup Responsibility Act (New Jersey)
EEI	Edison Electric Institute
EIL	Environmental Impairment Liability (insurance)
EM	Electromagnetic conductivity
EP	Extraction procedure
EPA	Environmental Protection Agency
EPRI	Electric Power Research Institute
FDER	Florida Department of Environmental Regulation
FID	Flame ionization detector
FIFRA	Federal Insecticide, Fungicide, and Rodenticide Act
FS	Feasibility studies

GC	Gas chromatography
GPR	Ground penetrating radar
HEA	Health effect assessment
HI	Hazard index
HP	Hewlett-Packard
HPLC	High performance liquid chromatography
HV	Hydrocarbon vapor
IDNR	Iowa Department of Natural Resources
IR	Infrared spectrophotometry
IRA	Initial remedial action
ISRTP	International Society of Regulatory Toxicology & Pharmacology
LUFT	Leaking underground fuel tank
MCL	Maximum contaminant level
MGL	Massachusetts General Law
MTBE	Methyl-tertiary-butyl ether
NAPL	Nonaqueous phase liquid
NCP	National Contingency Plan
NJDEP/OSR	New Jersey Department of Environment Protection/Office of Science & Research
NOAEL	No observable adverse effect level
NOR	Notice of Responsibility
NPL	National Priority List
NRC	National Research Council
NTP	National Toxicology Program
OER	Office of Environmental Response
OHA	Office of Health Assessment
OUST	Office of Underground Storage Tanks (of EPA)
PAH	Polynuclear aromatic hydrocarbon
PCB	Polychlorinated biphenol
PCP	Pentachlorophenol
PCS	Petroleum contaminated soils
PID	Photoionization detector
PNA	Polynuclear Aromatic
PRP	Potentially Responsible Party
RfD	Reference dose
RFP	Request for proposal
RCRA	Resource Conservation and Recovery Act
RI	Remedial investigation
RP	Responsible party
RSD	Relative Standard Deviation
SARA	Superfund Amendments and Reauthorization Act of 1986
SD	Standard deviations
SOP	Standard operating procedures

TCLP	Toxic characteristic leaching procedure
TCP	2,4,5-Trichlorophenol
TEF	Toxicity equivalency factor
THC	Total hydrocarbons
TOC	Total organic carbon
TOV	Total organic vapor
TPH	Total petroleum hydrocarbons
TSCA	Toxic Substances Control Act
UHL	University Hygienic Laboratory
UST	Underground storage tanks
USWAG	Utility Solid Waste Activities Group (of EEI)
VOC	Volatile organic compounds
WQC	Water quality criteria
WSSOM	Water-soluble soil organic material

List of Contributors

Barkach, John, The Dragun Corporation, 3240 Coolidge Highway, Berkley, MI 48072-1634

Barnes, Ramon, Department of Chemistry, University of Massachusetts, Amherst, MA 01003

Bauman, Bruce, American Petroleum Institute, 1220 L Street, NW, Washington, DC 20005

Bell, Charles E., Environmental and Health Sciences Program, University of Massachusetts, Amherst, MA 01003

Biehl, Francis J., Department of Civil Engineering and Environmental Institute, Drexel University, Philadelphia, PA 19104

Boeve, Lucas, Excalibur Enterprises, Inc., 314 West 53rd Street, New York, NY 10019

Boscardin, Marco, Department of Civil Engineering, Marston Hall, University of Massachusetts, Amherst, MA 01003

Bouchard, Robert J., Haley & Aldrich, Inc., 58 Charles Street, Cambridge, MA 02141 (NOTE: Work was performed while employed by BSC Engineering, Boston, MA)

Bozzelli, Joseph W., Department of Chemical Engineering, Chemistry, and Environmental Science, New Jersey Institute of Technology, Newark, NJ 07102

Burgher, Brian J., The Environmental Protection Systems, 3800 Concorde Parkway, Suite 1200, Chantilly, VA 22021

Burris, David R., Tyndall Air Force Base, HQAFESC/RDVW Florida, 32403

Calabrese, Edward J., Division of Public Health, University of Massachusetts, Amherst, MA 01003

Casana, John, Underground Storage Tank Branch, Versar, Inc., 6820 Versar Center, P.O. Box 1549, Springfield, VA 22151

Chemburkar, Arun, Department of Chemical Engineering, Chemistry, and Environmental Science, New Jersey Institute of Technology, Newark, NJ 07102

Conlon, Peter, Association of American Railroads; Environment, Facilities, and Security, P.O. Box 11130, Pueblo, CO 81001

Czarnecki, Raymond C., George Brox, Inc., 1471 Methuen Street, Dracut, MA 01826

Daugherty, Seth J., Orange County Health Care Agency, P.O. Box 355, Santa Ana, CA 92702

Dawsey, W. John, Senior Environmental Analyst, Environmental Department, San Diego Gas and Electric Company, San Diego, CA 92112

Del Pup, John, Texaco, Inc., Environmental Affairs, P.O. Box 509, Beacon, NY 12508

Dixon, Gina, Underground Storage Tank Branch, Versar, Inc., 6820 Versar Center, P.O. Box 1549, Springfield, VA 22151

Dragun, James, The Dragun Corporation, 3240 Coolidge Highway, Berkley, MI 48072-1634

Dykes, Russell S., EA Engineering, Science, and Technology, Inc., 8300 Esters Blvd., Suite 940, Irving, TX 75063

Edwards, Carolyn, School of Education, University of Massachusetts, Amherst, MA 01003

Fiering, Myron B., Division of Applied Sciences, Harvard University, Cambridge, MA 02138

Findlay, Margaret, CAA Bioremediation Systems, 1106 Commonwealth Avenue, Boston, MA 02215

Fitzgerald, John, Division of Hazardous Waste, Massachusetts Department of Environmental Quality Engineering, Northeast Region, 5 Commonwealth Avenue, Woburn, MA 01801

Fogel, Samuel, CAA Bioremediation Systems, 1106 Commonwealth Avenue, Boston, MA 02215

Gersberg, Richard M., Assistant Professor of Environmental Health, Graduate School of Public Health, San Diego State University, San Diego, CA 92182.

Gilbert, Charles E., Division of Public Health, University of Massachusetts, Amherst, MA 01003

Greenthal, John L., Nixon, Hargrave, Devans & Doyle, 1 Keycorp Plaza, Albany, NY 12207

Grubbs, Robert B., Solmar Corporation, 625 West Katella Avenue, Suite 5, Orange, CA 92667

Gulledge, William, Environmental Insurance Management, Inc., 7900 Westpark Drive, Suite A300, McLean, VA 22102

Haney, Jeanmarie, Arizona Department of Environmental Quality, Hydrology Section, 2005 N. Central Avenue, Phoenix, AZ 85004

Hansen, Michael E. F., Conoco, 600 North Dairy Ashford, Houston, TX 77079

Harris, Cynthia, Agency for Toxic Substances and Disease Registry, Office of Health Assessment, 1600 Clifton Road, NE, Atlanta, GA 30333

Henry, Evan C., Bank of America, Environmental Services, Unit #4122, One City Boulevard West, Second Floor, Orange, CA 92668

Hinchee, Robert E., Battelle Memorial Institute, 505 King Avenue, Columbus, OH 43201

Hornsby, Michael, Department of Chemical Engineering, Chemistry, and Environmental Science, New Jersey Institute of Technology, Newark, NJ 07102

Kerr, James M., Jr., Delta Environmental Consultants, Inc., 2637 Midpoint Drive, Suite F, Fort Collins, CO 80525

Knapp, James, The Earth Technology Corporation, 15481 Electronic Lane, Unit D, Huntington Beach, CA 92649

Kostecki, Paul T., Environmental and Health Sciences Program, University of Massachusetts, Amherst, MA 01003

Kostle, Pamela A., University Hygienic Laboratory, Oakdale Campus, Iowa City, IA 52242

Krueger, Andrew J., Mobil Oil Corporation, Toxicology Division, P.O. Box 1029, Princeton, NJ 08540

Leis, Walter M., Roy F. Weston, Inc., 1350 Treat Blvd., Suite 200, Walnut Creek, CA 94596

Lemley, Ann T., Dept. of Textiles and Apparel, Field of Environmental Toxicology, Cornell University, Ithaca, NY 14852

Lion, Leonard W., School of Civil and Environmental Engineering, Field of Environmental Toxicology, Cornell University, Ithaca, NY 14852

Mackerer, Carl R., Mobil Oil Corporation, Toxicology Division, P.O. Box 1029, Princeton, NJ 08540

Magee, Brian R., Roy F. Weston, Inc., Weston Way, West Chester, PA 19380 (NOTE: Work reported in Chapter 15 was performed at Cornell University, Field of Environmental Toxicology.)

Malot, James J., Terra Vac, Inc., P.O. Box 2199, Princeton, NJ 08543

Martin, Joseph P., Department of Civil Engineering and Environmental Institute, Drexel University, Philadelphia, PA 19104

McGriff, E. Corbin, Jr., 126 Olympia Fields, Jackson, MS 39211

McLearn, Mary E., Electric Power Research Institute, Coal Combustion Systems Division, 3412 Hillview Avenue, Palo Alto, CA 94303

Metelski, John, Environmental Insurance Management, Inc., 7900 Westpark Drive, Suite A300, McLean, VA 22102

Metzger, Bernhard H., Arthur D. Little, Inc., Acorn Park, Cambridge, MA 02140

Millspaugh, Mark P., Nixon, Hargrave, Devans & Doyle, 1 Keycorp Plaza, Albany, NY 12207

Molnaa, Barry A., Solmar Corporation, 625 West Katella Avenue, Suite 5, Orange, CA 92667

Moore, Alan, CAA Bioremediation Systems, 1106 Commonwealth Avenue, Boston, MA 02215

Neil, William, Mobil Oil Corporation, Toxicology Division, P.O. Box 1029, Princeton, NJ 08540

Ostendorf, David W., Department of Civil Engineering, Marston Hall, University of Massachusetts, Amherst, MA 01003

Pastides, Harris, Division of Public Health, University of Massachusetts, Amherst, MA 01003

Pavlick, Raphe, Roy F. Weston, Inc., 1350 Treat Blvd., Suite 200, Walnut Creek, CA 94596

Potter, Thomas L., Chenoweth Laboratory, University of Massachusetts, Amherst, MA 01003

Preslo, Lynne M., Roy F. Weston, Inc., 1350 Treat Blvd., Suite 200, Walnut Creek, CA 94596

Reisinger, H. James, EA Engineering, Science, and Technology, Inc., 1900 Lake Park Drive, Suite 350, Smyrna, GA 30080

Ridgway, Harry F., Orange County Water District, Fountain Valley, CA 92728-8300.

Robinson, W. Terry, Department of Civil Engineering and Environmental Institute, Drexel University, Philadelphia, PA 19104

Roy, Timothy A., Mobil Oil Corporation, Toxicology Division, P.O. Box 1029, Princeton, NJ 08540

Rubel, Fred N., Environmental Protection Systems, 21-00 Route 208, Fairlawn, NJ 07410

Schoeny, Rita, U.S. Environmental Protection Agency, ECAO, 26 W. Martin Luther King Drive, Cincinnati, OH 45268

Seeger, Dennis R., PACE Laboratories, Inc., 1710 Douglas Drive North, Minneapolis, MN 55422

Simpson, Garey L., EA Engineering, Science, and Technology, Inc., 1900 Lake Park Drive, Suite 350, Smyrna, GA 30080

Stanek, Edward J., III, Division of Public Health, University of Massachusetts, Amherst, MA 01003

Taylor, Byron, Environmental Realty Services, 240 Great Circle Road, Suite 342, Nashville, TN 37228

Varuntanya, C. Peter, 100 Magnolia Avenue, Apartment 3-R, Jersey City, NJ 07306

Veneman, Peter, Department of Plant and Soil Sciences, University of Massachusetts, Amherst, MA 01003

Wilhelm, Robert W., II, Haley & Aldrich, Inc., 58 Charles Street, Cambridge, MA 02141 (NOTE: Work reported in Chapter 27 was performed while employed by BSC Engineering, Boston, MA)

Yang, Joseph J., Mobil Oil Corporation, Toxicology Division, P.O. Box 1029, Princeton, NJ 08540

Index

T - #0059 - 101024 - C0 - 234/156/29 [31] - CB - 9780873712262 - Gloss Lamination